T0313704

FUNDAMENTAL SPACECRAFT DYNAMICS AND CONTROL

FUNDAMENTAL SPACECRAFT DYNAMICS AND CONTROL

Weiduo Hu

Beihang University, China

This edition first published 2015
©2015 John Wiley & Sons Singapore Pte. Ltd.

Registered office
John Wiley & Sons Singapore Pte. Ltd., 1 Fusionopolis Walk, #07-01 Solaris South Tower, Singapore 138628.

For details of our global editorial offices, for customer services and for information about how to apply for permission to reuse the copyright material in this book please see our website at www.wiley.com.

Library of Congress Cataloging-in-Publication Data applied for.

ISBN: 9781118753538

A catalogue record for this book is available from the British Library.

Typeset in 10/12pt TimesLTStd by SPi Global, Chennai, India

Printed and bound in Singapore by Markono Print Media Pte Ltd

1   2015

Contents

Preface

I have taught orbital mechanics for international students since 2006 at Beihang University, and have found that many Chinese students are also interested in learning this course in English. But a lot of these students whose mother language is not English have told me that most present English textbooks use long and complicated English descriptions, which means that they find it difficult to understand the basic materials of this course at the beginning.

One of the major objectives of this book is therefore to use simple and concise English to introduce orbital theory and applications, which other textbooks may already contain. Of course, some augmentations and extensions are also included, such as the orbital motion around asteroids, which are closely related to my Ph.D. dissertation at the University of Michigan; and satellite full attitude capture, which I studied in my work at the Beijing Institute of Control Engineering. Moreover many special simulation figures are made for this book. Finally, the organization of the content is different from similar books to allow it to be read easily, step by step.

To realize this special objective, detailed derivations of theories, a large number of illustrations and many practical application examples are given, making this textbook especially suitable for students whose mother language is not English when they begin to learn spacecraft dynamics in English.

Spacecraft dynamics in this book is concerned with the description of the motion of bodies in space. This subject is divided into two parts, one of which is termed orbital mechanics, while the other is referred to as attitude dynamics. As we know, spacecraft orbital motion is the motion of its mass center, whereas attitude motion is the spacecraft motion about its center of mass.

This book focuses mainly on the *orbital mechanics* course, which has been taught for the past 10 years by the author. It is also intended as a reference for other courses such as *spacecraft attitude dynamics and control*. Hence some fundamental information on attitude dynamics has also been included. In the first part of this book from Chapter 1 to 7, orbital related topics are discussed, such as Keplerian motion, orbital maneuvers, CW-equations, the Lambert problem, orbital determination and optimal control, the three body problem, perturbed orbits, asteroid exploration, orbital applications etc. In the second part, from Chapter 8 to 10, attitude kinematics, dynamics, stabilization and determination are introduced. The two parts can be treated independently, but they are related. For example, attitude determination and gravity gradient stabilization are based on knowledge of both orbital mechanics and attitude dynamics. Spacecraft control is not discussed independently, but general speaking, orbital maneuver, attitude stabilization, etc., are in the category of spacecraft control which are all included in this book. Although in this book, fundamental knowledge is mainly described, some relatively new definitions and methods or complicated problems are also introduced, such as periodic orbit, chaotic motion, f, g expansion, Lagrange bracket, resonant orbit, elliptic expansion, three body problem, orbital determination and optimization, CW equation, Lambert problem, Poincaré map, liquid sloshing model, two line element, GPS RINEX, inertia tensor, spherical geometry etc. All these materials could be interested not only by beginners but for experienced aerospace engineers.

Though I try to correct mistakes in the book very hard, there must be some errors left. If you find any questions, please tell me by email: weiduo.hu@buaa.edu.cn. All comments are grateful.

Weiduo Hu

Acknowledgments

Some of the material in this book originate from my class notes, handouts, and homework of different courses at the University of Michigan where I pursued my Ph.D. and also from discussions and exchanges with my friends in the USA, and also my students at Beihang. Here I wish to thank all of those who have contributed to this book, including Professors Scheeres, McClamours, Kambaber, Kuo, C. Hall, P. Lu, Chen, Bernstein, Greenwood, and also my friends FY Hsiao, B.F. Villac, Dan, David, Z. Shen, Z. Wu, L.Tian, as well as several reviewers such as K. Anupama whose suggestions have been helpful.

Many thanks are given to Beihang International School and Astronautics School for their support and encouragement.

Special thanks go to my family for their love, blessing, inspiration and patience.

About the Author

Weiduo Hu obtained his bachelor, masters and Ph.D. degrees in the Department of Automatic Control at Beihang in 1986, 1989, 1993 respectively. He then worked as an engineer and senior engineer at Beijing Institute of Control Engineering, where he was in charge of fault detection of a satellite control system. In 1998, he went to the USA, in pursuit of a second Ph.D. in the Department of Aerospace Engineering at the University of Michigan, supported by NASA JPL, focusing on research on spacecraft orbital motions around asteroids. In 2005, he joined the School of Astronautics at Beihang University (BUAA).

He has published many research papers on spacecraft dynamics and control in journals such as *Planetary and Space Science*, *AIAA Journal of GCD*, *Celestial Mechanics and Dynamical Astronomy*, *IEEE Trans. on Aero. and Elec. Sys.*, *Chinese Journal of Astronomy and Astrophysics*, *Acta Mechanica Sinica*, etc. He is currently an editor of the *Journal of Aerospace Science and Technology* at Hans Publishers Inc. USA. His research interests are spacecraft attitude and orbit determination, control and dynamics. He teaches courses such as Orbital Mechanics, Spacecraft Control, Integrated Navigation, Modern Control Theory etc. for undergraduate, graduate and international students.

Part One

Orbital Mechanics

1

Introduction

Spacecraft dynamics in this book means the study of motion of man-made objects in space, subject to both natural and artificially induced forces, which includes orbital mechanics and attitude dynamics.

Orbital mechanics is the science concerned with the trajectory motion of a spacecraft, whereas attitude dynamics is concerned with the orientation motion of a spacecraft.

This text book mainly focuses on trajectory motion, that is, the motion of mass center of a man-made spacecraft or a natural body. Natural body's motion is also within the scope of celestial mechanics. A basic introduction to attitude dynamics is given in this book.

Since dynamics are closely related to GNC (guidance, navigation and control), a brief discussion about its meaning is given here. Simply speaking, guidance answers the question where to go, navigation tells us where you are, and control is concerned with how to go. In this book, orbit and attitude determination belong to the category of navigation, while orbit maneuver, attitude stabilization and control are in the category of control. Dynamics form the foundation in analyzing a GNC system. Very little discussion is given to guidance in this book, and it can be simply regarded as a reference. Figure 1.1 shows the basic relations between these definitions using the classical feedback control diagram.

Many sub-systems of a spacecraft are closely related to GNC, including attitude control, communication, power supply, thermal control, structure, propulsion subsystem, etc.

1.1 History

The scientists who have made great contributions to the development of orbital mechanics include Aristotle, Ptolemy, Copernicus, Tycho Brahe, Kepler, Galileo, Newton, Euler, Lagrange, ...

1.1.1 Kepler's Laws

The Danish astronomer Tycho Brahe (1546–1601) gathered extremely accurate observational data on planetary motion. He developed and maintained detailed and precise records. But he

Fundamental Spacecraft Dynamics and Control, First Edition. Weiduo Hu.
© 2015 John Wiley & Sons (Asia) Pte Ltd. Published 2015 by John Wiley & Sons (Asia) Pte Ltd.

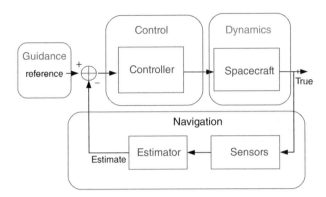

Figure 1.1　Basic relations between GNC and dynamics

didn't capitalize on observations, as he lacked the vision and mathematical skills. And he kept the Earth at the center of the planets.

Johannes Kepler (1571–1630), a gifted mathematician and astronomer, distilled Brahe's observation data to provide the first quantitative statements about orbital mechanics, which is known as Kepler's three Laws:

Law of ellipse: The orbits of the planets are ellipses with the Sun at one focus.
Law of equal area: The line joining a planet to the Sun sweeps out equal areas at equal times.
Law of harmony: The square of the orbital period (time to complete one orbit) is directly proportional to the cube of the average distance between the Sun and the planet.

Kepler's Laws are descriptive, which answer how the planets move around the Sun. But they can not explain why planetary motion satisfies these three laws. It was Isaac Newton (1642–1727), who established the mathematical foundation from which Kepler's Laws can be derived.

1.1.2　Newton's Laws

Newton is arguably the greatest physicist ever. His major discoveries and developments were differential and integral calculus, the gravitational laws, and also contributions in optics.

Isaac Newton (1642–1727) established the fundamentals of celestial mechanics based on the earlier work of Tycho and Kepler. Newton formulated the basic concepts of the Laws of Motion and Law of Gravitation around 1666, but it was not until 1687, at the urging of Edmund Halley, the Principia was published. It originally presented the three laws of motion.

Law of inertia: Every body continues in its state of rest or of uniform motion in a straight line unless it is compelled to change that state by forces impressed upon it.
Law of momentum: The rate of change of momentum ($m\mathbf{v}$) is proportional to the force ($\mathbf{F}$) impressed and is in the same direction as that force.

$$\mathbf{F} = \frac{\mathrm{d}}{\mathrm{d}t}(m\mathbf{v}).$$

This law is also sometimes called *the first principle*.

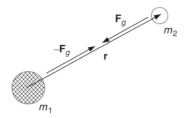

Figure 1.2   Universal gravity

Law of action and reaction: To every action there is always an opposed equal reaction.
Law of Universal Gravity: Any two bodies attract one another with a force ($\mathbf{F}_g$) proportional to the product of their masses (m_1, m_2) and inversely proportional to the square of the distance (r) between them.

Thus,

$$F_g = \left| \mathbf{F}_g \right| = \frac{G m_1 m_2}{r^2} \tag{1.1}$$

or in vector form:

$$\mathbf{F}_g = -\frac{G m_1 m_2}{r^2} \frac{\mathbf{r}}{\left| \mathbf{r} \right|}, \tag{1.2}$$

where $G = 6.67259 \times 10^{-20}$ km$^3 \cdot$ kg$^{-1} \cdot$ s^{-2} is universal gravitational constant, where $\mathbf{F}_g$ is the force on m_2 and $\mathbf{r}$ is a vector from m_1 to m_2. This assumes a spherically symmetric point mass. See Figure 1.2.

The subsequent major contributions in orbital mechanics are mainly in the sense of orbital determination. Newton needed three observations of position vectors; Halley mastered and refined Newton's methodology; Lambert used geometrical arguments; Lagrange developed mathematical basis for orbit calculations; Laplace developed new methodology; and Gauss summarized, simplified and completed the orbit determination work.

1.1.3   Space Missions

Some major figures in developing theory and practice in space technology should be mentioned. Konstantin Tsiolkovsky – Russia (1857–1935), who derived the fundamental equation of rocketry, is called the father of modern space technology. Robert Goddard – USA (1882–1945) built the laboratory experiments of rocket engine. Wernher Von Braun-Germany/ USA (1912–1977) brought the space technology to practical application, he designed the V2 rocket and Explore I satellite.

During the early space age, Sputnik was the first man-made object flying in space, having been launched in October 1957. Yuri Gagarin was the first man in space in April 1961. And Neil Armstrong was the first man to land on the Moon in July 1969. These are some of the milestones in human space exploration.

China's first satellite Dongfanghong-1 was successfully launched in 1970. In 2003, the Shenzhou-5 spaceship carried out manned spaceflight and Chang'e-1 in 2007 entered into lunar orbit. These have been the major breakthroughs in the history of Chinese space technology.

Table 1.1 Classification and examples of missions

Mission	Orbit/name	a (km)	Period (hr)	i (deg)	e
Communication	Geostationary	42 158	24	0	0
Remote sensing	Sun-synchronous	6500–7300	1.5	95	0
Navigation (GPS)	Semi-synchronous	26 610	12	55	0
Comm/Intel	Molniya	26 571	12	63.4	0.72 335
Space station	ISS (ZARYA)	6770.6	1.5401	51.643	0.0008
Space station	Tiangong 1	6731.5	1.5268	42.785	0.0008
Science	HST	6968	1.617	28.5	0.001
Weather	Fengyun 1D	7239.7	1.703	98.758	0.0016
Weather	Fengyun 2F	42 166	23.936	2.196	0.0007
Navigation	Beidou IGSO 2	42 167	23.937	54.625	0.0021
Navigation	Beidou G3	42 167	23.937	1.51	0.0002

At the present time, space technology is beging used everywhere, including communications, weather forecasting, science exploration, navigation etc. See examples in Table 1.1 and some of the figures and data in Chapter 7.

There are an extensive number of orbit types, for which a rough classification can be made. For a low Earth orbit (LEO), its height is normally less than 2000 km (altitude). Mid Earth Orbit (MEO)'s height is between 2000 and 30 000 km. Highly Elliptical Orbit (HEO) is an orbit with large eccentricity. Geosynchronous Orbit (GEO) (circular)'s period is the time it takes Earth to rotate once with respect to (wrt) stars, $R = 42\,164$ km. Polar orbit's inclination is nearly 90 degrees. Molniya's orbit is a highly eccentric orbit with 12 hr period (developed by Soviet Union to optimize coverage of the Northern hemisphere).

In Table 1.1, ISS (International Space Station), Tiangong 1 are space stations whose orbits belong to low Earth orbit (LEO). And HST (Hubble Space Telescope) is also in low Earth orbit. Fengyun 2F is a geostationary orbit (GEO) whose inclination is near zero. Beidou 2 is a geosynchronous orbit whose inclination is not zero, but its period is still near 24 hrs. A GPS satellite belongs to a MEO satellite. Molniya is a high elliptical orbit designed by the former Soviet Union. Most Sun-synchronous satellites, such as Fengyun 1, are polar orbit satellites whose inclinations are near 90 deg.

An excellent source of information on specific spacecraft is the *NASA JPL Missions*, located on the World Wide Web at http://www.jpl.nasa.gov/missions/.

1.2 Coordinate Systems

Before we discuss orbital motion, orbital transfer, station keeping, and interplanetary missions, a reference frame is needed to describe these motions. In this section we just give a very brief description; more detailed discussion and figures can be found later in the book [1, 6].

The frame must be inertial (nonaccelerating) for Newton's Laws to apply. To define a frame, an origin, a fundamental plane and a reference are needed.

1.2.1 Earth Reference Frame

As much work describes the Earth orbiting spacecraft, it is important that the coordinate system/reference frame is Earth-oriented.

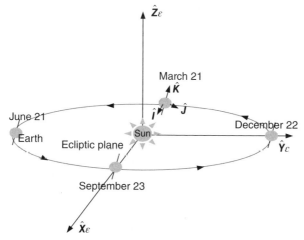

Vernal equinox direction is the direction for the
Earth to the Sun on the first day of spring.

Figure 1.3 Sun-centered HECS and Earth-centered ECI frames

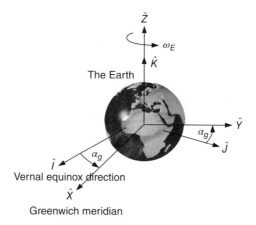

Figure 1.4 Earth centered ECI and ECEF frames

The origin of the coordinate system is at the center of the Earth. Its fundamental plane is the Earth's equatorial plane. Perpendicular to the plane is the North pole direction. The principal direction is the Vernal Equinox direction, which is from the Earth to the Sun at the first day of spring. That is the $\hat{X}_\varepsilon$ direction in Figure 1.3.

If the Vernal Equinox direction is $\hat{I}$, the North Pole direction is $\hat{K}$, $\hat{J}$ is in fundamental plane, the $\hat{I}\hat{J}\hat{K}$ compose a right-handed system. See Figure 1.4.

This $\hat{I}\hat{J}\hat{K}$ system is called Geocentric-Equatorial Coordinate System (GES). It is also known as (aka) Earth Centered Inertial (ECI), or the Conventional Inertial System (CIS), where J2000-Vernal equinox on January 1, 2000 is nonrotating.

If $\hat{X}$ is pointing to the Greenwich meridian from the Earth center, and $\hat{X}\hat{Y}\hat{Z}$ rotate with the Earth with angular velocity ω_E about the axis $\hat{K}$, $\hat{Z}$ axis, it is the well-known geographic frame, that is Earth centered Earth fixed frame (ECEF). α_g is the angle between Vernal Equinox and Greenwich meridian.

1.2.2   Sun-centered Frame

Its origin is at the center of the Sun. The fundamental plane is Earth's mean orbital plane about the Sun which is called the ecliptic plane. The fundamental direction is the Vernal Equinox direction.

Vernal Equinox direction is $\hat{X}_e$, $\hat{Y}_e$ is in ecliptic plane $90°$ from $\hat{X}_e$ in the direction of Earth's motion, $\hat{Z}_e$ is perpendicular to ecliptic plane. It is also a right-handed system.

This is called the Heliocentric-Ecliptic Coordinate System (HECS). See Figure 1.3.

The angle ϵ between the Earth's mean equator and the ecliptic is called the obliquity of the ecliptic, which approximates $23.5°$; see also Figure 2.16.

The intersection of the ecliptic and equatorial planes is called the line of nodes.

The Sun crosses the line of nodes twice a year at the equinoxes (Latin for equal day and night, Earth experiences equal day and night): ascending node is at the vernal equinox on March 21; Descending node is at the autumnal equinox on Sept 23.

Note that the Vernal Equinox occurs when the Sun's declination is $0°$ as it approaches the ascending node.

1.2.3   Right Ascension-Declination System

The origin is at the center of the Earth or point on the surface of the Earth, but it is not very important. The fundamental plane is also Earth's equatorial plane extended to the sphere of infinite radius, which is called the celestial equator.

The right ascension angle (α) is the angle measured eastward from vernal equinox in the plane of the celestial equator. The declination angle (δ) is the angle measured up (north) from the celestial equator. This frame is primary used to catalog stars, then used to help determine spacecraft position. See Figure 1.5. This system is also called the celestial sphere.

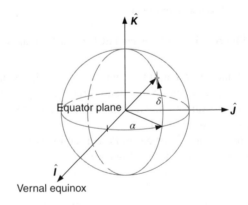

Figure 1.5   Celestial sphere

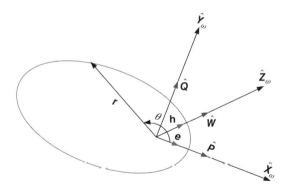

Figure 1.6   Perifocal coordinate system

The relationship between α and δ and latitude/longitude is the hour angle which is elapsed time in hours since the object was directly overhead. Greenwich Hour Angle (α_g, see Figure 1.3) is generally referred as the angle between vernal equinox and $0°$ longitude or the Greenwich Meridian.

1.2.4   Perifocal Coordinate System

The origin is at the center of gravity of a spacecraft. Its fundamental plane is the plane of the spacecraft's orbital plane. The direction perpendicular to the plane is aligned with angular momentum vector (**h**). The principal direction $\hat{X}_\omega$ is pointing towards the perigee which is denoted as vector $\hat{P}$. Its 2nd vector ($\hat{Y}_\omega$) is in the plane $90°$ from $\hat{X}_\omega$ in the direction of orbital motion which is denoted as vector $\hat{Q}$. $\hat{Z}_\omega$ is along angular momentum vector, denoted as vector $\hat{W}$. It is also a right-handed system. It is primarily used in the analysis associated with a spacecraft. This frame is also known as $\hat{P}\hat{Q}\hat{W}$ frame. See Figure 1.6.

1.2.5   Satellite Coordinate System

The Satellite Coordinate System is a so-called Radial Transverse Normal system, also known as Radial Tangential Normal system.

This system moves with the satellite. The $\hat{R}$ axis points the satellite, the $\hat{W}$ axis is normal to the orbital plane (and usually not aligned with the $\hat{K}$ axis), and the $\hat{S}$ axis is normal to the position vector. The $\hat{S}$ axis is continuously aligned with the velocity vector only for circular orbits. See Figure 1.7.

1.2.6   Topo-centric-horizon Coordinate System

The origin of the system is at a point on the surface of the Earth (topos) where a sensor such as radar is located. Its fundamental plane is the horizon at the sensor, and the coordinate directions are $\hat{X}$ points south, $\hat{S}$; $\hat{Y}$ points east, $\hat{E}$; $\hat{Z}$ points up, $\hat{Z}$. See Figures 1.8 and 2.19. In the $\hat{S}\hat{E}\hat{Z}$ system (topocentric), the sensor gives the satellite elevation and azimuth (Az, El). It is a noninertial system. This frame is also called THS (topcentric horizontal system), denoted as $\hat{X}_h\hat{Y}_h\hat{Z}_h$.

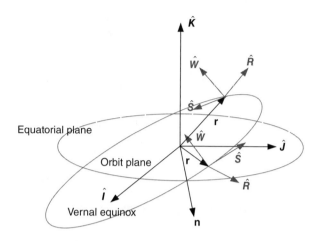

Figure 1.7   Satellite coordinate system

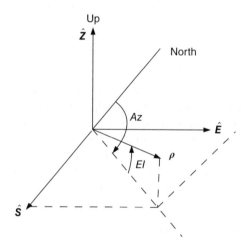

Figure 1.8   Top-centric-horizon coordinate system

1.3   Time System

There are a few time systems which are frequently used in astrodynamics, such as Julian day, Solar day, Sidereal day, Greenwich Mean Time, UTC, Zulu time. Time is important as the signal travel time of electromagnetic waves in altimetry, GPS, SLR and VLBI should be exactly known by some reference.

For positioning and orbit determination, one nanosecond (10^{-9}) seconded is 30 cm of distance of relative motion of celestial bodies or satellites.

1.3.1   Clocks

Before discussion of the time system, clock should first be introduced.

1.3.1.1 Astronomical Clocks

We commonly define time in terms of astronomical and geodetic periods, such as the rotation of the Earth (day), the revolution of the Earth around the Sun (year), the orbit of the Moon around the Earth (month), and the number of bodies in the solar system visible to the naked eye or 1/4 of a lunar cycle (week).

These astronomical clocks are not consistent. The current length of a year is 365.242 190 days; it was 365.242 196 days in 1990, and will be 365.242 184 days in 2100.

1.3.1.2 Mechanical Clocks

When mechanical clocks were less accurate than variations in astronomical clocks, astronomical clocks were used to correct mechanical clocks. The pendulum clock was invented in the 17th century.

Since 1928, with the invention of the quartz clock, it has been apparent that the uncertainty in the astronomical day was 10^{-7} due to irregularities in Earth rotation.

The idea of using hyperfine quantum states of atoms for a clock was first proposed by US physicist Isador Rabi in 1945, which lead to the invention of atomic clocks.

1.3.1.3 Atomic Clocks

Certain atoms in a magnetic field can exhibit one of two hyperfine states, that is, the spin of the outermost electron of an atom either points in the same direction as the magnetic field of the nucleus, or it points opposite. The laws of quantum physics forbid other orientations.

Generally, an atom remains in its hyperfine state. But when prodded by electromagnetic radiation at a specific frequency, it will switch to the other state, undergoing the so-called "hyperfine transition."

Essentially, an electronic clock selects atoms in one hyperfine state and exposes them to radiation which causes them to switch to the other state. The frequency of the radiation causing the transition becomes the regular beat that the clock counts to register time.

Hydrogen-masers, cesium, and rubidium standards are usually used. Each GPS satellite has 2 Cs and 2 Ru clocks.

1.3.2 Time

1.3.2.1 International Atomic Time (TAI)

Atomic Time, with the unit of duration being the Systems International (SI) second, was defined in 1967 as the duration of 9 192 631 770 cycles of radiation corresponding to the transition between two hyperfine levels of the ground state of cesium-133. The second, defined in 1967, corresponds to traditional measurement.

A uniform time-scale of high accuracy is provided by International Atomic Time (Temps Atomique International, TAI). TAI was chosen to start on January 1, 1958 0h. It is estimated by a large set (more than 200) of atomic clocks which are mostly cesium atomic clocks and a few hydrogen masers at 60 laboratories worldwide.

Time centers compare clocks, using GPS observations as time links. A weighted mean of local time centers results in the TAI.

1.3.2.2 Sidereal time

Sidereal time is directly related to the rotation of the Earth. Local Apparent (or true) Sidereal Time (LAST) refers to the observer's local meridian's hour angle to the true vernal equinox. Greenwich Sidereal Time (GST) is called Zulu time.

The vernal equinox is the intersection of the ecliptic and the equator, where the Sun passes from the southern to the northern hemisphere. The vernal equinox is affected by precession and nutation and experiences long and short period variations.

1.3.2.3 Solar Time and Universal Time

Solar time is used in everyday life. It is related to the apparent diurnal motion of the Sun about the Earth. See Figure 1.9.

This motion is not uniform, but it is assumed to be a constant velocity in the motion about the Sun. Mean solar time is equal to the hour angle of the mean sun plus 12 hours. If referred to the Greenwich mean astronomical meridian, it is termed Universal Time (UT). Its fundamental unit is the mean solar day, the interval between two transits of the sun through the meridian. Conversion of UT to GST is defined by convention, based on the orbital motion of the Earth (about $360°/365$ days):

1 mean sidereal day = 1 mean solar day - 3 min 55.90 sec = 86 164.10 s. Similarly one 1 sidereal month is about 27.5 days, one lunar month is 29.5 days.

1.3.2.4 Fundamentals of Time

Universal Time (UT) is the Greenwich hour angle of a fictitious sun uniformly orbiting in the equatorial plane, augmented by 12 hours (eliminates ecliptic motion of the Sun). UT0

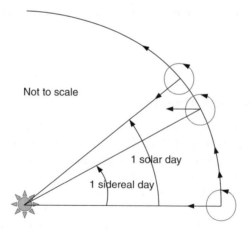

Not to scale

1 solar day

1 sidereal day

Figure 1.9   Sidereal day and solar day

is determined by the motions of the stars; thus it is a function of sidereal time. UT1 is UT0 corrected for polar motion. It closely approximates the mean diurnal motion of the Sun (Solar Time).

Dynamic Time (Ephemeris Time) is derived from planetary motions in the solar system (deduce time from position of planets and equations of motion) which is an independent variable in the equations of motion. Dynamic time is classified as two groups. Barycentric Dynamic Time (BDT) is based on planetary motions wrt (with respect to) the solar system barycenter. Terrestrial Dynamic Time (TDT) is derived from satellite motions around the Earth.

1.3.2.5 UTC

A practical time scale, as needed in navigation for instance, has to provide a uniform unit of time and maintain a close relationship with UT1. This led to the introduction of Coordinated Universal Time (UTC). Its time interval corresponds to atomic time TAI and its epoch differs by not more than 0.9 sec from UT1. In order to keep the difference, leap seconds are introduced to UTC when necessary. A specification of a UTC clock should always differ from TAI by an integer number of seconds. GPS provides easy access to UTC with an accuracy within 100 nanoseconds. GPS navigation data provides the integer offset for TAI.

1.3.2.6 Leap Seconds

Since the first leap second in 1972, all leap seconds have been positive.

Currently the Earth rotates slow at roughly 2 milliseconds per day, so a leap second is needed about every 500 days.

There have been 32 leap seconds in the 27 years to January 1999. The latest leap second happened on Jun. 30, 2015.

This pattern mostly reflects the general slowing trend of the Earth due to tidal braking.

Sometimes there is a misconception that the regular insertion of leap seconds every few years indicates that the Earth should stop rotating within a few millennia.

Leap seconds are not a measure of the rate at which the Earth is slowing. The 1 second increments are indications of the accumulated difference in time between the two systems. Usually, you would reset a slow clock to the accurate time. As we can't alter the Earth's rotation, so we reset the accurate clock.

As we have already introduced, Atomic Time – International Atomic Time (TAI) is based on vibrations of the Cesium atom. TDT = TAI + 32.184 sec. Coordinated Universal Time (UTC) is equal to TAI but augmented by leap seconds to keep it close to UT. TAI = UTC + 1s. 0n (n=integer=28 in early 1994). GPS Time needs to be uniform and has a constant offset of 19 secs wrt TAI, and was equal to UTC at the GPS standard epoch of January 6, 1980.

1.3.2.7 Present Time Differences

As of January 1, 1999, TAI is ahead of UTC by 32 seconds, TAI is ahead of GPS by 19 seconds, GPS is ahead of UTC by 13 seconds. The Global Positioning System (GPS) epoch is January 6, 1980 and is synchronized to UTC. No positive leap second was introduced at the end of June 2004.

1.3.2.8 Yet More Definitions

Terrestrial Time (TT) (or Terrestrial Dynamical Time, TDT), with unit of duration 86,400 SI seconds on the geoid, is the independent argument of apparent geocentric ephemerides. TDT = TAI + 32.184 seconds.

The epoch of TDT is defined as January 1, 1977 0h TAI.

Julian Day Number is a count of days elapsed since Greenwich mean noon on January 1, 4713 BC, Julian calendar. The Julian Date is the Julian day number followed by the fraction of the day elapsed since the preceding noon.

The Julian Day Number for February 7, 2002 is 2 452 313. The Modified Julian Day, was introduced by space scientists in the late 1950s. It is defined as MJD = JD -2 400 000.5.

Julian Date (JD) defines the number of mean solar days since 4713 BC, January 1, 0.5 (noon). Modified Julian Date (MJD) is obtained by subtracting 2 400 000.5 days from JD. Thus, MJD commences at midnight instead of noon. More discussion can be found later in the book [6].

1.3.3 Reference Motions

1.3.3.1 Earth Orientation Measurements

Length of Day must be measured, preferably using VLBI (very long baseline interferometry). Polar Motion must also be measured with space geodetic techniques. Nutation in both theory and modeling can generally be used (sometimes applying small corrections from VLBI). Precession can also generally be used.

Although in this book, we will generally ignore precession, nutation, and polar motion, a brief explanation of all three terms is given in the next two sections.

1.3.3.2 Precession and Nutation

Luni-Solar Precession is about 50 arcseconds per year whose period is 26 000 years (due to the torques of the Moon and the Sun on the Earth's equatorial bulge). Planetary Precession is 12 arcseconds per century and there is a decrease in the obliquity of the ecliptic of 47 arcseconds per century (due to the planetary perturbations on the Earth's orbit, causing changes in the ecliptic). 9 arcseconds nutation amplitude occurs at orbital periods of the Sun and the nutation periods due to the Moon are 13.7 days, 27.6 days, 6 months, 1 year, 18.6 years, etc., where the 18.6 year motion is largest – 20 arcseconds amplitude (0.5 km).

1.3.3.3 Polar Motion

Linear drift of the rotation pole is about 3–4 milli arc seconds/year in a direction between Greenland and Hudson Bay (due to post glacial rebound). Long period wobble (30 years) is about amplitude 30 milli arc seconds (cause unknown). 75% of the Annual Wobble (amplitude of 0.1 arc seconds, i.e., 3 meters on the Earth's surface) is caused by annual variation in the inertia tensor of the atmosphere; the rest is caused by mass variations in snow, ice, ground water, etc. Chandler Wobble (430 day period) is about 6 meters amplitude which is the normal mode of the Earth and is caused by atmospheric and oceanic effects (1 arcmin = 1/60 deg; 1 arcsec = 1/60 arcmin).

1.3.3.4   Motion of the Coordinate System

The relation of local sidereal time and Greenwich sidereal time is (see also Section 2.4.4)

$$\alpha_g = \alpha_{g0} + \omega_E \cdot \Delta UT1 \tag{1.3}$$

$$\omega_E = 7.2\,921\,158\,553 \times 10^{-5} \text{rad/s} \tag{1.4}$$

$$\alpha_{g0} = 1.753\,368\,560 + 628.3\,319\,706\,889\,T_{UT1} + 6.7707 \times 10^{-6} T_{UT1}^2$$
$$-4.5 \times 10^{-10} T_{UT1}^3, \tag{1.5}$$

where T_{U11} is number of Julian centuries elapsed from epoch J2000, which is only concerned with rotation of Earth about $\hat{Z}$. See Figure 1.3. The motion of the coordinate systems is

$$\mathbf{r}_{IJK} = \mathbf{R}_3(-\alpha_g)\mathbf{r}_{ECEF} \tag{1.6}$$

$$\mathbf{v}_{IJK} = \mathbf{R}_3(-\alpha_g)\mathbf{v}_{ECEF} + \dot{\mathbf{R}}_3(-\alpha_g)\mathbf{r}_{ECEF}$$
$$+ \boldsymbol{\omega}_\mathbf{E} \times \mathbf{R}_3(-\alpha_g)\mathbf{r}_{ECEF} \tag{1.7}$$

$$\mathbf{R}_3(-\alpha_g) = \begin{bmatrix} \cos\alpha_g & -\sin\alpha_g & 0 \\ \sin\alpha_g & \cos\alpha_g & 0 \\ 0 & 0 & 1 \end{bmatrix} \tag{1.8}$$

$$\dot{\mathbf{R}}_3(-\alpha_g) = \begin{bmatrix} \omega_E\sin\alpha_g & \omega_E\cos\alpha_g & 0 \\ -\omega_E\cos\alpha_g & \omega_E\sin\alpha_g & 0 \\ 0 & 0 & 1 \end{bmatrix}. \tag{1.9}$$

1.3.3.5   Variations of Length of Day (LOD)

There is a linear increase in the LOD of 1–2 msec/century. The increase is mainly due to tidal breaking. Some decrease is due to decreasing moment of inertia (J_2, e.g. a skater spins faster when their hands are drawn in). The decadal fluctuations is of 4–5 msec (due to transfer of angular momentum between the core and the mantle). Short period variability is of 2–3 msec at a period of less than 5 years (mainly 2 weeks, 1 month, 6 months, 1 year) (due to the Earth and ocean tides, the atmosphere, and wind-driven ocean circulation). Before modern times, ancient eclipse records have been used to determine variations in LOD.

Length of day changes are highly correlated with changes in atmospheric angular momentum (AAM).

1.3.3.6   Reference Frame Terminology

The Equator is the great circle on the surface of a body formed by the intersection of the surface with the plane passing through the center of the body perpendicular to the axis of rotation. The Celestial Equator is the projection on to the celestial sphere of the Earth's equator. The Ecliptic plane is the plane of the Earth's orbit about the Sun, affected by planetary precession.

The True Equator and Equinox of Date in the celestial coordinate system are determined by the instantaneous positions of the celestial equator and ecliptic (motion due to precession and nutation). The Mean Equator and Equinox of Date of the celestial reference system are determined by ignoring small variations of short-period (nutation) in the motions of the celestial equator (motion due to only precession). Mean Equator and Equinox of J2000.0 in the celestial reference system are at 12 hours, January 1, 2000.

1.3.3.7   Reference Frame Transformation

These frames are related,

$$\mathbf{r}_{J2000} = \mathbf{P} \cdot \mathbf{N} \cdot \mathbf{S} \cdot \mathbf{M} \cdot \mathbf{r}_{ECEF}, \tag{1.10}$$

where $\mathbf{P}$ is the precession transformation (moves state from mean equinox of epoch (J2000) to mean equinox of date), $\mathbf{N}$ is the nutation transformation (moves state from mean equinox to true equinox of date), and $\mathbf{S}$ is the Apparent Sidereal Time Transformation (referenced to true, equinox)

$$\mathbf{S} = \mathbf{R}_3(\alpha_g) + \mathbf{E}_q, \tag{1.11}$$

where α_g is Greenwich Mean Sidereal Time, $\mathbf{E}_q$ is the "equation of the equinoxes" (see reference [6] for more discussion), which is equal to the difference between mean and true equinoxes, and $\mathbf{M}$ is the polar motion transformation:

$$\mathbf{M} = \begin{bmatrix} 1 & 0 & x_p \\ 0 & 1 & -y_p \\ -x_p & y_p & 1 \end{bmatrix}. \tag{1.12}$$

The rotations here mean that $\mathbf{M}$ is about Polar Motion which rotates the terrestrial (Earth Center Earth Fixed) (ECEF) frame from the conventional pole to the celestial ephemeris pole; $\mathbf{S}$ is about Sidereal Time which rotates the terrestrial frame from the Greenwich Meridian to the true equinox of date; $\mathbf{N}$ is about Nutation which rotates the celestial frame from the true equinox of date to the mean equinox of date; and $\mathbf{P}$ is about Precession which rotates the celestial frame from the mean equinox of date to the mean equinox of J2000.0.

1.4   References and Further Reading

There are a few classical text book for spacecraft dynamics. The books by Bate [1], Kaplan [9], and Wertz [10] are good for beginners. Battin [7] is suitable for advanced readers. Kaula [3] focused on orbital perturbation. Szebehely [2] was the first author to discuss three body problem comprehensively. Goodwin [8] and Schaub [5] discuss space dynamics from the view of dynamics. Chobotov [4] and Vallado [6] add many modern applications. Some astrodynamics series also appear. Scheeres [11] focuses on asteroid exploration, Alfriend [12] mainly discusses formation flying, etc.

1.5   Summary and Keywords

This chapter provides a brief discussion on Kepler's three laws and Newton's three laws as well as some basic definitions such as GNC, a few important coordinates, time systems and related reference motions.

Keywords: orbital mechanics, attitude dynamics, guidance navigation and control (GNC), Kepler's Law, Newton's Law, universal gravity, Earth centered inertial frame (ECI), Earth centered Earth fixed frame (ECEF), heliocentric ecliptic frame, radial transverse normal system, topcentric frame, solar time, sidereal time, universal time (UT).

Problems

1.1 About orbit and attitude, answer the following questions:

(a) What is the major difference between orbital mechanics and attitude dynamics?
(b) What is the major reason that orbit and attitude are decoupled in most cases?
(c) Can you give a natural example where orbit and attitude are coupled?

1.2 What is GNC? Explain this in simple words.

1.3 Give some examples of sub-systems in a spacecraft.

1.4 Name some examples of orbit which are LEO, MEO, HEO, GEO in each case?

1.5 Show the difference between a Geostationary orbit and a Geosynchronous orbit? What are their periods?

1.6 List some coordinates which are inertial and noninertial in each case?

1.7 Answer the following regarding sidereal and solar time:

(a) Which is longer: a sidereal or solar day? Why?
(b) What causes an apparent solar day to be different from a mean solar day?
(c) What was the local sidereal time (in degrees) of Greenwich, England at 04:48 hours (local) on January 3, 2004 if α_{g0} for January 1, 2004 0-hr UT is 100.229°.
(d) What was the local sidereal time (in degrees) for Beijing at that same moment?
(e) Does it make a difference whether New York is on standard or daylight savings time?
(f) How long is one Lunar day? Is it one sidereal month or Lunar month? Sometimes one Lunar month is also called a synodic month.

References

[1] Bate, R., Mueller, D., and White, PJ. (1971) *Fundamentals of Astrodynamics*. New York: Dover Publications, Inc.
[2] Szebehely V. (1967) *Theory of Orbits*. New York: Academic Press Inc.
[3] Kaula, W.M. (1966), *Theory of Satellite Geodesy*. Boston: Blaisdell, p. 6.
[4] Chobotov, V. (2002) *Orbital Mechanics*, 3rd edn. AIAA Education Series, Inc. Reston, VA: AIAA.
[5] Schaub, H. and Junkins, J. (2003) *Analytical Mechanics of Space Systems*. AIAA Education Series, Reston, VA: AIAA.
[6] Vallado, D. (1997) *Fundamentals of Astrodynamics and Applications*. El Segundo, CA: Microcosm Press.
[7] Battin R.H. and Vaughan, R.J., (1987) *An Introduction to the Mathematics and Methods of Astrodynamics*. New York.
[8] Greenwood, D.T., (1988) *Principles of Dynamics*, 2nd edn. New York: Prentice-Hall.
[9] Kaplan. M., (1976) *Modern Spacecraft Dynamics and Control*. New York: John Wiley & Sons,
[10] Wertz J.R. (1985) *Spacecraft Attitude Determination and Control*. Boston: Reidel.
[11] Scheeres D.J., (2012) *Orbital Motion in Strongly Perturbed Environments: Applications to Asteroid, Comet and Planetary Satellite Orbiters*. London: Springer-Praxis.
[12] Alfriend, K., Vadali, S., and Gurfil, P., *et al.* (2010) *Spacecraft Formation Flying*. Oxford: Elsevier Astrodynamics Series.

2

Keplerian Motion

Keplerian motion means the relative conic motion of two bodies, which is fundamental in orbital mechanics. For the Keplerian problem in a nonlinear, inverse-square gravitational force field, much of the analysis can be performed in closed form in terms of a simple conic orbit. Only very few cases, such as Kepler's equation's calculation, need a numerical iterative solution, which is a little complicated. But almost all space applications are related to knowledge of Keplerian motion.

Based on kinematics, Newton's Second Law and Newton's Gravitational Law, equations of a particle's motion in the plane in vector and polar coordinate form can be obtained.

2.1 N-body System

Assume the gravitational forces are $\mathbf{F}_{g_1}, \mathbf{F}_{g_2}, ..., \mathbf{F}_{g_n}$. External forces are thrust, drag, solar, and other forces such as nonspherical bodies. The total forces are

$$\mathbf{F}_t = \sum \mathbf{F}_{g_i} + \mathbf{F}_t + \mathbf{F}_d + \mathbf{F}_s + \text{other forces} \tag{2.1}$$

See Figure 2.1. Let i denote a spacecraft of interest whose gravitational force is

$$\mathbf{F}_{g_i} = -\frac{Gm_i m_j}{r_{ji}^3} \cdot \mathbf{r}_{ji}, \quad \text{where} \quad \mathbf{r}_{ji} = \mathbf{r}_i - \mathbf{r}_j, \quad j \neq i \tag{2.2}$$

$$\mathbf{F}_t = \sum_{\substack{j=1 \\ j \neq i}}^{n} \mathbf{F}_{g_j} + \mathbf{F}_o. \tag{2.3}$$

Newton's Second Law (when mass changes very slowly):

$$\mathbf{F} = \frac{\mathrm{d}}{\mathrm{d}t}(m\mathbf{v}) \tag{2.4}$$

$$= \dot{m} \cdot \mathbf{v} + m \cdot \dot{\mathbf{v}} = \dot{m} \cdot \dot{\mathbf{r}} + m \cdot \ddot{\mathbf{r}} \tag{2.5}$$

$$\approx m \cdot \mathbf{a}. \tag{2.6}$$

Fundamental Spacecraft Dynamics and Control, First Edition. Weiduo Hu.
© 2015 John Wiley & Sons (Asia) Pte Ltd. Published 2015 by John Wiley & Sons (Asia) Pte Ltd.

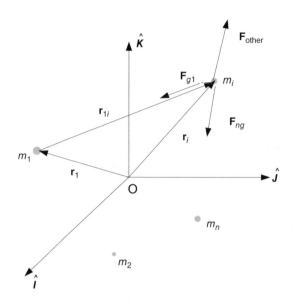

Figure 2.1   The forces to which a spacecraft is subjected

In general,

$$\dot{m}_i \cdot \mathbf{r}_i + m_i \cdot \ddot{\mathbf{r}}_i - \mathbf{F}_t = 0 \tag{2.7}$$

has no solution. For a pure N-body system, such as the Solar System and the Earth–Moon system, when only mutual gravity force is included

$$m_i \ddot{r}_i = G \sum_{\substack{j=1 \\ j \neq i}}^{n} \frac{m_i m_j}{r_{ij}} \mathbf{r}_{ij}. \tag{2.8}$$

The center of the masses,

$$\mathbf{r}_{CM} = \frac{1}{M} \sum_{i=1}^{n} m_i \mathbf{r}_i. \tag{2.9}$$

where $M = \sum_{i=1}^{n} m_i$. Total angular momentum

$$\mathbf{H} = \sum_{i=1}^{n} m_i \mathbf{r}_i \times \dot{\mathbf{r}}_i. \tag{2.10}$$

The kinetic energy of the system

$$T = \frac{1}{2} \sum_{i=1}^{n} m_i \dot{\mathbf{r}}_i^T \dot{\mathbf{r}}_i. \tag{2.11}$$

The gravitational potential energy of the collection of particles is

$$V = -\frac{G}{2} \sum_{i=1}^{n} \sum_{\substack{j=1 \\ j \neq i}}^{n} \frac{m_i m_j}{r_{ij}}. \tag{2.12}$$

2.2 The Two-body Problem

2.2.1 Geometry for Two Bodies in an Inertial Reference Frame

$\hat{I}\hat{J}\hat{K}$ is assumed to be an inertial coordinate system. $\hat{X}\hat{Y}\hat{Z}$ is displaced from $\hat{I}\hat{J}\hat{K}$ and normally rotates or accelerates with respect to $\hat{I}\hat{J}\hat{K}$. See Figure 1.3.

It is assumed that the mass of spacecraft is constant (unpowered flight), i.e. $\dot{m} = 0$ in Equation (2.6), all bodies are spherically symmetric, with mass concentrated at the center, and all external and internal forces other than gravity between Earth and spacecraft being zero.

2.2.2 Relative Motion of Two Bodies

See Figure 2.2; m_2 is a spacecraft's mass, m_1 is the Earth's mass, and $\mathbf{r}_{CM}$ is the center of the two masses. Based on Newton's Second Law,

$$m_2 \cdot \ddot{\mathbf{r}}_2 = -\frac{Gm_1 m_2}{r^2} \cdot \frac{\mathbf{r}}{r} \quad \text{and} \quad m_1 \cdot \ddot{\mathbf{r}}_1 = \frac{Gm_1 m_2}{r^2} \cdot \frac{\mathbf{r}}{r}, \tag{2.13}$$

where $\mathbf{r} = \mathbf{r}_2 - \mathbf{r}_1$. Then

$$\ddot{\mathbf{r}} = -\frac{G(m_1 + m_2)}{r^3} \cdot \mathbf{r}. \tag{2.14}$$

This is the basic equation of motion (EOM) of two-bodies, which is also called the equation of restricted relative motion for two bodies, where $\mu = G(m_1 + m_2)$ is the gravitational constant for two bodies.

Since a spacecraft is much smaller than the Earth, $m_1 + m_2 \approx m_1$, the two-body equation of relative motion becomes:

$$\ddot{\mathbf{r}} + \frac{\mu}{r^3} \cdot \mathbf{r} = 0, \tag{2.15}$$

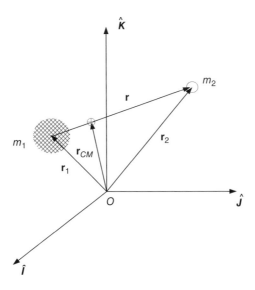

Figure 2.2 The relative motion of two bodies

where ($\mu \approx Gm_1 = 3.986 \times 10^5$ km^3 s^{-2}). Note that we have effectively moved the origin of our coordinate system to m_1 which is not inertial, but our "modified" form of Newton's Law accounts for this.

Note also that we assumed:

1. the mass of the satellite is negligible,
2. it is in an inertial coordinate system,
3. the bodies are spherically symmetric with uniform density,
4. no other forces act except for the gravitational attraction of the two bodies.

Let us give a rough justification of item (3) of the above assumptions.

When planets move around the Sun, the distance between the planets and the Sun is much larger than the radius of the planets. At this scenario, it is reasonable to assume that both the Sun and the planet are mass points. But when spacecraft is near to the Earth's surface, the altitude of spacecraft, for example 200 km, is much smaller than the Earth's radius, $R_E = 6378$ km, then the dimension of the Earth is significantly larger than infinitesimal. This apparent deficiency of the simple two-body model can be alleviated by analyzing the gravitation force due to a finite uniform sphere of mass.

This conclusion was first proofed by Newton. See Figure 2.3. Let us consider a hollow thin spherical shell and a particle m_2 is located outside the shell. To obtain the potential of the shell to the mass point, first calculate the potential of a band (or annulus) on the shell whose distance ρ to the particle is the same.

$$dV = -\frac{Gm_2 dm}{\rho} = -\frac{2\pi R_E^2 G\sigma m_2 \sin \eta d\eta}{\rho}. \tag{2.16}$$

Then integrate the annulus; its potential of the shell can be found as follows:

$$V_s = -2\pi R_E^2 G\sigma m_2 \int_0^\pi \frac{\sin \eta}{\rho} d\eta \tag{2.17}$$

$$= -\frac{2\pi R_E^2 G\sigma m_2}{R_E r} \int_{r-R}^{r+R} d\rho \tag{2.18}$$

$$= -\frac{4\pi R_E^2 G\sigma m_2}{r} = -Gm_s m_2 / r. \tag{2.19}$$

where m_2 is the particle's mass; r is the distance from the particle to the center of the shell; and m_s is the total mass of the shell. In the derivation, variable transform is performed by

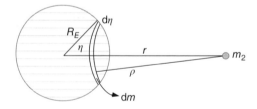

Figure 2.3   A hollow thin spherical shell

$2\rho d\rho = 2rR_E \sin \eta d\eta$ as shown in Figure 2.3, $\rho^2 = R_E^2 + r^2 - 2rR_E \cos \eta$, while a solid sphere is composed of concentric spherical shells. It is clear that any particle outside a sphere is the same as a mass point at the center of the sphere with the same mass of the solid sphere.

2.2.3 Constants of the Motion

Constants here mean integrals of motions for general two bodies motion and relative motion of two bodies.

$$a(t) = a_0 \tag{2.20}$$

$$v(t) = a_0 t + v_0 \tag{2.21}$$

$$r(t) = \frac{1}{2}a_0 t^2 + v_0 t + r_0. \tag{2.22}$$

Here we have two integrals v_0, r_0. For the general two-body problem, we need $3 \times 2 \times 2 = 12$ integrals as they are two vectors in second order differential equations.

Let's discuss the center of mass integral first. See Figure 2.2. Note that:

$$\mathbf{r}_{CM} = \frac{m_1 \mathbf{r}_1 + m_2 \mathbf{r}_2}{m_1 + m_2}. \tag{2.23}$$

Differentiate twice and by Equation (2.13)

$$\ddot{\mathbf{r}}_{CM} = \frac{m_1 \ddot{\mathbf{r}}_1 + m_2 \ddot{\mathbf{r}}_2}{m_1 + m_2} = 0. \tag{2.24}$$

Integrating the above equation twice

$$\mathbf{r}_{CM} = \mathbf{c}_1 t + \mathbf{c}_2. \tag{2.25}$$

Hence, the center of mass (COM) is moving in a straight line or motionless and is an integral point that we can use for our reference frame. $\mathbf{c}_1 = \mathbf{v}_0, \mathbf{c}_2 = \mathbf{r}_0$

2.2.3.1 Vis-Visa Equation

For the Polar coordinates, see Figure 2.4.

$$\hat{e}_r = \cos \theta \hat{P} + \sin \theta \hat{Q} \tag{2.26}$$

$$\hat{e}_t = -\sin \theta \hat{P} + \cos \theta \hat{Q} \tag{2.27}$$

$$\dot{\hat{e}}_r = (-\sin \theta \hat{P} + \cos \theta \hat{Q})\dot{\theta} = \dot{\theta}\hat{e}_t \tag{2.28}$$

$$\dot{\hat{e}}_t = (-\cos \theta \hat{P} - \sin \theta \hat{Q})\dot{\theta} = -\dot{\theta}\hat{e}_r. \tag{2.29}$$

Then

$$\mathbf{r} = r \cos \theta \hat{P} + r \sin \theta \hat{Q} = r\hat{e}_r \tag{2.30}$$

$$\dot{\mathbf{r}} = (\dot{r} \cos \theta - r \sin \theta \dot{\theta})\hat{P} + (\dot{r} \sin \theta + r \cos \theta \dot{\theta})\hat{Q} = \dot{r}\hat{e}_r + r\dot{\theta}\hat{e}_t \tag{2.31}$$

$$\ddot{\mathbf{r}} = \ddot{r}\hat{e}_r + \dot{r}\dot{\hat{e}}_r + \dot{r}\dot{\theta}\hat{e}_t + r\ddot{\theta}\hat{e}_t + r\dot{\theta}\dot{\hat{e}}_t \tag{2.32}$$

$$= (\ddot{r} - r\dot{\theta}^2)\hat{e}_r + (r\ddot{\theta} + 2\dot{r}\dot{\theta})\hat{e}_t. \tag{2.33}$$

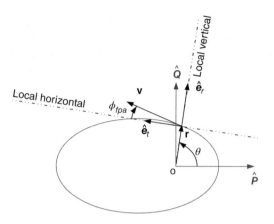

Figure 2.4   Polar coordinate and flight path angle

Then to get the energy integral, $\dot{\mathbf{r}}$ is multiplied into the two-body EOM (Equation (2.15)). As $\mathbf{r} = r\hat{e}_r$, $\dot{\mathbf{r}} = \dot{r}\hat{e}_r + r\dot{\theta}\hat{e}_t$, so

$$\mathbf{r} \cdot \dot{\mathbf{r}} = r\dot{r}. \tag{2.34}$$

Note

$$\mathbf{r} \times \dot{\mathbf{r}} = r^2\dot{\theta}\hat{e}_r \times \hat{e}_\theta = \mathbf{h} = h\hat{e}_n, \tag{2.35}$$

so

$$\dot{\mathbf{r}} \cdot \ddot{\mathbf{r}} + \dot{\mathbf{r}} \cdot \frac{\mu}{r^3}\mathbf{r} = 0 \tag{2.36}$$

$$\dot{\mathbf{r}} \cdot \ddot{\mathbf{r}} + \frac{\mu\dot{r}}{r^2} = 0. \tag{2.37}$$

Note that

$$\frac{\mathrm{d}}{\mathrm{d}t}\left(\frac{\dot{\mathbf{r}} \cdot \dot{\mathbf{r}}}{2}\right) = \dot{\mathbf{r}} \cdot \ddot{\mathbf{r}} \tag{2.38}$$

$$\frac{\mathrm{d}}{\mathrm{d}t}\left(\frac{\mu}{r}\right) = -\frac{\mu\dot{r}}{r^2}. \tag{2.39}$$

Thus

$$\frac{\mathrm{d}}{\mathrm{d}t}\left(\frac{\dot{\mathbf{r}} \cdot \dot{\mathbf{r}}}{2}\right) - \frac{\mathrm{d}}{\mathrm{d}t}\left(\frac{\mu}{r}\right) = 0. \tag{2.40}$$

So orbital motion occurs in a conservative gravitational field. Specific Mechanical Energy $\mathcal{E}$ is the sum of kinetic energy KE and potential energy PE which gives one integral about semi-major axis a:

$$E = KE + PE = \frac{1}{2}mv^2 - \frac{\mu m}{r}. \tag{2.41}$$

Let $\mathcal{E} = E/m$. Note that PE is defined as 0 when r approaches to infinity.

$$\Rightarrow \mathcal{E} = \frac{1}{2}v^2 - \frac{\mu}{r}, \tag{2.42}$$

where $\mathcal{E}$ is constant along an orbital path, known as energy integral. And $\mathcal{E} = -\mu/(2a)$ which we will show later. This equation is sometimes called the Vis-Viva Equation, where $v^2/2$ is relative kinetic energy per unit mass. $-\mu/r$ is relative potential energy per unit mass relative to one of the bodies. The units in this equation are: $\mathcal{E} = \text{km}^2/\text{s}^2$, $v = \text{km/s}$, $\mu = \text{km}^3/\text{s}^2$, $r = \text{km}$, $\mu = 3.986 \times 10^5 \text{ km}^3/\text{s}^2$ for the Earth. A useful expression is:

$$v = \sqrt{2(\mu/r + \mathcal{E})} = \sqrt{\dot{r}^2 + r^2\dot{\theta}^2}. \tag{2.43}$$

2.2.3.2 Angular Momentum Constant

For the angular momentum

$$\mathbf{H} = \mathbf{r} \times (m \cdot \mathbf{v}).$$

Specific angular momentum gives three integral ($\mathbf{h}$) or (Ω, i, h)
Let

$$\mathbf{h} = \frac{\mathbf{H}}{m} \Rightarrow \mathbf{h} = \mathbf{r} \times \mathbf{v}.$$

Note:

$$\frac{d}{dt}(\mathbf{r} \times \mathbf{v}) = \dot{\mathbf{r}} \times \dot{\mathbf{r}} + \mathbf{r} \times \ddot{\mathbf{r}} = 0 \tag{2.44}$$

$$= \dot{\mathbf{r}} \times \dot{\mathbf{r}} + \mathbf{r} \times (-\frac{\mu}{r^3}\mathbf{r}) = 0 \tag{2.45}$$

$$\Rightarrow \mathbf{h} = \text{constant} = r^2\dot{\theta}\hat{e}_n. \tag{2.46}$$

$\mathbf{h}$ is perpendicular to the orbital plane, constant in direction; thus $\mathbf{r}, \mathbf{v}$ lie in the same plane for all time.

Units of these variables are km^2/s for $\mathbf{h}$; km for $\mathbf{r}$; km/s for $\mathbf{v}$.

So, we have $3 + 3 + 3 + 1 = 10$ constants of the motion for $\mathbf{c}_1, \mathbf{c}_2, \mathbf{c}_3 = \mathbf{h}, c_4 = \mathcal{E}$ each.

A useful expression is:

$$h = r \cdot v \cdot \cos\phi_{fpa}, \tag{2.47}$$

where ϕ_{fpa} is the flight path angle; this is the angle from the local horizontal to the velocity vector. See Figure 2.4 and Figure 2.5. The angle is positive when the orbit is from periapsis to apoapsis and negative from apoapsis to periapsis. It is zero at periapsis and apoapsis. It is

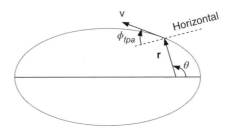

Figure 2.5 Flight path angle

always zero for circular orbits. Equations (2.48) and (2.49) can be derived later by Equations (2.98) and (2.99):

$$\cos \phi_{fpa} = \frac{h}{r \cdot v} = \frac{1 + e \cos \theta}{\sqrt{1 + 2e \cos \theta + e^2}} \tag{2.48}$$

$$\tan \phi_{fpa} = \frac{\dot{r}}{r\dot{\theta}} = \frac{e \sin \theta}{1 + e \cos \theta}. \tag{2.49}$$

Given an Earth satellite:

$$\mathbf{r} = 12\,500\hat{I} + 19\,000\hat{J} + 32\,000\hat{K} \text{ (km)}$$

$$\mathbf{v} = 8\hat{I} + 16\hat{J} - 3\hat{K} \text{ (km/s)}$$

$$r = 39\,258.76 \text{ km and } v = 18.14 \text{ km/s}$$

$$\mathcal{E} = \frac{1}{2}v^2 - \frac{\mu}{r} = (0.5)(18.14)^2 - (3.986 \times 10^5)/(39\,258.76)$$

$$= 154.38 \text{ (km}^2/\text{s}^2) > 0$$

$$\mathbf{h} = \mathbf{r} \times \mathbf{v} = \begin{vmatrix} \hat{I} & \hat{J} & \hat{K} \\ 12\,500 & 19\,000 & 32\,000 \\ 8 & 16 & -3 \end{vmatrix} = -569\,000\hat{I} - 293\,500\hat{J} + 48\,000\hat{K}$$

$$h = 642\,033.68 \text{ km}^2/\text{s}$$

$$\cos \phi_{fpa} = \frac{h}{r \cdot v} = 0.9015, \ \mathbf{r} \cdot \mathbf{v} > 0 \Rightarrow \phi_{fpa} = 25.64°.$$

2.2.3.3 Canonical Units

See Figure 2.6.

The reference orbit is the Earth's circular orbit around the Sun (AU), the left-hand figure in Figure 2.6, in the heliocentric frame. Generally the circular orbit near Earth's surface, is referenced, whose subscript is DU; see the right-hand figure in Figure 2.6, in the geocentric

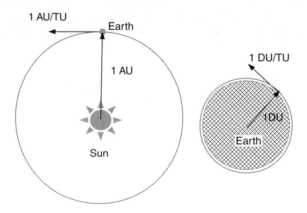

Figure 2.6 Canonical units

frame, while TU is normalized time unit. 1 TU = 7.905/1 DU.

In the geocentric frame:

$$1 \text{ DU}(R_\oplus) = 6378.145 \text{ km}, 1 \text{ TU} = 806.8\,118\,744 \text{ sec},$$

$$\mu_\oplus = 1 \text{ DU}^3/\text{TU}^2 = 3.986\,012 \times 10^5 \text{ km}^3/\text{s}^2, 1 \text{ SU}_\oplus = 1 \text{ DU/TU} = 7.905 \text{ km/s}.$$

In the heliocentric frame:

$$1 \text{ AU (or 1DU}(R_\odot) = 1.4\,959\,965 \times 10^8 \text{ km}, 1 \text{ TU} = 5.0\,226\,757 \times 10^6 \text{ sec},$$

$$\mu_\odot = 1 \text{ AU}^3/\text{TU}^2 = 1.3\,271\,544 \times 10^{11} \text{ km}^3/\text{s}^2, 1 \text{ SU}_\odot = 1 \text{ DU/TU} = 29.7859 \text{ km/s}.$$

Most common used units:

Distance: Kilometer, mile and nautical mile: 1 mile = 1.609 km, 1 nm = 1.852 km
Time: Seconds.

2.2.4 Orbital Path

Some facts are reviewed here. The two-body problem has only a circle, ellipse, parabola and hyperbola as possible orbital paths which are called conic sections as a whole; see Figure 2.12 for examples. The focus of a conic is located at the center of a central body (the Earth). Mechanical energy remains constant on an orbital path. Orbital motion occurs in a plane fixed in inertial space. Angular momentum remains constant along an orbital path.

2.2.4.1 Proof of Kepler's First Law

We have developed: $\ddot{\mathbf{r}} = -\mu/r^3 \cdot \mathbf{r}$, that is, Equation (2.15). But it can not be solved for $\mathbf{r}$ in closed form. Given $\mathbf{r}(0)$, we can solve it numerically using RK4.

After vector-differential math, for which we will provide a proof later in this section, we can have

$$r = \frac{h^2/\mu}{1 + (B/\mu)\cos\theta}, \tag{2.50}$$

where $\mathbf{B}$ is a constant vector and θ is the angle between $\mathbf{B}$ and $\mathbf{r}$ where $\mathbf{B} = \mu\mathbf{e}$, see Equation (2.93).

For analytical geometry:

$$r = \frac{p}{1 + e\cos\theta}. \tag{2.51}$$

This is the Polar Equation of a conic section, where p is the semi-latus rectum, and e is eccentricity. When $e = 0$ the orbit is a circle; when $0 < e < 1$ the orbit is an ellipse; when $e = 1$ the orbit is a parabola; and when $e > 1$, it is a hyperbola.

Actually this is Kepler's First Law which can be proofed using Newton's Law. For this problem

$$-\frac{\mu}{r^3}\mathbf{r} = -\frac{\mu}{r^2}\hat{e}_r \tag{2.52}$$

$$(\ddot{r} - r\dot{\theta}^2)\hat{e}_r + (r\ddot{\theta} + 2\dot{r}\dot{\theta})\hat{e}_t = -\frac{\mu}{r^2}\hat{e}_r. \tag{2.53}$$

Thus, in polar coordinates:

$$\ddot{r} - r\dot{\theta}^2 = -\frac{\mu}{r^2} \tag{2.54}$$

$$r\ddot{\theta} + 2\dot{r}\dot{\theta} = 0. \tag{2.55}$$

Equation (2.55) is equivalent to

$$\frac{\mathrm{d}}{\mathrm{d}t}\left(r^2\dot{\theta}\right) = 0, \tag{2.56}$$

which is our angular momentum integral $h = r^2\dot{\theta}$. We will see that this is actually Kepler's Second law. See Equation (2.70).

To solve $\ddot{r} - r\dot{\theta}^2 = -\mu/r^2$, change independent variable t to θ and change dependent variable r to $1/u$. Then

$$\frac{\mathrm{d}r}{\mathrm{d}t} = \frac{\mathrm{d}r}{\mathrm{d}\theta}\frac{\mathrm{d}\theta}{\mathrm{d}t}, \tag{2.57}$$

or in general

$$\frac{\mathrm{d}}{\mathrm{d}t} = \dot{\theta}\frac{\mathrm{d}}{\mathrm{d}\theta}. \tag{2.58}$$

But $r^2\dot{\theta} = h$, then

$$\frac{\mathrm{d}}{\mathrm{d}t} = \frac{h}{r^2}\frac{\mathrm{d}}{\mathrm{d}\theta} = hu^2\frac{\mathrm{d}}{\mathrm{d}\theta}; \tag{2.59}$$

hence,

$$\dot{r} = \frac{\mathrm{d}r}{\mathrm{d}t} = hu^2\frac{\mathrm{d}}{\mathrm{d}\theta}\left(\frac{1}{u}\right) = -\frac{hu^2}{u^2}\frac{\mathrm{d}u}{\mathrm{d}\theta} = -h\frac{\mathrm{d}u}{\mathrm{d}\theta} \tag{2.60}$$

$$\ddot{r} = \frac{\mathrm{d}}{\mathrm{d}t}\left(\frac{\mathrm{d}r}{\mathrm{d}t}\right) = hu^2\frac{\mathrm{d}}{\mathrm{d}\theta}\left(-h\frac{\mathrm{d}u}{\mathrm{d}\theta}\right) = -h^2u^2\frac{\mathrm{d}^2u}{\mathrm{d}\theta^2}; \tag{2.61}$$

also $r\dot{\theta}^2 = r(h/r^2)^2 = h^2/r^3 = h^2u^3$. So $\ddot{r} - r\dot{\theta}^2 = -\mu/r^2$ becomes

$$-h^2u^2\frac{\mathrm{d}^2u}{\mathrm{d}\theta^2} - h^2u^3 = -\mu u^2 \tag{2.62}$$

$$\Rightarrow \frac{\mathrm{d}^2u}{\mathrm{d}\theta^2} + u = \frac{\mu}{h^2} = \text{constant}, \tag{2.63}$$

which has the form of an undamped spring-mass (harmonic oscillator) system with with a constant forcing term.

The general solution to this equation is:

$$u = \frac{\mu}{h^2} + A\cos(u_t - \omega), \tag{2.64}$$

where A and ω are two more constants of motion.

Substituting $u = 1/r$

$$r = \frac{1}{\mu/h^2 + A\cos(u_t - \omega)} = \frac{h^2/\mu}{1 + (Ah^2/\mu)\cos(u_t - \omega)}. \tag{2.65}$$

Let $\theta = u_t - \omega$ as true anomaly, ω as argument of periapse, $p = h^2/\mu$ as semi-latus rectum or semi-parameter, and $e = Ah^2/\mu$ eccentricity where $e = B/\mu$ in Equation (2.66). Thus

$$r = \frac{p}{1 + e\cos\theta} = \frac{a(1 - e^2)}{1 + e\cos\theta}. \tag{2.66}$$

This is exactly the equation of a conic section. This is also sometimes called the "trajectory equation." At the same time, it proofs and extends Kepler's First Law from ellipses to parabolas and hyperbolas.

Conic sections are formed by slicing a cone with a plane. When the plane is parallel to the base, a circle results. Planes which are perpendicular to the base yield hyperbolas, and planes parallel to the outer surface yield parabolus. All other sections are ellipses, except for special cases in which the plane is on the surface (a line or rectilinear orbit) or only through the vertex (a point) (Vallado [3]).

2.2.4.2   Proof of Kepler's Second and Third Laws

Horizontal component of velocity. See Figure 2.7.

$$v \cdot \cos\phi_{fpa} = r \cdot \frac{d\theta}{dt}$$

$$h = rv\cos\phi_{fpa} = r^2\frac{d\theta}{dt}$$

$$\Rightarrow d\theta = \frac{h}{r^2}dt$$

$$dA = \frac{1}{2}r^2 d\theta = \frac{1}{2}r^2\frac{h}{r^2}dt = \frac{h}{2}dt$$

$$\Rightarrow dt = \frac{2}{h}dA.$$

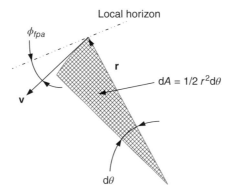

Figure 2.7   Differential element of area in an ellipse

Integrate for one period:

$$\int dt = TP, \quad \int dA = \pi ab \tag{2.67}$$

$$TP = \frac{2\pi ab}{h}, \quad b = \sqrt{ap}, \quad h = \sqrt{\mu p} \tag{2.68}$$

$$\Rightarrow TP = \frac{2\pi}{\sqrt{\mu}} \cdot a^{3/2}. \tag{2.69}$$

This is the so-called Kepler's Third Law.

Actually in the middle we already have

$$\frac{dA}{dt} = \frac{1}{2} r^2 \frac{d\theta}{dt} = \frac{h}{2}, \tag{2.70}$$

which is the proof of Kepler's Second Law.

2.2.4.3 Elliptic Orbit

Ellipse is a set of all points whose sum of distances from two fixed points is constant. See Figure 2.8, $r + r' = 2a$. For elliptical orbit $0 < e < 1$, $e = c/a$, $b = a\sqrt{1 - e^2}$, a is a semi-major axis, b is a semi-minor axis. Also, $e = \sqrt{a^2 - b^2}/a$. Sometimes flattening f is also used:

$$f = (a - b)/a \tag{2.71}$$

$$e^2 = 2f - f^2. \tag{2.72}$$

The semi-latus rectum is

$$p = b^2/a = a(1 - e^2) = h^2/\mu. \tag{2.73}$$

The names of periapsis and apoapsis for the Earth, the Sun and the Moon are perigee, apogee; perihelion, aphelion; periselenium, aposelenium respectively.

Let $2c$ = distance between foci ($c = 0$ for circle, $c = \infty$ for parabola). Let $2a$ = length of chord through foci (a = radius for circle, $a = \infty$ for parabola, $a < 0$ for hyperbola);

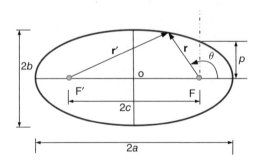

Figure 2.8   Ellipse geometry

e = eccentricity = c/a, $p = a(1 - e^2)$ and $r = p/(1 + e \cdot \cos \theta)$:

$$r_{min} = r_{perigee} = \frac{p}{1 + e \cdot \cos 0°} = \frac{p}{1 + e} = a(1 - e) \qquad (2.74)$$

$$r_{max} = r_{apogee} = \frac{p}{1 + e \cdot \cos 180°} = \frac{p}{1 - e} = a(1 + e) \qquad (2.75)$$

$$p = h^2/\mu, \quad \mu \text{ is the gravity constant,} \qquad (2.76)$$

$$h = r_p \cdot v_p = r_a \cdot v_a, \qquad (2.77)$$

$$\mathcal{E} - \frac{1}{2}v^2 \frac{\mu}{r} = -\frac{\mu}{2a}, \qquad (2.78)$$

$$e = \sqrt{1 + 2\mathcal{E} \cdot (h/\mu)^2}, \qquad (2.79)$$

$$p = a(1 - e^2), \qquad (2.80)$$

while $\mathcal{E} = (1/2)v^2 - \mu/r = -\mu/2a$ is the Vis-Viva Equation (2.42).

Given an elliptical orbit: $\mathcal{E} = -15(km/s)^2$, and $e = 0.2$, then

$$a = -\frac{\mu}{2\mathcal{E}} = \frac{3.986 \times 10^5 \text{ km}^3/s^2}{30 \text{ km}^2/s^2} = 13\,287 \text{ km}$$

$$p = a(1 - e^2) = 12\,756 \text{ km}$$

$$h = \sqrt{p \cdot \mu} = 71\,306 \text{ km}^2/s$$

$$r_a = c + a = 15\,944 \text{ km}$$

$$r_p = \frac{p}{1 + e} = 10\,630 \text{ km}$$

Ellipse summary: $r_a + r_p = 2a$, a = semi-major axis, $r_a - r_p = 2c$, b = semi-minor axis, $e = c/a = (r_a - r_p)/(r_a + r_p)$, $a^2 = b^2 + c^2$.

2.2.4.4   Circular Orbit

Circle: Set of all points whose distance from a fixed point is constant. Semi-major axis = a = radius = r. Period: $TP = 2\pi r^{3/2}/\sqrt{\mu}$. Orbital speed:

$$\mathcal{E} = \frac{1}{2}v^2 - \frac{\mu}{r} = -\frac{\mu}{2a} \Rightarrow \frac{1}{2}v^2 = \frac{\mu}{r} - \frac{\mu}{2r} = \frac{\mu}{2r}$$

$$\Rightarrow v_{cir} = \sqrt{\mu/r}. \qquad (2.81)$$

Note: v square and r are inversely proportional related here.

2.2.4.5   Parabolic Orbit

Parabola: Set of all points equidistant from a fixed point and a fixed line. Parabolic orbits are considered open because they don't repeat. The semi-major axis is infinite. The second focus is at infinite. See Figure 2.9. For a parabola, its eccentricity $e = 1 \Rightarrow r_p = p/2$, and $r_a = \infty$,

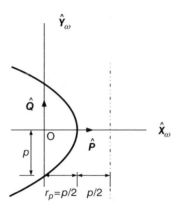

Figure 2.9   Geometry of a parabola

$\mathcal{E} = 0, r = p/(1 + \cos\theta)$. Though $e = \sqrt{1 + 2\mathcal{E}h^2/\mu^2} = 1$, this is not a sufficient condition for a parabola, for rectilinear orbit, $e = 1$ could be an ellipse or hyperbola. The sufficient condition for a parabola is $\mathcal{E} = 0$. $\mathcal{E} = 0 = v^2/2 - \mu/r, v = \sqrt{2\mu/r}$; as $r \to \infty$, $v \to 0$, and also as $\theta \to 180°, r \to \infty$.

Parabolic trajectory is a kind of escape trajectory. Escape speed is the speed necessary to overcome gravity effects and coast to infinity ($v_\infty = 0$, and $r_\infty = \infty$). The escape speed is v_{esc}

$$\mathcal{E} = \frac{1}{2}v^2 - \frac{\mu}{r} = \frac{1}{2}v_{esc}^2 - \frac{\mu}{r_{esc}} = \frac{1}{2}v_\infty^2 - \frac{\mu}{r_\infty} = 0$$

$$\Rightarrow v_{esc} = \sqrt{2\mu/r}. \tag{2.82}$$

See Table 2.1 for the planets' escape velocities and circular velocities near the planets' surfaces with some useful parameters.

Table 2.1   Escape velocities for the planets, the Moon and the Sun near their surfaces with some useful planetary parameters [3]. Comparison can be made with Table 3.1, Table 3.2 and Table 3.3

Planet	Mass $(10^{24}$ kg$)$	μ $(10^3$ km^3/s$^2)$	Equatorial radius $(10^3$ km$)$	v_{cir} km/s	v_{esc} km/s
Mercury	0.3302	22.032	2.436	3.01	4.26
Venus	4.8690	325.7	6.153	7.33	10.4
Earth	5.9742	398.6	6.378	7.9	11.2
Mars	0.6419	43.05	3.393	3.54	5.0
Jupiter	1898.8	126 800	71.491	42.1	60.0
Saturn	568.5	37 940	60.527	25.0	35.4
Uranus	86.625	5794	27.167	14.6	20.6
Neptune	102.78	6809	22.428	17.4	24.6
Pluto	0.015	0.9	1.148	0.885	1.25
Moon	0.0735	4.903	1.738	1.68	2.38
Sun	1.9891×10^6	1.327×10^8	6.9599×10^2	436.65	717.52

A parabolic orbit is a borderline case between an open hyperbolic orbit and a closed elliptic orbit.

2.2.4.6 Hyperbolic Orbits

Hyperbola: Set of all points whose difference of distances from two fixed points is constant, $r' - r = -2a$. For a hyperbolic orbit, the semi-major axis is negative, and the eccentricity is larger than 1.0. Notice the rapid departure from the Earth with a small eccentricity. The orbit approaches the asymptotes. Hyperbolas occur in pairs called branches. The second branch is a mirror image of the first branch about the conjugate axis. Some meteors and space probes' trajectories are hyperbolic. See Figure 2.10. Their trajectory has parabolic speed + hyperbolic excess speed (v_∞). See Figure 2.11. At burnout

$$\mathcal{E} = \frac{1}{2}v_{bo}^2 - \frac{\mu}{r_{bo}}, \quad bo = \text{burnout}.$$

At infinity

$$\mathcal{E} = \frac{1}{2}v_\infty^2 - \frac{\mu}{r_\infty}, \quad r_\infty = \infty.$$

Then

$$\Rightarrow \frac{1}{2}v_\infty^2 = \frac{1}{2}v_{bo}^2 - \frac{\mu}{r_{bo}} \quad \Rightarrow \quad v_\infty^2 = v_{bo}^2 - \frac{2\mu}{r_{bo}}.$$

So,

$$v_\infty^2 = v_{bo}^2 - v_{esc}^2. \tag{2.83}$$

$e > 1$, $\mathcal{E} > 0$, $\mathcal{E} = v^2/2 - \mu/r$, $v^2 = 2\mathcal{E} + 2\mu/r$, as $r \to \infty$, $v^2 \to 2\mathcal{E} = v_0^2 - 2\mu/r_0$ for any other point on the hyperbola. Or $v_\infty^2 = v^2/2 \to \mathcal{E}$, excess velocity at $r = \infty$, and since $\mathcal{E} = -\mu/(2a)$, $v_\infty^2 = -\mu/a$ is hyperbolic excess velocity. One important application is planetary approach. $p/r = 1 + e\cos\theta$, as $r \to \infty$, $e\cos\theta_\infty \to -1$, so for the true anomaly of the asymptotes,

$$\cos\theta = -\frac{1}{e}. \tag{2.84}$$

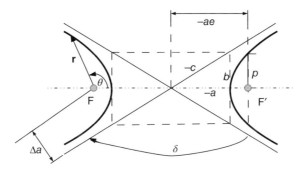

Figure 2.10 Hyperbolic geometry

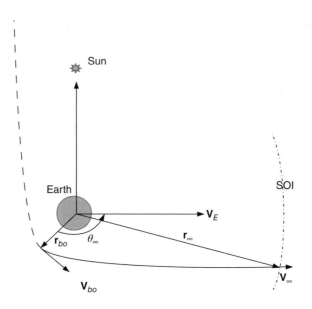

Figure 2.11   A hyperbolic escape trajectory

The turning angle is the deflection angle of v_∞, i.e. δ, $\theta_\infty = (\pi + \delta)/2$, $\cos\theta_\infty = -\sin(\delta/2)$

$$\sin\frac{\delta}{2} = \frac{1}{e}. \tag{2.85}$$

The smaller the e, the greater the turning angle. Aiming distance Δ_a is the distance between the focus and the asymptotes of the hyperbola. See Figure 2.10.

$$\Delta_a = \sin\theta_\infty(-ae) = -a\sqrt{e^2-1} = r_p\sqrt{\frac{e+1}{e-1}}. \tag{2.86}$$

2.2.4.7   Examples and Summary

Given: $r_p = 7000$ km and $r_a = 10\,000$ km
Eccentricity:

$$e = \frac{r_a - r_p}{r_a + r_p} = \frac{3000}{17\,000} = 0.1765.$$

Semi-major axis:

$$a = \frac{r_a + r_p}{2} = \frac{17\,000}{2} = 8500\,(\text{km}).$$

Altitude when $\theta = 90°$:

$$R = \frac{a(1-e^2)}{1+e\cdot\cos\theta} = 8235.3\,\text{km}, \quad Alt = R - R_{\text{earth}} = 1857.1\,(\text{km}).$$

Specific mechanical energy:

$$\mathcal{E} = -\frac{\mu}{2a} = -\frac{3.986 \times 10^5}{2(8500)} = -23.45 \, (\text{km/s})^2$$

Velocity when $\theta = 90°$

$$v = \sqrt{2\left(\frac{\mu}{r} + \mathcal{E}\right)} = \sqrt{2\left(\frac{3.986 \times 10^5}{8235.2} - 23.45\right)} = 7.065 \, (\text{km/s}).$$

Period:

$$TP = 2\pi\sqrt{a^3/\mu} = 7799 \text{ seconds} \approx 130 \text{ minutes.}$$

Figure 2.12 shows some typical examples of conics; the data for the orbits are listed in Table 2.2. A summary of this section is given in Table 2.3.

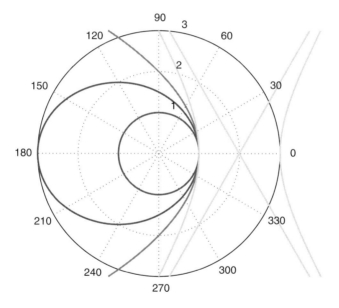

Figure 2.12   Examples of conic orbits whose data are listed in Table 2.2

Table 2.2   Examples of the conics in Figure 2.12

	Circle	Ellipse	Parabola	Hyperbola
r_p	1	1	1	1
e	0	0.5	1	2
p	1	1.5	2	3
a	1	2	∞	−1

Table 2.3 Comparison of the conics. As a special case when $e = 1$, the orbit could be a rectilinear orbit where $h = p = 0$, which might be an ellipse, a parabola, or a hyperbola

	a	$\mathcal{E}$	e
Circle	> 0	< 0	0
Ellipse	> 0	< 0	≤ 1
Parabola	∞	0	1
Hyperbola	< 0	> 0	≥ 1

2.3 Orbital Elements

By position and velocity vectors $\mathbf{r}, \mathbf{v}$, orbital elements $(a, e, i, \Omega, \omega, \theta)$ can be determined.

The Classic Orbital Elements (COEs), that is, Orbital Element Sets, are six quantities needed to describe orbit and spacecraft's position within an orbit.

For airplane, they are: Latitude, Longitude, Altitude (Range), Horizontal Velocity, Heading, Vertical Velocity (Range rate = velocity).

For satellite, they are Range (3-component vector), Velocity (3-component vector).

2.3.1 Kepler's COEs

See Figure 2.13 and Figure 2.14; see also Figure 2.8 with the following definition.

1. Orbit size: Semi-major axis, a
   Circle: a = radius, Ellipse: a = semi-major axis, Parabola: $a = \infty$, Hyperbola: $a < 0$.
2. Orbit shape: Eccentricity e (out-of-roundness)
   Circle: $e = 0$, Ellipse: $0 < e < 1$, Parabola: $e = 1$, Hyperbola: $e > 1$.
3. Orbit plane orientation: Inclination i (tilt of orbital plane wrt equatorial plane)
   For the Earth satellite:
   $i = 0$ or $180°$: equatorial orbit, $i = 90°$: polar orbit, $0 < i < 90°$: direct or prograde orbit, $90° < i < 180°$: indirect or retrograde orbit.
4. Orbit plane orientation: Right ascension (longitude) of the ascending node, Ω $[0, 360°]$.
   Ascending node: Point where orbit crosses equatorial plane from below to above.
   Descending node: point where orbit crosses equatorial plane from above to below.
   Right ascension: Angle between vernal equinox direction and the ascending node (measured eastward).
5. Orbit orientation within plane: Argument of perigee, ω $(0, 360°)$
   Angle along orbital path in direction of motion between ascending node and perigee.
6. Satellite location: Time of perigee passage, T or True Anomaly, θ $(0, 360°)$ which is the angle between perigee and position vector at time T called "epoch."

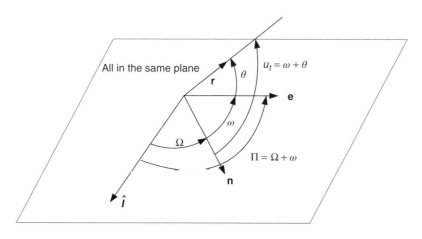

Figure 2.13   The elements in orbital plane

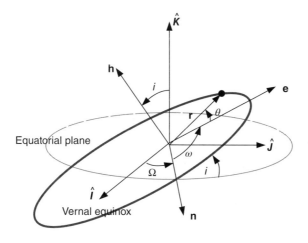

Figure 2.14   The elements in three dimensions in ECI frame

See Figure 2.13 and Figure 2.14 for the geometry of the elements. In the figures

e	eccentricity vector
i	inclination
Ω	longitude of ascending node
$\Pi = \Omega + \omega$	longitude of periapsis
ω	argument of periapsis
θ	true anomaly

$u_t = \omega + \theta$	argument of latitude
h	the angular momentum vector
n	the vector of ascending node.
r	the position vector of spacecraft.

2.3.2   Alternate Orbital Element Quantities

See also Figure 2.13 and figure 2.14; instead of argument of perigee (ω), longitude of perigee (Π) is sometimes used, especially when $i = 0$.

$$\Pi = \Omega + \omega. \tag{2.87}$$

In the previous example, $\Pi = 395.25° - 360° = 35.25°$
Instead of true anomaly θ, any of the following can be used: $\theta = \theta(t)$
Time T of perigee passage
Argument of latitude or sometimes called argument of orbit (u_t):

$$u_t = \omega + \theta(t). \tag{2.88}$$

True longitude (l_t):

$$l_t = \Omega + \omega + \theta(t) = \Pi + \theta(t) = \Omega + u_t. \tag{2.89}$$

Equinoctial elements are

$$a, e\cos(\Pi), e\sin(\Pi), \tan(i/2)\sin\Omega, \tan(i/2)\cos\Omega, \Pi + M. \tag{2.90}$$

Delaunay elements are

$$\sqrt{\mu a}, h = \sqrt{\mu a(1 - e^2)}, h\cos i, \Omega, \omega, M. \tag{2.91}$$

Delaunay elements are one kind of canonical elements. These different sets of elements can be used for different types of problems.

2.3.3   A Calculation Example

Radar has picked up an NFL (new foreign launch) spacecraft. Its initial orbit data is:

$$\mathbf{r} = 8250\hat{I} + 390\hat{J} + 6900\hat{K} \text{ (km)}$$

$$\mathbf{v} = -0.70\hat{I} + 6.6\hat{J} - 0.6\hat{K} \text{ (km/sec)}.$$

Develop the initial orbital element set 3 by fundamental vectors: **h**, **n**, **e**.
   Angular momentum $\mathbf{h} = \mathbf{r} \times \mathbf{v}$.

$$\mathbf{h} = \begin{vmatrix} \hat{I} & \hat{J} & \hat{K} \\ 8250 & 390 & 6900 \\ -0.70 & 6.6 & -0.60 \end{vmatrix} = -45\,774\hat{I} + 120\hat{J} + 54\,723\hat{K} \text{ (km}^2/\text{s)}.$$

r and **v** in orbital plane $\Rightarrow$ **h** orthogonal to orbital plane.

Ascending node vector,

$$
\mathbf{n} = \hat{K} \times \mathbf{h} =
\begin{vmatrix}
\hat{I} & \hat{J} & \hat{K} \\
0 & 0 & 1 \\
-45\,774 & 120 & 54\,723
\end{vmatrix}
= -120\hat{I} - 45\,774\hat{J}.
$$

As $\mathbf{n}$ is orthogonal to $\hat{K} \Rightarrow \mathbf{n}$ is in the equatorial plane and also $\mathbf{n}$ is orthogonal to $\mathbf{h} \Rightarrow \mathbf{n}$ is the orbital plane. So $\mathbf{n}$ vector is along the line of nodes in the direction of the ascending node.

Eccentricity vector, by crossing Equation (2.15) into $\mathbf{h}$, then integrating:

$$
\mathbf{e} = \frac{1}{\mu}\mathbf{v} \times \mathbf{h} - \frac{\mathbf{r}}{r} \tag{2.92}
$$

$$
= \frac{1}{\mu}\left[\left(v^2 - \frac{\mu}{r}\right)\mathbf{r} - (\mathbf{r} \cdot \mathbf{v})\,\mathbf{v}\right]. \tag{2.93}
$$

$\mathbf{e}$ from focus (center of earth) toward perigee. $|\mathbf{e}|$ = eccentricity:

$$
v = |\mathbf{v}| = \sqrt{(-0.70)^2 + (6.6)^2 + (-0.6)^2} = 6.664 \ (\text{km/s})
$$

$$
r = |\mathbf{r}| = \sqrt{(8250)^2 + (390)^2 + (6900)^2} = 10\,762.184 \ (\text{km})
$$

$$
\mathbf{e} = \frac{1}{398\,600}
\left\{
\begin{array}{l}
[(6.664)^2 - 398\,600/10\,762.184]\,(8250\hat{I} + 390\hat{J} + 6900\hat{K}) - \\
(8250\hat{I} + 390\hat{J} + 6900\hat{K}) \cdot (-0.70\hat{I} + 6.6\hat{J} - 0.6\hat{K}) \\
\hspace{4.5cm} \cdot(-0.70\hat{I} + 6.6\hat{J} - 0.6\hat{K})
\end{array}
\right\}
$$

$$
= 0.1397\hat{I} + 0.1288\hat{J} + 0.1166\hat{K}.
$$

Semi-major axis:

$$
\mathcal{E} = \frac{v^2}{2} - \frac{\mu}{r} = (0.5)(6.664)^2 - (398\,600)/(10\,762.184) = -14.8326 \ (\text{km}^2/\text{s}^2)
$$

$$
\mathcal{E} = -\frac{\mu}{2a} \Rightarrow a = -\frac{\mu}{2\mathcal{E}} = 13\,436.62 \ \text{km}.
$$

Eccentricity: $e = |\mathbf{e}| = 0.2229 \to 0.22$

Inclination:

$$
\mathbf{A} \cdot \mathbf{B} = AB\cos\theta
$$

$$
\Rightarrow i = \cos^{-1}\left[\frac{\hat{K} \cdot \mathbf{h}}{Kh}\right]
$$

$$
\hat{K} \cdot \mathbf{h} = 54\,734, \quad K = 1, \quad h = 71\,343.39
$$

$$
i = \cos^{-1}(0.767) = 39.91° \text{ or } i = 360 - 39.91 = 320.09°
$$

$$
0 \le i \le 180 \Rightarrow i = 39.91°.
$$

Right ascension of ascending node:

$$\Omega = \cos^{-1}\left[\frac{\hat{I} \cdot \mathbf{n}}{In}\right], \quad \hat{I} \cdot \mathbf{n} = -120, \quad I = 1, \quad n_j = -45\,774.16$$

$$\Omega = \cos^{-1}(-0.00\,262) = 90.15° \text{ or } 269.85°$$

$n_j \geq 0$, then $0° \leq \Omega \leq 180°$, $n_j < 0$, then $180° < \Omega < 360°$, $n_j = -45\,774 \Rightarrow \Omega = 269.85°$.

Argument of perigee, ω

$$\omega = \cos^{-1}\left[\frac{\mathbf{n} \cdot \mathbf{e}}{ne}\right]$$

$$\mathbf{n} \cdot \mathbf{e} = -5912.455, \quad n = 45\,774.16, \quad e = 0.2229$$

$$\omega = \cos^{-1}(-0.5795) = 125.4° \text{ or } 234.6°.$$

If $e_k \geq 0$, then $0° \leq \omega \leq 180°$; If $e_k < 0$, then $180° < \omega < 360°$

$$e_k = +0.1166 \Rightarrow \omega = 125.4°.$$

True anomaly, θ

$$\theta = \cos^{-1}\left[\frac{\mathbf{e} \cdot \mathbf{r}}{er}\right],$$

$$\mathbf{e} \cdot \mathbf{r} = 2007.297, \quad e = 0.2229, \quad r = 10\,762.184$$

$$\theta = \cos^{-1}(0.8368) = 33.2° \text{ or } 326.8°.$$

If $\mathbf{r} \cdot \mathbf{v} \geq 0$, then $0 \leq \theta \leq 180°$; If $\mathbf{r} \cdot \mathbf{v} < 0$, then $180° < \theta < 360°$. So,

$$\mathbf{r} \cdot \mathbf{v} = -7341 \Rightarrow \theta = 326.8°.$$

Here is the summary:

Given: $\mathbf{r} = 8250\hat{I} + 390\hat{J} + 6900\hat{K}$, $\mathbf{v} = -0.70\hat{I} + 6.6\hat{J} - 0.60\hat{K}$.
Determined: $a = 13\,436.62$ km, $e = 0.22$, $i = 39.91°$, $\Omega = 269.85°$, $\omega = 125.4°$, $\theta = 326.8°$.

Figure 2.15 visualizes the orbit.

2.4 Coordinate Transformations

A vector may be expressed in any coordinate frame. It is necessary to know how to transform among coordinate frames.

Coordinate transformation means to change the basis of a vector, while both its magnitude and direction remain the same. And also what it represents remains the same. The transforms use rotation (rigid body). Positive rotation is with right-hand rule: thumb + curl fingers.

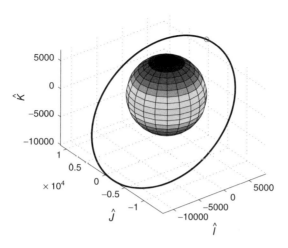

Figure 2.15   The 3D simulation of this orbit. The point "o" is the initial position

2.4.1   Rotation

x-axis ($\mathbf{R}_1$)

$$
\begin{bmatrix} x' \\ y' \\ z' \end{bmatrix} = \begin{bmatrix} 1 & 0 & 0 \\ 0 & \cos\alpha & \sin\alpha \\ 0 & -\sin\alpha & \cos\alpha \end{bmatrix} \begin{bmatrix} x \\ y \\ z \end{bmatrix}. \tag{2.94}
$$

An example of this rotation is the coordinate change from ECI to HECS; here $\alpha = \epsilon \sim 23.5°$. See Figure 2.16 and Figure 1.4.

y-axis ($\mathbf{R}_2$)

$$
\begin{bmatrix} x' \\ y' \\ z' \end{bmatrix} = \begin{bmatrix} \cos\beta & 0 & -\sin\beta \\ 0 & 1 & 0 \\ \sin\beta & 0 & \cos\beta \end{bmatrix} \begin{bmatrix} x \\ y \\ z \end{bmatrix}, \tag{2.95}
$$

z-axis ($\mathbf{R}_3$)

$$
\begin{bmatrix} x' \\ y' \\ z' \end{bmatrix} = \begin{bmatrix} \cos\gamma & \sin\gamma & 0 \\ -\sin\gamma & \cos\gamma & 0 \\ 0 & 0 & 1 \end{bmatrix} \begin{bmatrix} x \\ y \\ z \end{bmatrix}. \tag{2.96}
$$

An example of this rotation is the coordinate change from ECI to ECEF; here $\gamma = \alpha_g$ is the Greenwich sidereal time angle. See Figure 2.17 and Figure 1.3; also see Equation (2.117) Note: Order is important as matrix multiplication is not commutative. Euler angles are the angles through which one frame must be rotated from another frame. More information about Euler angle can be found in Chapter 8. Max of three rotations is sufficient to align any two frames.

2.4.2   From $\hat{P}\hat{Q}\hat{W}$ to $\hat{I}\hat{J}\hat{K}$

The transformation of the position and velocity in the perifocal system to the Earth center inertial frame is simple but very useful. The perifocal system here means, $\hat{x}$ i.e. $\hat{P}$ goes through

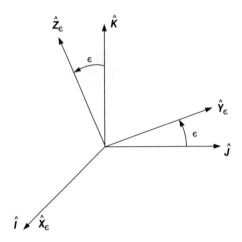

Figure 2.16   Earth-centered inertial and heliocentric ecliptic coordinates

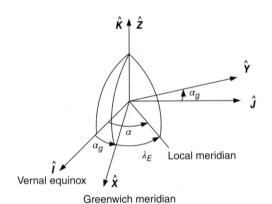

Figure 2.17   ECI and ECEF coordinate

periapse, $\hat{z}$ i.e. $\hat{W}$ in the direction of $\mathbf{h}$, $\hat{y}$ i.e. $\hat{Q}$ perpendicular to $\hat{x}$ and $\hat{z}$ in the orbital plane. See Figure 2.18.

$$\mathbf{r}_{PQW} = r\cos\theta\hat{P} + r\sin\theta\hat{Q} = \begin{bmatrix} p\cos\theta/(1+e\cos\theta) \\ p\sin\theta/(1+e\cos\theta) \\ 0 \end{bmatrix}. \tag{2.97}$$

By Equation (2.46), Equation (2.66) and Equation (2.31)

$$r\dot{\theta} = \frac{\sqrt{\mu p}(1+e\cos\theta)}{p} = \sqrt{\frac{\mu}{p}}(1+e\cos\theta) \tag{2.98}$$

$$\dot{r} = \frac{r\dot{\theta}e\sin\theta}{1+e\cos\theta} = \sqrt{\frac{\mu}{p}}(e\sin\theta); \tag{2.99}$$

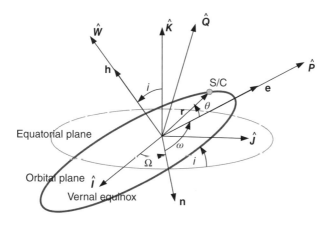

Figure 2.18 Earth-centered inertial and perifocal coordinates

thus

$$\mathbf{v}_{PQW} = \begin{bmatrix} \dot{r}\cos\theta - r\dot{\theta}\sin\theta \\ \dot{r}\sin\theta + r\dot{\theta}\cos\theta \\ 0 \end{bmatrix} = \sqrt{\frac{\mu}{p}} \begin{bmatrix} -\sin\theta \\ e + \cos\theta \\ 0 \end{bmatrix}. \tag{2.100}$$

Now we simply need to rotate into the Geocentric Equatorial System. The order of the rotations does matter

$$\mathbf{r}_{IJK} = \mathbf{R}_3(-\Omega)\mathbf{R}_1(-i)\mathbf{R}_3(-\omega)\mathbf{r}_{PQW} = \begin{bmatrix} \hat{I}\hat{J}\hat{K} \\ \hat{P}\hat{Q}\hat{W} \end{bmatrix} \mathbf{r}_{PQW} \tag{2.101}$$

$$\mathbf{v}_{IJK} = \mathbf{R}_3(-\Omega)\mathbf{R}_1(-i)\mathbf{R}_3(-\omega)\mathbf{v}_{PQW} = \begin{bmatrix} \hat{I}\hat{J}\hat{K} \\ \hat{P}\hat{Q}\hat{W} \end{bmatrix} \mathbf{v}_{PQW}, \tag{2.102}$$

where

$$\begin{bmatrix} \hat{I}\hat{J}\hat{K} \\ \hat{P}\hat{Q}\hat{W} \end{bmatrix} = \begin{bmatrix} c\Omega c\omega - s\Omega s\omega ci & -c\Omega s\omega - s\Omega c\omega ci & s\Omega si \\ s\Omega c\omega + c\Omega s\omega ci & -s\Omega s\omega + c\Omega c\omega ci & -c\Omega si \\ s\omega si & c\omega si & ci \end{bmatrix}. \tag{2.103}$$

Here letters c and s are used as abbreviations for cos and sin trigonometric function respectively.

2.4.3 From $\hat{I}\hat{J}\hat{K}$ to $\hat{S}\hat{E}\hat{Z}$

$\hat{I}\hat{J}\hat{K}$ is the Earth Center Inertial (ECI) frame, $\hat{S}\hat{E}\hat{Z}$ is the so-called topocentric horizon coordinate system in Chapter 1.

2.4.3.1 From $\hat{I}\hat{J}\hat{K}$ to $\hat{S}\hat{E}\hat{Z}$

Letus discuss how to transform from $\hat{I}\hat{J}\hat{K}$(aka GES)$\rightarrow \hat{S}\hat{E}\hat{Z}$ (aka THS) and its application. See Figure 2.19. Assume α is the local sidereal time in Equation (2.117), δ is the geodetic latitude:

Step 1: $\mathbf{R}_3\,(\alpha)$
Step 2: $\mathbf{R}_2(90° - \delta)$

$$
\begin{bmatrix} a_S \\ a_E \\ a_Z \end{bmatrix} = \begin{bmatrix} \cos(90° - \delta) & 0 & -\sin(90° - \delta) \\ 0 & 1 & 0 \\ \sin(90° - \delta) & 0 & \cos(90° - \delta) \end{bmatrix} \begin{bmatrix} \cos\alpha & \sin\alpha & 0 \\ -\sin\alpha & \cos\alpha & 0 \\ 0 & 0 & 1 \end{bmatrix} \begin{bmatrix} a_I \\ a_J \\ a_K \end{bmatrix}.
$$

Note: $\sin(90° - \delta) = \cos(\delta)$ and $\cos(90° - \delta) = \sin\delta$.

$$
\begin{bmatrix} a_S \\ a_E \\ a_Z \end{bmatrix} = \begin{bmatrix} \sin\delta & 0 & -\cos\delta \\ 0 & 1 & 0 \\ \cos\delta & 0 & \sin\delta \end{bmatrix} \begin{bmatrix} \cos\alpha & \sin\alpha & 0 \\ -\sin\alpha & \cos\alpha & 0 \\ 0 & 0 & 1 \end{bmatrix} \begin{bmatrix} a_I \\ a_J \\ a_K \end{bmatrix}
$$

$$
= \begin{bmatrix} \sin\delta\cos\alpha & \sin\delta\sin\alpha & -\cos\delta \\ -\sin\alpha & \cos\alpha & 0 \\ \cos\delta\cos\alpha & \cos\delta\sin\alpha & \sin\delta \end{bmatrix} \begin{bmatrix} a_I \\ a_J \\ a_K \end{bmatrix}. \tag{2.104}
$$

R is Direction Cosine Matrix (DCM) (single product matrix) whose properties are:

1. any row (column) is orthogonal to any other row (column),
2. any row (column) has magnitude 1,
3. its determinant is 1,
4. $[\mathbf{R}(\hat{I}\hat{J}\hat{K} \rightarrow \hat{S}\hat{E}\hat{Z})]^{-1} = [\mathbf{R}(\hat{I}\hat{J}\hat{K} \rightarrow \hat{S}\hat{E}\hat{Z})]^{T} = \mathbf{R}(\hat{S}\hat{E}\hat{Z} \rightarrow \hat{I}\hat{J}\hat{K})$.

In this example: $\mathbf{R}(\hat{S}\hat{E}\hat{Z} \rightarrow \hat{I}\hat{J}\hat{K}) = \mathbf{R}_3^T(\alpha)\mathbf{R}_2^T(90° - \delta)$.

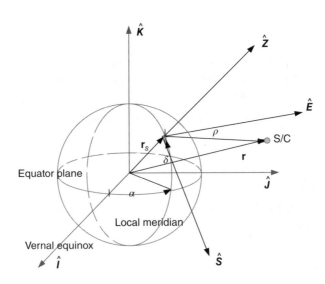

Figure 2.19   Earth centered inertial and topocentric horizon coordinates

2.4.3.2 From $\hat{N}\hat{E}\hat{D}$ to $\hat{X}\hat{Y}\hat{Z}$

Similarly we can translate from frame $\hat{N}\hat{E}\hat{D}$ (North-East-Down) to $\hat{X}\hat{Y}\hat{Z}$ (geographic frame, that is, ECEF in Chapter 1). Assume δ is the geodetic latitude which is used to approximate to its local geographic latitude, geographic longitude is λ_E in Equation (2.117)
Step 1: $\mathbf{R}_2(90° + \delta)$
Step 2: $\mathbf{R}_3(-\lambda_E)$.

$$\mathbf{R} = \begin{bmatrix} \cos(-\lambda_E) & \sin(-\lambda_E) & 0 \\ -\sin(-\lambda_E) & \cos(-\lambda_E) & 0 \\ 0 & 0 & 1 \end{bmatrix} \begin{bmatrix} \cos(90° + \delta) & 0 & -\sin(90° + \delta) \\ 0 & 1 & 0 \\ \sin(90° + \delta) & 0 & \cos(90° + \delta) \end{bmatrix},$$

as $\sin(-\lambda_E) = -\sin(\lambda_E)$, $\cos(-\lambda_E) = \cos(\lambda_E)$ and $\sin(90° + \delta) = \cos(\delta)$ and $\cos(90° + \delta) = -\sin(\delta)$ then

$$\mathbf{R}(\hat{N}\hat{E}\hat{D} \to \hat{X}\hat{Y}\hat{Z}) = \begin{bmatrix} -\cos(\lambda_E)\sin(\delta) & -\sin(\lambda_E) & -\cos(\lambda_E)\cos(\delta) \\ -\sin(\lambda_E)\sin(\delta) & \cos(\lambda_E) & -\sin(\lambda_E)\cos(\delta) \\ \cos(\delta) & 0 & -\sin(\delta) \end{bmatrix}. \tag{2.105}$$

2.4.3.3 Position and Velocity from COEs

With preceding discussions, it is easy to obtain $\mathbf{r}, \mathbf{v}$ from $a, e, i, \Omega, \omega, \theta$. First use polar Equation (2.66) to calculate the radius, then use Equations (2.97)–(2.100) to obtain the position and velocity in $\hat{P}\hat{Q}\hat{W}$ frame. Finally using Equations (2.101)–(2.102) the $\mathbf{r}, \mathbf{v}$ in ECI frame can be obtained.

2.4.4 Single Radar Observation

Consider a radar site in a THS(the Topocentric-Horizon Coordinate System) frame. More detailed discussion about this section can be found in reference [1].

If the measures are range and direction to satellite, range $= \rho$, radar axis gives angles, Azimuth (Az) and elevation (El).

For Doppler radar, the measures are frequency shift in return signal which measures range rate $\dot{\rho}$. Radar axis measures rate of change of $\dot{Az}$ and $\dot{El}$. See Figure 1.8 and Figure 2.19.

In THS: $\boldsymbol{\rho} = \rho_S \hat{S} + \rho_E \hat{E} + \rho_z \hat{Z}$; see Figure 1.8. Using spherical trig:

$$\rho_S = -\rho\cos(El)\cos(Az) \tag{2.106}$$

$$\rho_E = \rho\cos(El)\sin(Az) \tag{2.107}$$

$$\rho_Z = \rho\sin(El). \tag{2.108}$$

2.4.4.1 Position and Velocity

$\mathbf{v} = \dot{\boldsymbol{\rho}}$ is the relative velocity to radar site:

$$\mathbf{v} = \dot{\boldsymbol{\rho}} = \dot{\rho}_S \hat{S} + \dot{\rho}_E \hat{E} + \dot{\rho}_z \hat{Z}$$

$$\dot{\rho}_S = -\dot{\rho}\cos(El)\cos(Az) + \rho\sin(El)\cos(Az)\dot{El} + \rho\cos(El)\sin(Az)\dot{Az}$$

$$\dot{\rho}_E = \dot{\rho}\cos(El)\sin(Az) - \rho\sin(El)\sin(Az)\dot{El} + \rho\cos(El)\cos(Az)\dot{Az}$$

$$\dot{\rho}_Z = \dot{\rho}\sin(El) + \rho\cos(El)\dot{El}.$$

Then we can have the position relative to Geocentric-Equatorial Coordinate System (GES). Let $\mathbf{r}_s$ be the vector from the center of the Earth to the radar site, $\mathbf{r} = \mathbf{r}_s + \rho$. Assume the Earth is a perfect sphere, then the velocity relative to GES can be found. As GES is in inertial space, THS is noninertial-rotating. The relation between inertial and rotating frame should be clear.

Let $\mathbf{A} = \mathbf{A}(t)$, in fixed coordinate frame: $\mathbf{A} = A_i\hat{I} + A_j\hat{J} + A_k\hat{K}$. in rotating coordinate frame: $\mathbf{A} = A_u\hat{U} + A_v\hat{V} + A_w\hat{W}$ with angular velocity $\boldsymbol{\omega}$,

$$\frac{d\mathbf{A}}{dt} = \mathbf{A}' = \dot{A}_i\hat{I} + \dot{A}_j\hat{J} + \dot{A}_k\hat{K} + A_i\dot{\hat{I}} + A_j\dot{\hat{J}} + A_k\dot{\hat{K}}[\dot{\hat{I}} = \dot{\hat{J}} = \dot{\hat{K}} = 0] \tag{2.109}$$

$$= \dot{A}_u\hat{U} + \dot{A}_v\hat{V} + \dot{A}_w\hat{W} + A_u\dot{\hat{U}} + A_v\dot{\hat{V}} + A_w\dot{\hat{W}} \tag{2.110}$$

$$\dot{\hat{U}} = \boldsymbol{\omega}\times\hat{U}, \dot{\hat{V}} = \boldsymbol{\omega}\times\hat{V}, \dot{\hat{W}} = \boldsymbol{\omega}\times\hat{W} \tag{2.111}$$

$$A_u\dot{\hat{U}} + A_v\dot{\hat{V}} + A_w\dot{\hat{W}} = \boldsymbol{\omega}\times\mathbf{A} \tag{2.112}$$

$$\frac{d\mathbf{A}}{dt}\Big|_{Fixed} = \frac{d\mathbf{A}}{dt}\Big|_{Rotating} + \boldsymbol{\omega}\times\mathbf{A}[\text{Coriolis}]. \tag{2.113}$$

Coriolis as an operator:

$$\frac{d(\bullet)}{dt}\Big|_{Fixed} = \frac{d(\bullet)}{dt}\Big|_{Rotating} + \boldsymbol{\omega}\times(\bullet). \tag{2.114}$$

Review:

$$\frac{d\mathbf{r}}{dt}\Big|_{Fixed} = \frac{d\mathbf{r}}{dt}\Big|_{Rotating} + \boldsymbol{\omega}\times\mathbf{r}.$$

or

$$\mathbf{v}\Big|_{Fixed} = \mathbf{v}\Big|_{Rotating} + \boldsymbol{\omega}\times\mathbf{r} \quad\Rightarrow\quad \mathbf{v}\Big|_F = \mathbf{v}\Big|_R + \boldsymbol{\omega}\times\mathbf{r}.$$

Acceleration:

$$\frac{d(\bullet)}{dt}\Big|_F = \frac{d(\bullet)}{dt}\Big|_R + \boldsymbol{\omega}\times(\bullet)$$

$$\mathbf{a}_F = \frac{d\mathbf{v}_F}{dt}\Big|_F = \frac{d\mathbf{v}_F}{dt}\Big|_R + \boldsymbol{\omega}\times\mathbf{v}_F$$

$$= \mathbf{a}_R + \left[\frac{d\boldsymbol{\omega}}{dt}\Big|_R\times\mathbf{r} + \boldsymbol{\omega}\times\frac{d\mathbf{r}}{dt}\Big|_R\right] + \boldsymbol{\omega}\times\mathbf{v}_R + \boldsymbol{\omega}\times(\boldsymbol{\omega}\times\mathbf{r})$$

$$= \mathbf{a}_R + \left[\frac{d\boldsymbol{\omega}}{dt}\Big|_R\times\mathbf{r} + 2\boldsymbol{\omega}\times\frac{d\mathbf{r}}{dt}\Big|_R\right] + \boldsymbol{\omega}\times(\boldsymbol{\omega}\times\mathbf{r}). \tag{2.115}$$

Example

Shemya, the Alaska radar site, observes a space object. Its data are:

$$\rho = 0.4\text{DU}, Az = 90°, El = 30°$$

$$\dot{\rho} = 0, \dot{Az} = 10 \text{ rad/TU}, \dot{El} = 5 \text{ rad/DU},$$

the in THS frame:

$$\rho_s = -\rho \cos(El)\cos(A_z) = 0$$

$$\rho_e = \rho \cos(El)\sin(Az) = 0.3464$$

$$\rho_z = \rho \sin(El) = 0.2000$$

$$\rho = 0.3464\hat{E} + 0.2000\hat{Z}$$

$$\Rightarrow \mathbf{r} = 0.3464\hat{E} + 1.2000\hat{Z}$$

$$\dot{\rho}_s = 3.4641, \quad \dot{\rho}_E = -1.0000, \quad \dot{\rho}_z = 1.7321$$

$$\dot{\rho} = 3.4641\hat{S} - 1.000\hat{E} + 1.7321\hat{Z}.$$

We need the transformation matrix $\hat{S}\hat{E}\hat{Z} \rightarrow \hat{I}\hat{J}\hat{K}$

$$\mathbf{R}(\hat{S}\hat{E}\hat{Z} \rightarrow \hat{I}\hat{J}\hat{K}) = \begin{bmatrix} \sin\delta\cos\alpha & -\sin\alpha & \cos\delta\cos\alpha \\ \sin\delta\sin\alpha & \cos\alpha & \cos\delta\sin\alpha \\ -\cos\delta & 0 & \sin\delta \end{bmatrix}. \tag{2.116}$$

$\mathbf{R}(\hat{S}\hat{E}\hat{Z} \rightarrow \hat{I}\hat{J}\hat{K})$ is the inverse or transpose of $\mathbf{R}(\hat{I}\hat{J}\hat{K} \rightarrow \hat{S}\hat{E}\hat{Z})$ in Equation (2.104).
Radar site's Latitude $\delta = 60°$ N, and Longitude $= 150°$W; let us find its δ and α.

δ is the measure of angular distance(°) above or the below equatorial plane. The Geodetic vs Geocentric latitude transform can easily be derived. Use δ as geocentric latitude; α is the measure of angle from the vernal equinox ($\hat{I}$). There are parts in the angle $\alpha = \alpha_g + \lambda_E$, where α_g is angle between $\hat{I}$ and Greenwich meridian, λ_E is the longitude of interest, and α is called local sidereal time. See Figure 2.17.

For solar time, the Sun's successive passages over a given longitude is also called the apparent solar day. Since the Earth's orbit is elliptical ($e = 0.017$), the length of the apparent solar day varies slightly throughout a year. The average over the year is the mean solar day (clock time); 24 mean solar hours is one mean solar day. See Figure 1.9.

Sidereal time is the successive passages of vernal equinox direction over a specific longitude. A sidereal day equals 23 hours, 56 minutes, 4 seconds. One mean solar day equals 1.0027 sidereal days.

Time reference point is at Greenwich meridian; see Figure 1.3 and Figure 1.4 (Chapter 1, above)

$$\alpha = \alpha_g + \lambda_E \tag{2.117}$$

$$\alpha_g = \alpha_{g_0} + offset, \tag{2.118}$$

where α_{g0} is the angle between vernal equinox and Greenwich meridian at 0 hours, January 1 (reference [1]). *offset* is the # (number) days since 0 hours, January 1.

$$\alpha_g = \alpha_{g0} + 1.0027 \cdot 360° \cdot offset, \tag{2.119}$$

$1.0027 \approx 24/(23 + 56/60)$, λ_E is the $\pm$ longitude of interest. Roughly speaking, α_g equals $0°$ on September. 23, $90°$ on December 22, $180°$ on March 21, and $270°$ on June 21st respectively at 0 hours at Greenwich.

Radar site observes object at 0600 Zulu (Greenwich sidereal time), its sidereal time is $\alpha = (6)\cdot(15)\text{-}150 = -60°$.

By Equation (2.116), $\mathbf{R}(\hat{S}\hat{E}\hat{Z} \rightarrow \hat{I}\hat{J}\hat{K})$ with $\delta = 60°$ and $\alpha = -60°$

$$\mathbf{R} = \begin{bmatrix} 0.433 & 0.866 & 0.25 \\ -0.75 & 0.5 & -0.433 \\ -0.50 & 0.0 & 0.866 \end{bmatrix}$$

$$\mathbf{r} = \mathbf{R} \cdot \mathbf{r} = 0.6\hat{I} - 0.346\hat{J} + 1.04\hat{K}$$

$$\dot{\rho} = \mathbf{R} \cdot \dot{\rho} = 1.06\hat{I} - 3.84\hat{J} - 0.232\hat{K}$$

$$\mathbf{v} = \dot{\rho} + (0.0588\hat{K}) \times \mathbf{r} = 1.08\hat{I} - 3.8\hat{J} - 0.232\hat{K}[0.0588\hat{K} = \omega_\oplus].$$

The orbital elements can be found $a = -0.0695$, $e = 17.8096$, $i = 1.9977$, $\Omega = 1.8677$, $\omega = 1.6151$, $\theta = 0.3735$. This is a hyperbolic orbit.

Figure 7.9 and Figure 7.10 show a similar simulation example of the material in this section.

2.4.5 Summary of the Transformation

To convert from the ECI ($\hat{I}\hat{J}\hat{K}$) system to ECEF ($\hat{X}\hat{Y}\hat{Z}$), we simply rotate around Z by the GHA (Greenwich Hour Angle) α_g; see Figure 2.17:

$$\mathbf{r}_{XYZ} = \mathbf{R}_3(\alpha_g)\mathbf{r}_{IJK} \tag{2.120}$$

or

$$\mathbf{r}_{IJK} = \mathbf{R}_3(-\alpha_g)\mathbf{r}_{XYZ}, \tag{2.121}$$

where α_g is also called Greenwich Sidereal Time, ignoring precession, nutation, polar motion, motion of equinoxes.

To convert from $\hat{X}\hat{Y}\hat{Z}$ to $\hat{S}\hat{E}\hat{Z}$:

$$\mathbf{r}_{SEZ} = \mathbf{R}_2(90° - \phi)\mathbf{R}_3(\lambda_E)\mathbf{r}_{XYZ} \tag{2.122}$$

$$= \mathbf{R}_2(90° - \phi)\mathbf{R}_3(\alpha - \alpha_g)\mathbf{r}_{XYZ}. \tag{2.123}$$

To convert between $\hat{I}\hat{J}\hat{K}$ and $\hat{P}\hat{Q}\hat{W}$:

$$\mathbf{r}_{IJK} = \mathbf{R}_3(-\Omega)\mathbf{R}_1(-i)\mathbf{R}_3(-\omega)\mathbf{r}_{PQW} \tag{2.124}$$

$$\mathbf{r}_{PQW} = \mathbf{R}_3(\omega)\mathbf{R}_1(i)\mathbf{R}_3(\Omega)\mathbf{r}_{IJK}. \tag{2.125}$$

To convert between $\hat{P}\hat{Q}\hat{W}$ and $\hat{R}\hat{S}\hat{W}$:

$$\mathbf{r}_{RSW} = \mathbf{R}_3(\theta)\mathbf{r}_{PQW} \tag{2.126}$$

$$\mathbf{r}_{PQW} = \mathbf{R}_3(-\theta)\mathbf{r}_{RSW}. \tag{2.127}$$

Thus, $\hat{R}\hat{S}\hat{W} \rightarrow \hat{I}\hat{J}\hat{K}$ is:

$$\mathbf{r}_{IJK} = \mathbf{R}_3(-\Omega)\mathbf{R}_1(-i)\mathbf{R}_3(-u_t)\mathbf{r}_{RSW}. \tag{2.128}$$

where $u_t = \theta + \omega$. The Latitude/Longitude in $\hat{X}\hat{Y}\hat{Z}$ (ECEF) can be obtained by

$$\mathbf{r}_{XYZ} = \mathbf{R}_3(\alpha_g)\mathbf{r}_{IJK} \tag{2.129}$$

$$\lambda_E = \arctan(\frac{r_y}{r_x}) \tag{2.130}$$

$$\phi_{gc} = \arcsin(\frac{r_z}{r}) \tag{2.131}$$

$$r = \sqrt{r_x^2 + r_y^2 + r_z^2} \tag{2.132}$$

$$\mathbf{r}_{XYZ} = \begin{bmatrix} r\cos\phi\cos\lambda_E \\ r\cos\phi\sin\lambda_E \\ r\sin\phi \end{bmatrix} = \begin{bmatrix} r_x \\ r_y \\ r_z \end{bmatrix}. \tag{2.133}$$

Geocentric latitude, for geodetic latitude use:

$$\tan\phi_{\text{ground}} = \frac{\tan\phi_{\text{geocentric}}}{1 - e_\oplus^2} \tag{2.134}$$

where $e_\oplus = 0.081\,819\,221\,456$ in Equation (2.72) for the oblate Earth shape.
Right Ascension/Declination

$$\mathbf{r}_{IJK} = \begin{bmatrix} r\cos\delta\cos\alpha \\ r\cos\delta\sin\alpha \\ r\sin\delta \end{bmatrix} = \begin{bmatrix} r_I \\ r_J \\ r_K \end{bmatrix}; \tag{2.135}$$

$$r = \sqrt{r_I^2 + r_J^2 + r_K^2} \tag{2.136}$$

thus,

$$\delta = \arcsin(r_K/r) \tag{2.137}$$

$$\alpha = \arctan(r_J/r_I) \tag{2.138}$$

$$\lambda_E = \alpha - \alpha_g. \tag{2.139}$$

Compute Azimuth-Elevation, slant-range vector from site to satellite:

$$\boldsymbol{\rho}_{IJK} = \mathbf{r}_{IJK} - \mathbf{r}_{\text{site}IJK}. \tag{2.140}$$

Rotate in $\hat{S}\hat{E}\hat{Z}$

$$\boldsymbol{\rho}_{IJK} = \mathbf{R}_2(90° - \phi)\mathbf{R}_3(\lambda_E)\mathbf{R}_3(\alpha_g)\boldsymbol{\rho}_{IJK}, \tag{2.141}$$

since

$$\boldsymbol{\rho}_{SEZ} = \begin{bmatrix} -\rho\cos El\cos AZ \\ \rho\cos El\sin AZ \\ \rho\sin El \end{bmatrix} = \begin{bmatrix} \rho_S \\ \rho_E \\ \rho_Z \end{bmatrix} \tag{2.142}$$

$$\sin(El) = \frac{\rho_Z}{\rho}, \quad \cos(El) = \frac{\sqrt{\rho_S^2 + \rho_E^2}}{\rho} \tag{2.143}$$

$$\sin(Az) = \frac{\rho_E}{\sqrt{\rho_S^2 + \rho_E^2}}, \quad \cos(Az) = -\frac{\rho_S}{\sqrt{\rho_S^2 + \rho_E^2}}. \tag{2.144}$$

Alternatively:

$$\mathbf{r}_{XYZ} = \mathbf{R}_3(\alpha_g)\mathbf{r}_{IJK} \tag{2.145}$$

$$\boldsymbol{\rho}_{XYZ} = \mathbf{r}_{XYZ} - \mathbf{r}_{siteXYZ} \tag{2.146}$$

$$\boldsymbol{\rho}_{SEZ} = \mathbf{R}_2(90° - \phi)\mathbf{R}_3(\lambda_E)\boldsymbol{\rho}_{XYZ}. \tag{2.147}$$

2.4.6   Three Position Vectors (Gibbs Method)

In the previous section, we introduced the detailed method to determine the orbit by its position and velocity vectors. There are some other methods, such as J.W. Gibbsan's method [1], which uses three coplanar position vectors: $\mathbf{r}_1, \mathbf{r}_2, \mathbf{r}_3$ in a perifocal $(\hat{P}\hat{Q}\hat{W})$ system; see Figure 2.20. Procedure:

1. Test $\mathbf{r}_1 \cdot \mathbf{r}_2 \times \mathbf{r}_3 = 0$; ensure vectors are coplanar.
2. Develop the following vectors:

$$\mathbf{D} = \mathbf{r}_1 \times \mathbf{r}_2 + \mathbf{r}_2 \times \mathbf{r}_3 + \mathbf{r}_3 \times \mathbf{r}_1 \tag{2.148}$$

$$\mathbf{n} = r_3(\mathbf{r}_1 \times \mathbf{r}_2) + r_1(\mathbf{r}_2 \times \mathbf{r}_3) + r_2(\mathbf{r}_3 \times \mathbf{r}_1) \tag{2.149}$$

$$\mathbf{S} = (r_2 - r_3)\mathbf{r}_1 + (r_3 - r_1)\mathbf{r}_2 + (r_1 - r_2)\mathbf{r}_3 \tag{2.150}$$

$$\mathbf{B} = \mathbf{D} \times \mathbf{r}_2. \tag{2.151}$$

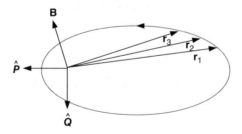

Figure 2.20   Gibbs method

3. Test $\mathbf{D} \neq 0$, $\mathbf{n} \neq 0$, $\mathbf{D} \cdot \mathbf{n} > 0$ ensure 2 - body orbit.
4. Let $L = \sqrt{\mu/(DN)}$.
5. $\mathbf{v}_2 = L/r_2\mathbf{B} + L\mathbf{S}$.
   Have position and velocity vectors.

For a circular polar orbit $\mathbf{r}_1 = [1, 0, 0]^T$, $\mathbf{r}_2 = [1/\sqrt{2}, 0, 1/\sqrt{2}]^T$, $\mathbf{r}_3 = [0, 0, 1]^T$, then $\mathbf{n} = [0, -0.4142, 0]^T$, $\mathbf{S} = [0, 0, 0]^T$, $\mathbf{B} = [-0.2929, 0, 0.2929]^T$, $L = 2.4142$, finally, $\mathbf{v}_2 = [-1/\sqrt{2}, 0, 1/\sqrt{2}]^T$.
Detailed discussion of this method can be found in reference [1].

2.5 Time of Flight (*TOF*)

In Section 2.5.1, Kepler's equation is discussed and later (Section 2.5.4) f and g expansion is considered. The key to this problem is the Prediction Problem of Position and Velocity.

Kepler's work gave rise to two classes of problems. One is to find the time to travel between two known points on an orbit, using Kepler's Equation. The other is to find a S/C future location given the last known position and velocity vectors at time t, which is Kepler's Problem or Propagation.

2.5.1 Kepler's Equation (Elliptical Orbits)

Assume T is time of perigee passage, t is any time after $(t > T)$, TP is orbital period, a is semi-major axis, b is semi-minor axis, θ is true anomaly. Then, during time period $t - T$ the radius vector sweeps out area A_1 which is the sector PFV. See Figure 2.21. And by Kepler's Second Law:

$$\frac{t - T}{A_1} = \frac{TP}{\pi ab}.$$ (2.152)

2.5.1.1 Circle vs. Ellipse

In Figure 2.21, a is the semi-major axis, b is the semi-mini axis and $OF = c = ae$. The ellipse equation is:

$$\frac{x^2}{a^2} + \frac{y^2}{b^2} = 1.$$

Its circumscribed circle is:

$$\frac{x^2}{a^2} + \frac{y^2}{a^2} = 1.$$

Given an x coordinate, the y coordinates are respectively
for the circle:

$$y_c = \left(a^2 - x^2\right)^{1/2} = QS,$$

for the ellipse:

$$y_e = \frac{b}{a}\left(a^2 - x^2\right)^{1/2} = PS,$$

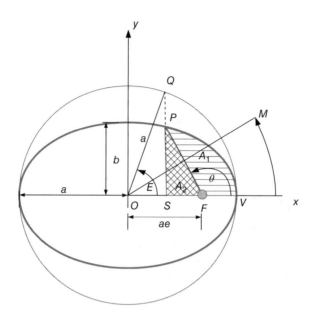

Figure 2.21   True anomaly θ is the angle xFP, mean anomaly M is the angle xOM and eccentric anomaly E is the angle xOQ, all in the orbital plane, xOy

hence the magnitude of the y coordinate of any point on the ellipse is smaller than the corresponding point on the circle by the factor, i.e. $PS/QS = b/a$ in Figure 2.21, where $a = a\cos E$, $b = \sin E$.

2.5.1.2   Derivation Outline

$$A_1 = \text{Area }(PSV) - A_2$$

$$\text{Area }(PSV) = \frac{b}{a} \cdot \text{Area }(QSV) = \frac{b}{a}\left[\frac{1}{2}a^2 E - \frac{1}{2}a\cos E \cdot a\sin E\right]$$

$$= \frac{ab}{2}\left[E - \sin E \cdot \cos E\right]$$

$$A_2 = \text{Area }(\Delta PSF) = \frac{1}{2}(ae - a\cos E)\left(\frac{b}{a}a\sin E\right)$$

$$= \frac{ab}{2}(e\sin E - \sin E \cdot \cos E)$$

$$A_1 = \frac{ab}{2}(E - e\sin E)$$

$$TP = 2\pi\sqrt{a^3/\mu} \Rightarrow t - T = \sqrt{a^3/\mu}\,(E - e\sin E),$$

where E is the eccentric anomaly in radians, the mean anomaly $M = E - e\sin E$, the mean motion $n = \sqrt{\mu/a^3}$, then we have Kepler's Equation:

$$M = n(t - T) = E - e \cdot \sin E. \tag{2.153}$$

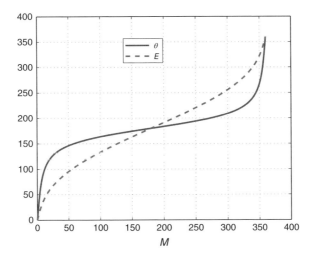

Figure 2.22 True anomaly θ and eccentric anomaly E as a function of mean anomaly M when $e = 0.8$

To know the position θ must be related to E. In Figure 2.21

$$OS = a \cos E = ae + r \cos \theta, \quad r = \frac{a(1 - e^2)}{1 + e \cos \theta}. \tag{2.154}$$

Then

$$\cos E = \frac{e + \cos \theta}{1 + e \cos \theta}. \tag{2.155}$$

Since

$$\tan^2 \left(\frac{E}{2}\right) = \frac{1 - \cos E}{1 + \cos E} = \left(\frac{1 - e}{1 + e}\right) \left(\frac{1 - \cos \theta}{1 + \cos \theta}\right) = \left(\frac{1 - e}{1 + e}\right) \tan^2 \left(\frac{\theta}{2}.\right) \tag{2.156}$$

Then,

$$\tan \frac{\theta}{2} = \left(\frac{1 + e}{1 - e}\right)^{1/2} \tan \frac{E}{2}. \tag{2.157}$$

Finally,

$$r = \frac{a(1 - e^2)}{1 + e \cos \theta} = a(1 - e \cos E). \tag{2.158}$$

It can be seen from Figure 2.22 that θ and E are monotonous functions of increase with M, and when $M < 180°, \theta > E$, when $M > 180°, \theta < E$. At the points $M = 0, 180°, \theta = E = M$.

2.5.2 Numerical Solution

Newton-Raphson Iteration can be used to solve $f(y) = 0$ for y. Assume $y = x + \Delta x$, where x is an approximate guess, and Δx is a small correction. This method is based on the expanding of Taylor Series. See Appendix A.1.

Example 1

Given: A S/C in orbit with $a = 7000$ km, $e = 0.05$ and $\theta = 270°$

Find: How long until S/C reaches $\theta = 50°$

Mean motion: $n = \sqrt{\mu/a^3} = 0.001\,078$ rad/sec $= 14.82$ rev/day which is often in TLE (see Chapter 7). E_i at initial:

$$E_i : \quad \cos(E_i) = \frac{e + \cos(\theta_i)}{1 + e\cos(\theta_i)} = 0.05 \Rightarrow E_i = 87.13° \text{ or } 272.87°.$$

E and θ always same half-plane: $\theta = 270° \Rightarrow E_i = 272.87° = 4.762$ rad E_f at final.

$$E_f : \quad \cos(E_f) = \frac{e + \cos(\theta_f)}{1 + e\cos(\theta_f)} = 0.6712 \Rightarrow E_f = 47.84 = 0.835 \text{ rad.}$$

$M_i : \quad M_i = E_i - e\sin(E_i) = 4.812$ rad
$M_f : \quad M_f = E_f - e\sin(E_f) = 0.798$ rad

Time: $t_f - t_i = (M_f - M_i + 2k\pi)/n$, ($k = $ # perigee passages)
$= 2105.00 \text{ sec}(k = 1) = 35.08$ min.

Example 2

Given: S/C with $a = 25\,512$ km and $e = 5/8$

Find: θ and r at 4 hours after perigee passage

Mean anomaly: $M = nt = \sqrt{\mu/a^3} \cdot 4 = 2.231$ rad (watch units $\mu = $ km^3/s^2)

Need eccentric anomaly $E : \quad M = E - e \cdot \sin E$

Kepler's equation: $f(E) = E - e \cdot \sin E - M = E - 5/8 \sin E - 2.231$

want E so that $f(E) = 0$

Newton's method: $x_{i+1} = x_i - f(x_i)/f'(x_i)$

$f(E) = E - 5/8 \sin E - 2.231, \quad f'(E) = 1 - 5/8 \cos E \quad$ then $\quad E = 2.5694, \quad \theta = 2.8608,$

$r = 38\,917$ km.

2.5.2.1 Nonelliptical orbits

For circles it is straightforward. For parabolic orbits, define

$$D = \sqrt{p}\tan(\theta/2), \tag{2.159}$$

as a parabolic eccentric anomaly. By $\dot{\theta} = \sqrt{\mu p}/r^2$ and $r = p/2(1 + \tan^2(\theta/2)) = (p + D^2)/2$
then

$$t - T = \int dt = \frac{1}{\sqrt{\mu p}} \int r^2 d\theta = \frac{1}{2\sqrt{\mu}}\left[pD + \frac{1}{3}D^3\right] \tag{2.160}$$

and

$$t - t_0 = \frac{1}{2\sqrt{\mu}} \left[(pD + \frac{1}{3}D^3) - (pD_0 + \frac{1}{3}D_0^3) \right] \tag{2.161}$$

For hyperbolic orbits, let F as hyperbolic eccentric anomaly, $E = \pm iF$:

$$\cosh F == \cos E = \frac{e + \cos\theta}{1 + e\cos\theta}, \tag{2.162}$$

or

$$F = \ln\left[y + \sqrt{y^2 - 1} \right], \tag{2.163}$$

where $y = \cosh F$, $0 \le \theta \le \pi$, $F \ge 0$, $\pi < \theta < 2\pi$, $F < 0$, then

$$t - T = \sqrt{\frac{(-a)^3}{\mu}} (e\sinh F - F), \tag{2.164}$$

or

$$t - t_0 = \sqrt{\frac{(-a)^3}{\mu}} [(e\sinh F - F) - (e\sinh F_0 - F_0)]. \tag{2.165}$$

Then, we plot the relation between true anomaly θ and mean anomaly M for circle($e=0$), ellipse($e=0.5$), parabola($e=1$), and a hyperbola($e=2$). See Figure 2.23.

Here we give a review of some **hyperbolic trigonometry**. The Euler's Formulas in complex domain is ($z = x + iy$)

$$\sin z = \frac{e^{iz} - e^{-iz}}{2i} \quad \text{and} \quad \cos z = \frac{e^{iz} + e^{-iz}}{2}.$$

Define hyperbolic sine and cosine:

$$\sinh z = \frac{e^z - e^{-z}}{2} \quad \text{and} \quad \cosh z = \frac{e^z + e^{-z}}{2}.$$

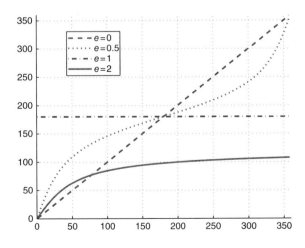

Figure 2.23   True anomaly θ as a function of mean anomaly M for different eccentricities

There relationships are

$$\sin z = \sin x \cdot \cosh y + i \cos x \cdot \sinh y \tag{2.166}$$

$$\cos z = \cos x \cdot \cosh y - i \sin x \cdot \sinh y \tag{2.167}$$

$$\sinh z = \sinh x \cdot \cos y + i \cosh x \cdot \sinh y \tag{2.168}$$

$$\cosh z = \cosh x \cdot \cos y + i \sinh x \cdot \sin y. \tag{2.169}$$

Note:$(\sinh z)' = \cosh z$, $(\cosh z)' = \sinh z$.

2.5.2.2   Kepler's Problem – Propagation

The problem: Determine a satellite's future position and velocity given the position and velocity vectors at a specified time. There are three general approaches to solving this problem. In this section, orbital element approach is skipped, f and g series are concentrated on, and classical formulation is introduced lightly.

2.5.3   Universal Variable X

Derivation outline: as we have $h = r^2\dot{\theta} = \sqrt{\mu p}$ and $\mathcal{E} = (1/2)v^2 - \mu/r = -\mu/(2a)$ resolve $\mathbf{v}$ into radial and transverse components, $v^2 = \dot{r}^2 + r^2\dot{\theta}^2$, then we have the differential equation:

$$\dot{r}^2 = -\frac{\mu p}{r^2} + \frac{2\mu}{r} - \frac{\mu}{a}. \tag{2.170}$$

Let $\dot{X} = \sqrt{\mu}/r$ and consider $\dot{r}^2/\dot{X}^2 = (dr/dx)^2$, then $dX = dr/\sqrt{-p + 2r - r^2/a}$
So,

$$X + C_0 = \sqrt{a}\sin^{-1}\left(\frac{r/a - 1}{\sqrt{1 - p/a}}\right) = \sqrt{a}\sin^{-1}\left(\frac{r/a - 1}{e}\right) \tag{2.171}$$

$$\Rightarrow r = a\left[1 + e\sin\left(\frac{X + C_0}{\sqrt{a}}\right)\right] \tag{2.172}$$

$$\sqrt{\mu}dt = a(1 + e\sin\left(\frac{X + C_0}{\sqrt{a}}\right)dX \tag{2.173}$$

$$\sqrt{\mu}t = aX - ae\sqrt{a}\left[\cos\left(\frac{X + C_0}{\sqrt{a}}\right) - \cos\left(\frac{C_0}{\sqrt{a}}\right)\right]. \tag{2.174}$$

Assume $X = 0$ when $t = 0$, we have r and t in terms of new variable X, which is called the universal variable instead of E, D, F for ellipse, parabola and hyperbola respectively.

2.5.3.1 Prediction Problem

Key equation:

$$r = a \left[1 + e \sin \left(\frac{X + C_0}{\sqrt{a}} \right) \right] = a(1 - e \cos E). \tag{2.175}$$

Then

$$X = \sqrt{a}(E - E_0). \tag{2.176}$$

At $t = 0$ (assume $X = 0$),

$$e \sin \left(\frac{C_0}{\sqrt{a}} \right) = \frac{r_0}{a} - 1. \tag{2.177}$$

Differentiate key equation and recall:

$$\sin(a + b) = \sin a \cdot \cos b + \cos a \cdot \sin b, \text{ and} \tag{2.178}$$

$$\cos(a + b) = \cos a \cdot \cos b - \sin a \cdot \sin b. \tag{2.179}$$

Obtain:

$$\sqrt{\mu} t = a \left[X - \sqrt{a} \sin \left(\frac{X}{\sqrt{a}} \right) \right] + \frac{\mathbf{r}_0 \cdot \mathbf{v}_o}{\sqrt{\mu}} a \left[1 - \cos \left(\frac{X}{\sqrt{a}} \right) \right]$$

$$+ r_0 \sqrt{a} \sin \left(\frac{X}{\sqrt{a}} \right) \tag{2.180}$$

$$r = a + ae \left[\sin \left(\frac{X}{\sqrt{a}} \right) \cos \left(\frac{C_0}{\sqrt{a}} \right) + \cos \left(\frac{X}{\sqrt{a}} \right) \sin \left(\frac{C_0}{\sqrt{a}} \right) \right]. \tag{2.181}$$

Define another variable $z = X^2 / a$

$$\sqrt{\mu} t = \left[\frac{\sqrt{z} - \sin \sqrt{z}}{\sqrt{z^3}} \right] X^3 + \frac{\mathbf{r}_0 \cdot \mathbf{v}_0}{\sqrt{\mu}} X^2 \cdot \frac{1 - \cos \sqrt{z}}{z} + \frac{r_0 X \sin \sqrt{z}}{\sqrt{z}} \tag{2.182}$$

$$r = \frac{X^2}{z} + \frac{\mathbf{r}_0 \cdot \mathbf{v}_0}{\sqrt{\mu}} \cdot \frac{X}{\sqrt{z}} \sin \sqrt{z} + r_0 \cos \sqrt{z} - \frac{X^2}{z} \cos \sqrt{z}. \tag{2.183}$$

If $z = 0$, both expressions are indeterminate and need series expansion:

$$S(z) = \frac{\sqrt{z} - \sin \sqrt{z}}{\sqrt{z^3}} = \frac{1}{3!} - \frac{z}{5!} + \frac{z^2}{7!} - \frac{z^3}{9!} + \dots \tag{2.184}$$

$$C(z) = \frac{1 - \cos \sqrt{z}}{z} = \frac{1}{2!} - \frac{z}{4!} + \frac{z^2}{6!} - \frac{z^3}{8!} + \dots \tag{2.185}$$

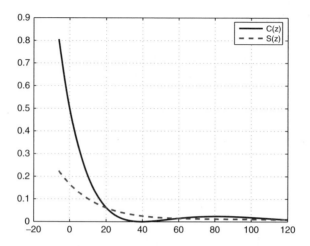

Figure 2.24   The S(z) and C(z) are functions of variable z

Figure 2.24 shows numerically the relation between S(z) and C(z) versus z. We can see $S(0) = 1/6$, $C(0) = 1/2$, $C((2\pi)^2) = 0$, etc.

Forms similar to S(z) and C(z) occur in applied math. The Euler formulas in complex domain are $\sin z = (e^{iz} - e^{-iz})/(2i)$ and $\cos z = (e^{iz} + e^{-iz})/2$. Then hyperbolic functions: $z = X^2/a$, $a < 0$, hyperbolic $\sinh z = (e^z - e^{-z})/2$ and $\cosh z = (e^z + e^{-z})/2$. Then

$$S(z) = \frac{\sqrt{z} - \sin\sqrt{z}}{\sqrt{z^3}} = \frac{\sinh\sqrt{-z} - \sqrt{-z}}{\sqrt{(-z)^3}} \tag{2.186}$$

$$C(z) = \frac{1 - \cos\sqrt{z}}{z} = \frac{1 - \cosh\sqrt{-z}}{z}. \tag{2.187}$$

Finally, we have the universal Kepler's equation:

$$\sqrt{\mu}t = X^3 S + \frac{\mathbf{r}_0 \cdot \mathbf{v}_0}{\sqrt{\mu}} X^2 C + r_0 X(1 - zS) \tag{2.188}$$

$$r = X^2 C + \frac{\mathbf{r}_0 \cdot \mathbf{v}_0}{\sqrt{\mu}} X(1 - zS) + r_0(1 - zC). \tag{2.189}$$

For the prediction problem, that is, given t to determine the universal variable X, X can not be isolated directly, just by the equation for $\sqrt{\mu}t$. Newton's iteration should be used

$$X_{n+1} = X_n + \frac{t - t_n}{dt/dX}, \quad \text{where } \frac{dt}{dX} = \frac{1}{\dot{X}} = \frac{r}{\sqrt{\mu}} \Rightarrow \sqrt{\mu}\frac{dt}{dX} = r \tag{2.190}$$

$$\sqrt{\mu}\frac{dt}{dX} = X_n^2 C + \frac{\mathbf{r}_0 \cdot \mathbf{v}_0}{\sqrt{\mu}} X_n(1 - zS) + r_0(1 - zC). \tag{2.191}$$

Example

Given: Position and velocity at a specific time:

$$\mathbf{r}_0 = 1131.34\hat{I} - 2282.343\hat{J} + 6672.423\hat{K} \text{ km}$$

$$\mathbf{v}_0 = -5.64\,305\hat{I} + 4.30\,333\hat{J} + 2.42\,879\hat{K} \text{ km/sec.}$$

Find the value of the universal variable X forty minutes later.

Problem set-up:

$$\mathbf{r}_0 = 0.17\,738\hat{I} - 0.35\,784\hat{J} + 1.04\,614\hat{K} \text{ DU}, \quad r_0 = 1.11\,978\,418$$

$$\mathbf{v}_0 = -0.71\,383\hat{I} + 0.54\,436\hat{J} + 0.30\,723\hat{K} \text{ DU/TU}, \quad v_0^2 = 0.90\,026\,11$$

$$t = 40\,\text{min} = \frac{40 \cdot 60}{806.8\,118\,744} = 2.97\,467\,114 \text{ TU}, \quad \mathbf{r}_0 \cdot \mathbf{v}_0 = 0.00\,000\,062.$$

Find X:

$$\mathcal{E} = \frac{1}{2}v_0^2 - \frac{\mu}{r_0} = -0.44\,289\,865, \Rightarrow a = -\frac{\mu}{2\mathcal{E}} = 1.12\,892\,645. \tag{2.192}$$

As the physical meaning of X is $\sqrt{a}(E - E_0)$ for an ellipse, the initial guess:

$$X_0 = \frac{\sqrt{\mu} \cdot t}{a} = \frac{(1)(2.97\,467\,114)}{1.12\,892\,645} = 2.63\,495\,566$$

$$z = \frac{X_0^2}{a} = 6.15\,008\,297, \quad \sqrt{z} = 2.47\,993\,608(142.0\,898\,709^\circ)$$

$$C = \frac{1 - \cos\sqrt{z}}{z} = 0.2\,908\,864, \text{ and } S = \frac{\sqrt{z} - \sin\sqrt{z}}{\sqrt{z^3}} = 0.12\,231\,408$$

$$r = \sqrt{\mu}\frac{dt}{dX} = X_0^2 C + \frac{\mathbf{r}_0 \cdot \mathbf{v}_0}{\sqrt{\mu}}X_0(1 - zS) + r_0(1 - zC) = 1.13\,613\,987$$

$$\sqrt{\mu}t_0 = \frac{\mathbf{r}_0 \cdot \mathbf{v}_0}{\sqrt{\mu}}X_0^2 C + \left(1 - \frac{r_0}{a}\right)X_0^3 S + r_0 X_0 = 3.01\,460\,215$$

$$X_1 = X_0 + \frac{t - t_0}{dt/dX} = 2.59\,980\,945$$

$$z = 5.9\,585\,711\,207 \Rightarrow \sqrt{z} = 2.44\,685\,759(140.194\,613^\circ) \tag{2.193}$$

$$C = 0.29\,533\,827, \quad S = 0.12\,332\,585$$

$$r = \sqrt{\mu}\frac{dt}{dX} = 1.13\,595\,018, \quad \sqrt{\mu}t = 2.92\,877\,632 \tag{2.194}$$

$$X_2 = 2.6\,4021\,159, z = 6.1\,746\,425 \Rightarrow \sqrt{z} = 2.48\,488\,279(142.3\,732\,967^\circ)$$

$$C = 0.290\,220\,072, \quad S = 0.12\,216\,231$$

$$r = \sqrt{\mu}\frac{dt}{dX} = 1.13\,616\,733, \quad \sqrt{\mu}t = 2.97\,467\,556 \tag{2.195}$$

$$X_3 = 2.6\,402\,077. \tag{2.196}$$

This is accurate to 4 decimal at which point the recursive calculation can be stopped;

Three iterations of Newton's Method converge: $X = 2.6\,402\,077$. The orbit elements of this example are $a = 7200$ km, $e = 0.0081$, $i = 98.6°$, $\Omega = 319.7°$, $\omega = 70.88°$, $\theta = 0.0°$. It is a near circular orbit, $r_p = 7142.1$ km, this is why the algorithm converges quickly. For more discussion for this cubic expression $t = f(X)$ see reference [1], More iterations are needed to have a more accurate final X to update z, C and S.

2.5.4 f and g Expansion

Keplerian motion is in a plane, that is, orbital plane. This discussion is in the Perifocal Coordinate System.

$$\mathbf{r} = x\hat{P} + y\hat{Q} \quad \text{and} \quad \mathbf{v} = \dot{x}\hat{P} + \dot{y}\hat{Q}, \tag{2.197}$$

$\mathbf{r}_0$ and $\mathbf{v}_0$ in plane

$$\mathbf{r} = f\mathbf{r}_0 + g\mathbf{v}_0 \quad \text{and} \quad \mathbf{v} = \dot{f}\mathbf{r}_0 + \dot{g}\mathbf{v}_0, \tag{2.198}$$

$$\mathbf{r} \times \mathbf{v}_0 \Rightarrow f = \frac{x\dot{y}_0 - \dot{x}_0 y}{h}. \tag{2.199}$$

As before h is the magnitude of angular momentum, and

$$\dot{f} = \frac{\dot{x}\dot{y}_0 - \dot{x}_0\dot{y}}{h} \tag{2.200}$$

$$\mathbf{r}_0 \times \mathbf{r} \Rightarrow g = \frac{x_0 y - x y_0}{h} \quad \text{and} \quad \dot{g} = \frac{x_0\dot{y} - \dot{x}y_0}{h}. \tag{2.201}$$

By $re\cos\theta = a(1 - e^2) - r$ and Equation (2.172) and $r^2 = x^2 + y^2$ Then

$$\mathbf{h} = \mathbf{r} \times \mathbf{v} \Rightarrow 1 = f\dot{g} - \dot{f}g. \tag{2.202}$$

As

$$x = r\cos\theta = -a\left(e + \sin\left[\frac{X + c_0}{\sqrt{a}}\right]\right), \quad y = a\sqrt{1 - e^2}\cos\left[\frac{X + c_0}{\sqrt{a}}\right] \tag{2.203}$$

$$\dot{x} = -\frac{\sqrt{\mu a}}{r}\cos\left[\frac{X + c_0}{\sqrt{a}}\right] \quad \text{and} \quad \dot{y} = -\frac{h}{r}\sin\left[\frac{X + c_0}{\sqrt{a}}\right]. \tag{2.204}$$

Then

$$f = 1 - \frac{X^2}{r_0}C, \quad g = t - \frac{X^3}{\sqrt{\mu}}S \tag{2.205}$$

$$\dot{f} = \frac{\sqrt{\mu}}{r_0 r}X(zS - 1), \quad \dot{g} = 1 - \frac{X^2}{r}C. \tag{2.206}$$

For example if $X = 2.64\,02\,077$ (in previous example), $z = 6.17\,462\,431$, $\Rightarrow \sqrt{z} = 2.48\,487\,913$ ($142.373\,087°$). Then $S = 0.12\,216\,242$ and $C = 0.29\,022\,057$,

$$f = 1 - \frac{X^2}{r_0}C = -0.80\,663\,346, \quad g = t - \frac{X^3}{\sqrt{\mu}}S = 0.72\,638\,774$$

$$\dot{f} = \frac{\sqrt{\mu}}{r \cdot r_0}X(zS - 1) = -0.50\,986\,379, \quad \dot{g} = 1 - \frac{X^2}{r}C = -0.78\,058\,242$$

Check: $1 = f\dot{g} - \dot{f}g = 1.00\,000\,271$

$$\mathbf{r}_{IJK} = f\mathbf{r}_0 + g\mathbf{v}_0 = -0.661\,598\hat{I} + 0.68\,406\,215\hat{J} - 0.62\,068\,342\hat{K} \text{ (DU)}$$

$$= -4219.77\hat{I} + 4363.05\hat{J} - 3958.81\hat{K} \text{ (km)}$$

$$\mathbf{v}_{IJK} = \dot{f}\mathbf{r}_0 + \dot{g}\mathbf{v}_0 = 0.4\,667\,351\hat{I} - 0.24\,246\,819\hat{J} - 0.77\,320\,725\hat{K} \text{ (DU/TU)}$$

$$= 3.6899\hat{I} - 1.9168\hat{J} - 6.1125\hat{K} \text{ (km/sec)}.$$

2.5.4.1   Kepler's Problem – Solution

1. From $\mathbf{r}_0$ and $\mathbf{v}_0$ determine r_0 and a.
2. Given $t - t_0$ (t_0 usually assumed $= 0$) solve for the universal variable X using Newton's scheme Equation (2.190).
3. Develop f and g then predicted $\mathbf{r}(t)$ by Equation (2.205).
4. Develop $\dot{f}$ and $\dot{g}$ then predicted $\mathbf{v}(t)$ by Equation (2.206).

2.5.4.2   Nonelliptical Orbits

Elliptical orbits: $X = \sqrt{a}(E - E_0)$.
Parabolic orbits: $a = \infty \Rightarrow z = 0$.
Then $C = 1/2$ and by Equation (2.189).

$$r = \frac{1}{2}X^2 + \frac{\mathbf{r}_0 \cdot \mathbf{v}_0}{\sqrt{\mu}}X + r_0$$

$$D_0 = \frac{\mathbf{r}_0 \cdot \mathbf{v}_0}{\sqrt{\mu}}, \quad r = \frac{1}{2}(p + D^2)$$

$$\frac{1}{2}(p + D^2) = \frac{1}{2}X^2 + D_0 X + \frac{1}{2}(p_0 + D_0^2)$$

$$\Rightarrow X = D - D_0.$$

For hyperbolic orbits: $a < 0 \Rightarrow z < 0$, let $F = iE$, then $X = \sqrt{-a}(F - F_0)$ and $-z = (F - F_0)^2$.

2.5.4.3 Classical Formulation

Elliptic eccentric anomaly E:

$$e \cdot \sin E = \frac{\mathbf{r} \cdot \mathbf{v}}{\sqrt{\mu a}}. \tag{2.207}$$

Parabolic eccentric anomaly D:

$$D = \frac{\mathbf{r} \cdot \mathbf{v}}{\sqrt{\mu}} = \frac{r \cdot v}{\sqrt{\mu}} = \sqrt{p} \tan \frac{\theta}{2}. \tag{2.208}$$

Hyperbolic eccentric anomaly F:

$$e \cdot \sinh F = \frac{\mathbf{r} \cdot \mathbf{v}}{\sqrt{-\mu a}}, \quad E = \pm iF. \tag{2.209}$$

The f and g – eccentric anomaly algorithm is:

1. Let $\Delta U = U - U_0$.
2. By being given $\mathbf{r}_0$ and $\mathbf{v}_0$, determine r_0, a, e, p, and θ_0.
3. Solve Kepler's time-of-flight: D, E, or F.
4. Then solve for r:

$$r = 1/2(p + D^2), r = a(1 - e \cdot \cos E), r = a(1 - e \cdot \cosh F).$$

5. Solve for f and g:

$$f = 1 - \frac{a}{r_0}(1 - \cos \Delta U), \quad \dot{f} = -\frac{\sqrt{\mu a}}{r \cdot r_0} \sin \Delta U \tag{2.210}$$

$$g = (t - t_0) - \sqrt{\frac{a^3}{\mu}}(\Delta U - \sin \Delta U), \quad \dot{g} = 1 - \frac{a}{r}(1 - \cos \Delta U). \tag{2.211}$$

Example, $\mathbf{r}_0 = [1, 0, 0]^T$, $\mathbf{v}_0 = [0, 0.9, 0]$, $\mu = 1$, $\Delta t = 1$, then following the above procedure, orbital elements are: $a = 0.84034$, $e = 0.19$, $\theta = \pi$; f, g expansions are: $f = 0.5208$, $g = 0.82773$, $\dot{f} = -0.91064$, $\dot{g} = 0.4728$. The final position and velocity are: $\mathbf{r}_f = [0.5208, 0.74496, 0]^T$, $\mathbf{v}_f = [-0.91064, 0.42552, 0]^T$.

2.6 Summary and Keywords

This chapter introduces Keperian motions, including basic two-body equations, the proofs of Kepler's laws, orbital elements, coordinate transformations. Many detailed examples have been given.

Keywords: Keplerian motion, polar equation, conic section, flight path angle, classic orbital elements, canonical units, Vis-Visa equation, universal variable, f, g function, Kepler's equation.

Problems

2.1 An Earth satellite has a semi-major axis of 20 000 km. The orbit eccentricity is 0.3. Find

(a) the periapsis and apoapsis radii and velocities,
(b) the specific energy,
(c) the semi-latus rectum and angular momentum magnitude,
(d) the orbital period and
(e) the radius and velocity magnitudes and flight-path angle for a true anomaly of 135 and 300 deg.

2.2 Sketch the following 5 orbits. For each orbit list the type of conic section and compute the following quantities: a, $\mathcal{E}$, r_p, r_a, and h when $\mu = 1$. See Figure 2.12.

(a) $p = 6$ and $e = 0$,
(b) $p = 12$ and $e = 1$,
(c) $p = 8$ and $e = 1/3$,
(d) $p = 18$ and $e = 2$,
(e) $p = 10$ and $e = 2/3$.

2.3 Prove Equations (2.79), (2.92) and (2.93) by the identity Equation (9.39)

2.4 Calculate the third cosmic velocity, that is, the minimum velocity needed to send a probe out of our solar system.

2.5 Show by means of differential calculus that the position vector is an extremum (maximum or minimum) at the apses of the orbit.

2.6 Analytically demonstrate that a flight-path angle reaches an extremum for two-body elliptic motion when the S/C crosses the minor axis of its orbit. Derive an expression for the true anomaly, θ, of this point that is solely a function of eccentricity. Make a qualitative sketch of flight-path angle as a function of θ.

2.7 Kepler's Second law states that a line joining a planet and the Sun sweeps out equal areas in equal times. For an ellipse, show that if Kepler's Second Law holds, angular momentum must be constant.

2.8 Two unit mass particles m_1 and m_2 move under the influence of gravitational force. Their initial conditions in inertial, Cartesian xyz coordinates are:

$$\mathbf{r}_1(0) = [0, 0, 0]^T, \dot{\mathbf{r}}_1(0) = [0, 0, 0]^T;$$
$$\mathbf{r}_2(0) = [2, 0, 0]^T, \dot{\mathbf{r}}_2(0) = [0, 1, 0]^T.$$

(a) Determine values of the constants of the motion $\mathbf{c}_1$, $\mathbf{c}_2$, $\mathbf{c}_3$, and c_4, assuming that the units are chosen so that $G = 1$.
(b) Form the 10 scalar algebraic equations that relate the time-varying Cartesian coordinates and velocities of each mass for $t > 0$ to the values of the 10 constants. Determine how many unknown quantities exist.

(c) Demonstrate that the motion lies entirely in the xy plane by showing that the equations of part (b) are satisfied by $z_i(t) = \dot{z}_i(t) \equiv 0$ for $i = 1, 2$. Determine how many unknowns and equations remain.

(d) Show that the solution to the equations obtained in part (b) is given by:

$$\mathbf{r}_1(t) = [1 - \cos(t/2), t/2 - \sin(t/2), 0]^T \text{ and}$$

$$\mathbf{r}_2(t) = [1 + \cos(t/2), t/2 + \sin(t/2), 0]^T.$$

(e) Interpret the solution in part (d) geometrically by determining the position of each mass relative to the center of mass of the system and sketching the result.

2.9 A space vehicle destined for Mars begins its flight in a 200-km circular Earth parking orbit. After coasting in orbit for a period of time to allow system checks, the vehicle's rocket propulsion system is utilized to instantaneously increase its velocity to 12 km/sec, placing it on an interplanetary transfer toward Mars.

(a) What is the velocity of the spacecraft while it is in its parking orbit?

(b) Find e, h and a relative to the Earth for the escape orbit.

(c) What is the hyperbolic excess velocity?

(d) Compare the velocity at a radius of 1.0e+06 km from the Earth with the hyperbolic excess velocity. Why are the two so nearly alike?

2.10 Derive from first principles that the speed of a satellite in circular orbit is: $v_{cir} = \sqrt{\mu/r}$, and that $\phi_{fpa} = 0$. Compare this speed to escape velocity (v_{esc}) at the same radius. With the Earth as the central body, calculate v_{cir} and v_{esc} at $r = 6578$ km (low Earth orbit), $r = 42\,166$ km (GEO) and $r = 385\,000$ km (lunar orbit).

2.11 Halley's comet last passed perihelion on February 9, 1986. It has a semi-major axis, $a = 17.9564$ AU and eccentricity, $e = 0.967\,298$. Calculate the period of Halley's comet and predict the date (month, day and year) of the next return (perihelion).

2.12 NASA has sent a space probe on a fly by mission of a newly discovered planet that is in a circular orbit about the Sun. The Deep Space Network precisely tracks the spacecraft's position and velocity relative to the new planet during the fly by event. Explain how you would determine the planet's mass from knowledge of v_∞, δ and: (a) r_p or (b) Δ_a; see Figure 2.10.

2.13 The center of the Moon is 60.27 DU from the center of the Earth and has a relative mass of 0.0123.

(a) Locate the center of mass of the Earth–Moon system with respect to the center of the Earth.

(b) Find the ratio of orbital velocities of the Earth and Moon about the center of mass of the system.

(c) Compute the orbital velocity of the Moon about the Earth (in DU/TU) assuming two-body motion and that the Moon moves about the Earth in a circular orbit.

(d) Given this velocity, what is the approximate orbital velocity of the Earth about the center of mass of the Earth–Moon system (in DU/TU and in km/s)?

(e) Are there any challenges associated with assuming two-body motion for the Earth–Moon system?

2.14 The following is given by a student taking the Ph.D. qualifying exams as a general derivation of the two-body equation of motion. The final result is mathematically equivalent to Equation (2.15) derived in class. Is this student's derivation valid? Explain your reasoning.

I shall begin by assuming:

(a) only two bodies, each with spherical mass distribution;

(b) no forces acting on the system other than gravity;

(c) no mass change.

Assume an inertial coordinate system with origin at the center of mass M, and radius $\mathbf{r}_m$ from this origin to mass m (position of body m relative to body M). Applying Newton's Second Law to body m yields:

$$m\ddot{\mathbf{r}}_m = \mathbf{F}_g = -\frac{GMm}{r^3}\mathbf{r} \Rightarrow \ddot{\mathbf{r}}_m = -\frac{GM}{r^3}\mathbf{r} \Rightarrow \ddot{\mathbf{r}}_m = -\frac{\mu}{r^3}\mathbf{r}.$$

2.15 A comet makes a close approach to Earth on a hyperbolic passage. The fly by induces a change in direction of the comet's velocity vector with respect to Earth of 60 deg and its velocity at infinity with respect to the Earth is 2.0 km/sec. Determine:

(a) eccentricity,

(b) true anomaly when r at infinity,

(c) semi-major axis,

(d) periapsis altitude,

(e) periapsis velocity.

2.16 The US early warning defense system detects an unknown object headed to Earth. Ground tracking stations determine that at one point in its flight: $v = 9.0$ km/sec, $r = 7500.0$ km, $\phi_{fpa} = 25$ deg.

(a) Determine whether this is an ICBM, and Earth science satellite, or an interplanetary probe.

(b) If it is an ICBM, calculate its range over the Earth's surface from launch to impact; otherwise, determine the geocentric or heliocentric period.

2.17 By inspection, determine the orbital elements for an object crossing the +Y axis in a retrograde, equatorial, circular orbit at an altitude of 1 DU. Sketch this orbit.

2.18 In the planet-centric equator and equinox coordinate frame (ECI), a vector is defined by:

$$\mathbf{r} = 3\hat{I} + 2\hat{I} + \hat{K}(\mathrm{DU}).$$

For a ground station located at 30 deg N latitude and 60 deg East of the reference direction ($\hat{I}$ axis), transform this vector into the topocentric-horizon and South spherical coordinate frame.

2.19 Given the following orbital elements (inclination, longitude of ascending node, longitude of periapsis and true longitude) for objects A, B, C and D in the table below:

Object	i (deg)	Ω (deg)	Π (deg)	l_t (deg)
A	0	undefined	210	30
B	4	180	260	90
C	110	90	110	140
D	23	60	260	160

Complete the following statements:
(a) Object _____ is in retrograde motion.
(b) Object _____ has a true anomaly at epoch (time of this observation) of 180 deg.
(c) Object _____ has its periapsis south of the reference plane.
(d) Object _____ has a line of nodes which coincides with the reference axis direction.
(e) Object _____ has an argument of periapsis of 200 deg.

2.20 (a) A radar site on the Earth determines the following position and velocity vector in the Earth centered Earth equator and equinox coordinate system. Determine the orbital elements.

$$\mathbf{r} = -(\hat{I} + \hat{J} + \hat{K})\mathrm{DU}, \mathbf{v} = \frac{1}{3}(\hat{I} - \hat{J} + \hat{K})\mathrm{DU/TU}.$$

(b) For the following orbital elements: $p = 0.23$ DU, $e = 0.82$, $i = 90$ deg, $\Omega = 180$ deg, $\omega = 260$ deg, $\theta = 190$ deg.
Determine the Cartesian position and velocity vectors in the perifocal-eccentricity coordinate system and the Earth-centric, Earth-equator and equinox coordinate system (ECI).

2.21 A radar tacking site located at 30 deg N, 97.5 deg W obtains the following data at 0930 GST for a satellite passing directly overhead (in Earth canonical units): Range $= 0.1$ DU, Azimuth $= 30$ deg, Elevation $= 90$ deg, Range rate $= 0.0$, Azimuth rate $= 0.0$, Elevation rate $= 10$ rad/TU.

(a) Determine the Cartesian position and velocity vectors in $\hat{X}_h\hat{Y}_h\hat{Z}_h$ frame.
(b) Assume a spherical Earth and express the Earth-centric position in the $\hat{X}_h\hat{Y}_h\hat{Z}_h$ frame.
(c) Compute the Earth-centric position and velocity vectors in the the inertia $\hat{I}\hat{J}\hat{K}$ frame.
(d) Compute the orbital elements (a, e, i, Ω, ω and θ).

2.22 Determine the true anomaly of a parabolic escape orbit from the Earth three hours after leaving a circular 200-km orbit. What are the position and velocity vectors at this point in the $\hat{X}_\omega\hat{Y}_\omega\hat{Z}_\omega$ reference frame?

2.23 A S/C is in orbit about Saturn with eccentricity of 0.6 and a semi-major axis of 9×10^5 km. Find the time of flight required to transit from $\theta = 300$ deg to $\theta = 190$ deg.

2.24 A comet is observed to have a position of $\mathbf{r} = 1.4\hat{X}_e$ AU and velocity of $\mathbf{v} = -4.0\hat{X}_e + 25.0\hat{Y}_e$ km/sec in the heliocentric-ecliptic and equinox Cartesian coordinate frame. Determine the comet's heliocentric position and velocity vector six months later in the $\hat{X}_e \hat{Y}_e \hat{Z}_e$ frame.

2.25 For an elliptical orbit, prove the following relationship:

$$\tan \phi_{fpa} = \frac{e \sin E}{\sqrt{1 - e^2}}.$$

where ϕ_{fpa} : flight-path angle, e: eccentricity, E: eccentric anomaly.

2.26 At a position, the S/C's velocity $v = 6$ km/s, its $\phi_{fpa} = 8°$. What is its $\dot{r}$, $\dot{r}$ and $|\dot{\mathbf{r}}|$ at this time?

2.27 Provide proofs for Equation (2.160) and Equation (2.208).

2.28 If an orbit's eccentricity is 1, must it be a parabola? Explain why?

2.29 Obtain the equation of motion of a mass particle in a frictionless straight-line tunnel connecting two arbitrary points on the surface of a large solid homogeneous sphere.

References

[1] Bate, R, Mueller, D, and White, J. (1971) *Fundamentals of Astrodynamics*. New York: Dover Publications, Inc.

[2] Chobotov, V. (2002) *Orbital Mechanics*, 3rd edn. AIAA Education. Series, Reston, VA: AIAA.

[3] Vallado, D. (1997) *Fundamentals of Astrodynamics and Applications*. El Segundo, CA: Mircocosm Press.

[4] Battin R.H., and Vaughan, R.J., (1987) *An Introduction to the Mathematics and Methods of Astrodynamics*. New York: AIAA.

3

Orbit Maneuver

This chapter discusses some orbital maneuver problems. The term orbit maneuver, as used here, implies the general class of problems that includes satellite interception, ballistic missile launch, gravity assist, Earth departure, Lunar exploration, etc. All maneuvers apply impulsive thrust to transfer between specific initial and terminal orbits, mainly based on the two-body problem using the basic knowledge in the previous chapter. The idea behind the impulsive thrust is that the burn times are much smaller than the time interval between burns.

Generally speaking, orbital maneuver belongs to the subject of spacecraft control.

3.1 Basic Orbital Transfer

In this section, most commonly used impulse transfers are discussed, typically the Hohmann transfer, which is the most economical solution for a large class of transfer problems, such as from LEO to GEO, lunar mission or interplanetary transfer.

3.1.1 In-plane (Coplanar) Changes

In-plane orbit maneuver can be used for acquiring precise orbit, station keeping, obtaining new orbit and rendezvous activities.

We have to change δv to realize the change of orbital size, shape (eccentricity), rotating major axis, but this coplanar maneuver cannot change orbital plane orientation.

One well-known in-plane orbital maneuver is **Hohmann transfer**. In this transfer, it is assumed that initial and final orbits are circular and coplanar; engine burn is instantaneous.

Its approach is to do a two-tangent transfer to realize minimum speed change, minimum burn, minimum fuel and minimum weight. See Figure 3.1.

3.1.1.1 Example

A communications satellite in an initial circular parking orbit with $R_1 = 6570$ km. Move the satellite to geosynchronous final orbit $R_2 = 42{,}160$ km. As we know

$$\mathcal{E} = \frac{v^2}{2} - \frac{\mu}{r} = -\frac{\mu}{2a} \Rightarrow v^2 = \mu \left(\frac{2}{r} - \frac{1}{a} \right). \tag{3.1}$$

Fundamental Spacecraft Dynamics and Control, First Edition. Weiduo Hu.
© 2015 John Wiley & Sons (Asia) Pte Ltd. Published 2015 by John Wiley & Sons (Asia) Pte Ltd.

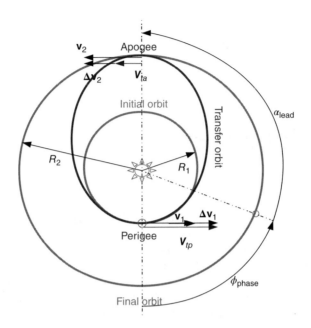

Figure 3.1 Hohmann transfer

Assume r constant at the engine burn instance, then

$$2v\,dv = \frac{\mu}{a^2}\,da \;\Rightarrow\; da = \frac{2a^2}{\mu}\,v\,dv. \tag{3.2}$$

We see the velocity change by increment Δv causes a change in a by Δa. So we can add or subtract velocity to change specific mechanical energy, and hence the orbit size (semi-major axis).

Apply Δv at perigee; this changes apogee height as $r_a + r_p = 2a$:

$$\Delta h_a \approx 2\Delta a \approx \frac{4a^2}{\mu}v_p\Delta v_p. \tag{3.3}$$

Apply Δv at apogee; this changes perigee height:

$$\Delta h_p \approx \frac{4a^2}{\mu}v_a\Delta v_a. \tag{3.4}$$

Note that as v_p is larger than the velocity at any other place, the best place to change semi-axis and height of an orbit is at its periapsis.

Then for the transfer orbit:

$$a_t = \frac{R_1 + R_2}{2} = \frac{6570 + 42\,160}{2} = 24\,365 \;\text{km} \tag{3.5}$$

$$\mathcal{E}_t = -\frac{\mu}{2a_t} = -\frac{398\,601.2}{(2)(24\,365)} = -8.1798 \;\text{km}^2/\text{s}^2 \tag{3.6}$$

$$v_{tp} = \sqrt{2\left(\frac{\mu}{R_1} + \mathcal{E}_t\right)} = 10.2460 \tag{3.7}$$

$$v_{ta} = \sqrt{2\left(\frac{\mu}{R_2} + \mathcal{E}_t\right)} = 1.5967. \tag{3.8}$$

For the initial orbit R1:

$$\mathcal{E}_1 = -\frac{\mu}{2a_1} = -\frac{\mu}{2R_1} = -30.3349 \tag{3.9}$$

$$v_1 = \sqrt{2\left(\frac{\mu}{R_1} + \mathcal{E}_1\right)} = \sqrt{\frac{\mu}{R_1}} = 7.7891 \tag{3.10}$$

$$\Delta v_1 = |v_{tp} - v_1| = 2.4569. \tag{3.11}$$

For the final orbit R2:

$$\mathcal{E}_2 = -\frac{\mu}{2R_2} = -4.7272 \tag{3.12}$$

$$v_2 = \sqrt{2\left(\frac{\mu}{R_2} + \mathcal{E}_2\right)} = \sqrt{\frac{\mu}{R_2}} = 3.0748 \tag{3.13}$$

$$\Delta v_2 = |v_{ta} - v_2| = 1.4781. \tag{3.14}$$

The total velocity needed is:

$$\Delta v_{\text{total}} = \Delta v_1 + \Delta v_2 = 3.9350 \, \text{km/sec}. \tag{3.15}$$

One orbit $TOF = 2\pi\sqrt{a^3/\mu}$, transfer requires 1/2 orbit period and as $a = 24\,365$.

$$TOF = \pi\sqrt{a^3/\mu} = 18\,924.75 \, \text{sec} = 315.4 \, \text{min} = 5 \, \text{hr}, 15 \, \text{min}. \tag{3.16}$$

For Hohmann transfer, it is the most efficient in terms of Δv, but not most efficient in terms of time.

3.1.2 Out-of-plane (Non-coplanar) Changes

For a simple plane change, it is just a change of direction to velocity vector. A combined plane change alters magnitude and direction of velocity vector simultaneously. See Figure 3.2.

As a simple plane change example, a satellite is in an orbit with inclination $i = 28°$. We want the satellite to be in an equatorial orbit whose $i = 0$. It is required to change the orbit's orientation, not its size, so that the velocity vector magnitude can be the same but the direction must be changed.

3.1.2.1 Simple Change: $v_1 = v_2$

This is an isosceles triangle relation. $\Delta v = 2 \cdot v_1 \cdot \sin(\theta/2)$. A simple method to change inclination is to change velocity at ascending or descending node. In our example: $\Delta v = 2 \cdot v_1 \cdot \sin(14°) = 0.4838 \cdot v_1 = 3.77 \, \text{km/s}$.

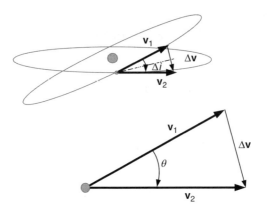

Figure 3.2   Inclination change

3.1.2.2   Combined Plane Change: $v_1 \neq v_2$

A satellite is in a low altitude parking orbit, $i = 28°$ and $R = 6570$ km, to transfer it to a geostationary orbit ($R = 42\,160$) and $i = 0°$, the Law of Cosines can be used:

$$\Delta v = \sqrt{v_1^2 + v_2^2 - 2 \cdot v_1 \cdot v_2 \cos \theta}. \tag{3.17}$$

Let's do some comparisons with the example:
Given circular orbit: $R_1 = 6570$ km, $R_2 = 42\,160$ km, $i_1 = 28°$, $i_2 = 0°$.
We want Δv_{total}. There are four options:

Case 1: first a simple 28° inclination change at R_1: $\Delta v = 3.77$; then a Hohmann transfer from orbit 1 to orbit 2: $\Delta v = 3.94$; total $\Delta v = 7.71$.

Case 2: first a Hohmann transfer from orbit 1 to orbit 2: $\Delta v = 3.94$; then a simple 28° inclination change at R_2: $\Delta v = 1.49$; total $\Delta v = 5.43$.

Case 3: first a combined plane change at perigee(R_1) of transfer orbit by Equation (3.17): $\Delta v = 4.98$; then a Δv_2 of Hohmann transfer at R_2: $\Delta v = 1.47$; total $\Delta v = 6.46$.

Case 4: first Δv_1 of Hohmann transfer: $\Delta v = 2.46$ at R_1, then a combined plane change at apogee(R_2) of transfer orbit: $\Delta v = 1.82$; total $\Delta v = 4.29$.

By comparison, it can be concluded that inclination change at high orbit can save fuel. A simple way to change inclination is at the ascending or descending node. And a good plane to change longitude of ascending is at the highest latitude of the orbit. See Figure 3.3(a).

3.1.3   The Phase Angle

For Hohmann transfer, the key is timing. Given initial orbit R_1 and final orbit R_2 (see Figure 3.1):

1. Compute a_t and *TOF*.
2. Compute angular velocity $\omega_2 = \sqrt{\mu/R_2^3}$.
3. Compute lead angle $\alpha_{\text{lead}} = \omega_2 \cdot TOF$.
4. Compute phase angle $\phi_{\text{phase}} = \pi - \alpha_{\text{lead}}$.

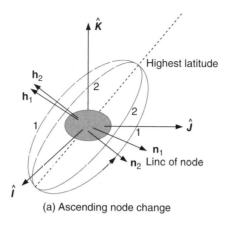

(a) Ascending node change

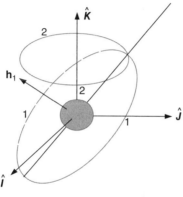

(b) Impossible transfer

Figure 3.3 Longitude change

For a rendezvous problem, the target and chaser are all in the same circular orbit at the beginning, but the target is ahead of the chaser. The chaser is slowed down to enter a phasing orbit first. Then the chaser travels 2π, while the target travels $\phi_{\text{travel}} = 2\pi - \phi_{\text{initial}}$. $\omega_{\text{target}} = \sqrt{\mu/a_{\text{target}}^3}$ and $TOF = \phi_{\text{travel}}/\omega_{\text{target}}$. The chaser's TOF is $2\pi\sqrt{a_{\text{phasing}}^3/\mu}$. Thus,

$$a_{\text{phasing}} = \left[\mu \left(\frac{\phi_{\text{travel}}}{2\pi\omega_{\text{target}}} \right)^2 \right]^{1/3}. \tag{3.18}$$

See Figure 3.4, where $r_t = r_c = a_{\text{target}}$ at the beginning.

3.2 Ballistic Missiles

An intercontinental ballistic missile (ICBM) can also be regarded as a kind of orbit maneuver. In powered flight phase, that us from launch to burnout, with energy continuously added during this phase, it is now no longer a two-body problem now. Guidance, navigation and control system are used.

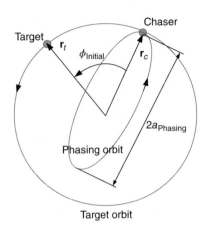

Figure 3.4 Rendezvous problem

In free-flight phase, that is, form burnout to reentry, generally it is an elliptical trajectory. Two-body analysis applies.

For the reentry phase, atmospheric drag is significant to impact. Energy continuously lessens due to friction. Phase design is required.

Detailed discussion appears in reference [1], a summary of which is given in this section.

3.2.1   Ballistic Missile Trajectory

At first some assumptions are made. Earth does not rotate. it is relaxed in later analysis. Burnout altitude is the start of reentry altitude. Then, free-flight trajectory is symmetrical, making this problem simpler. Position and velocity magnitudes at burnout, r_{bo} v_{bo} and θ_{bo}, are needed to perform this analysis.

Here are some notations:

Γ – powered flight range angle, R_p – ground range of powered flight,
Ψ – free-flight range angle, R_{ff} – ground range of free-flight,
Ω – reentry range angle, R_{re} – ground range of reentry,
Λ – total range angle, R_t – total ground range,
$\Lambda = \Gamma + \Psi + \Omega, R_t = R_p + R_{ff} + R_{re}$

Detailed description can be found in reference [1]. For convenience, its figure has been reproduced with a few changes here. See Figure 3.5.

Let $Q = v^2 r / \mu$ and $v_{cs} = \sqrt{\mu/r}, \Rightarrow Q = (v/v_{cs})^2$. For

$Q = 1$: the missile has circular speed;
$Q = 2$: the missile has escape speed and is on a parabolic orbit;
$Q > 2$: the missile is on a hyperbolic orbit.

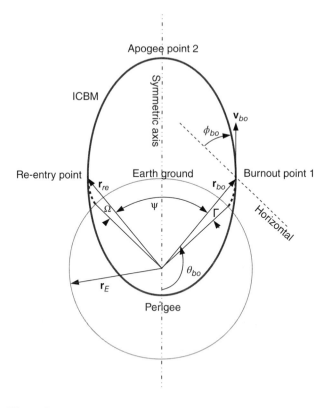

Figure 3.5 Ballistic missile trajectory adapted from reference [1]

These conclusions can be proved easily as

$$\mathcal{E} = \frac{v^2}{2} - \frac{\mu}{r} = -\frac{\mu}{2a}, \quad v^2 = \frac{\mu Q}{r}$$

$$\Rightarrow a = \frac{r}{2 - Q}, \text{ or } Q = 2 - \frac{r}{a}.$$

3.2.1.1 Free-flight Range

$$r_{bo} = \frac{p}{1 + e \cdot \cos \theta_{bo}} \Rightarrow \cos \theta_{bo} = \frac{p - r_{bo}}{e r_{bo}}. \tag{3.19}$$

Free-flight trajectory symmetric between burnout and reentry: $h_{bo} = r_{bo} - r_E = h_{re}$

$$\Rightarrow \cos\left(\frac{\Psi}{2}\right) = -\cos \theta_{bo}, \ (\theta_{bo} > 90°)$$

$$p = \frac{h^2}{\mu}, \ h = rv\cos\phi, \ e^2 = 1 - \frac{p}{a};$$

free-flight range equation:

$$\cos\left(\frac{\Psi}{2}\right) = \frac{1 - Q_{bo}\cos^2\phi_{bo}}{\sqrt{1 + Q_{bo}(Q_{b0} - 2)\cos^2\phi_{bo}}}. \tag{3.20}$$

3.2.1.2 Flight Path Angle

From the algebra and geometry of ellipse relation, the flight path angle equation becomes

$$\sin\left(2\phi_{bo} + \frac{\Psi}{2}\right) = \frac{2 - Q_{bo}}{Q_{bo}}\sin\left(\frac{\Psi}{2}\right). \tag{3.21}$$

Readers can examine when free-flight range is 90°; the missile is capable of $Q_{bo} = 0.9$. Then,

$$\sin\left(2\phi_{bo} + \frac{\Psi}{2}\right) = \frac{2 - 0.9}{0.9}\sin(45°) = 0.8624$$

$$\Rightarrow 2\phi_{bo} + \frac{\Psi}{2} = \sin^{-1}(0.8624) = 60° \text{ and } 120°$$

$$\Rightarrow \phi_{bo} = 7.5° \text{ and } 37.5°.$$

These are the low trajectory and high trajectory orbits.

3.2.1.3 Maximum Range

This occurs when the flight-path angle gives one trajectory, then,

$$\sin\left(2\phi_{bo} + \frac{\Psi}{2}\right) = \frac{2 - Q_{bo}}{Q_{bo}}\sin\left(\frac{\Psi}{2}\right) = 1$$

$$\Rightarrow 2\phi_{bo} + \frac{\Psi}{2} = 90° \Rightarrow \phi_{bo} = \frac{1}{4}\left(180° - \Psi\right).$$

Maximum range angle:

$$\sin\left(\frac{\Psi}{2}\right) = \frac{Q_{bo}}{2 - Q_{bo}}. \tag{3.22}$$

For maximum range:

$$Q_{bo} = \frac{2\sin(\Psi/2)}{1 + \sin(\Psi/2)}. \tag{3.23}$$

3.2.1.4 Free-flight

TOF is from burnout point to reentry, which is $2 \times TOF$ from burnout to apogee. Eccentric anomaly at apogee is 180°. Let point 1 be burnout point. We have:

$$\theta_1' = 180° - \Psi/2 \tag{3.24}$$

$$\cos E_1 = \frac{e + \cos\theta_1}{1 + e\cos\theta_1} = \frac{e - \cos(\Psi/2)}{1 - e\cos(\Psi/2)} \tag{3.25}$$

$$\cos(180° - \Psi/2) = \cos 180° \cdot \cos(\Psi/2) + \sin 180° \cdot \sin(\Psi/2) \tag{3.26}$$

$$t_{ff} = 2\sqrt{a^3/\mu}\left(\pi - E_1 + e\sin E_1\right), \tag{3.27}$$

where its semi-major axis is:

$$a = \frac{r}{2 - Q} \qquad (3.28)$$

and eccentricity:

$$e = \sqrt{1 + Q(Q - 2)\cos^2 \phi}. \qquad (3.29)$$

Then for given r, v, ϕ, ψ, we can have t_{ff}.

3.2.1.5 Error Effects

There are different types of error effects. In-plane errors result in "down-range" errors. Out-of-plane errors result in "cross-range" errors.

Cross-range errors are the lateral displacement of the burnout point. Assume a cutoff point displaced by Δx, with cross-range error being $\Delta c \approx \Delta x \cdot \cos \Psi$ [1].

If launch azimuth error is $\Delta \beta$, cross-range error being $\Delta c \approx \Delta \beta \cdot \sin \Psi$ [1].

Down-range errors are the burnout point displacement. If the downrange error at cutoff/burnout is Δx, then the downrange error at impact is Δx.

At burnout, if the flight path angle error is $\Delta \phi_{bo}$, then the downrange error at impact is

$$\Delta \Psi = \frac{2 \sin \left(\Psi + 2\phi_{bo} \right) - 2}{\sin 2\phi_{bo}} \cdot \Delta \phi_{bo}. \qquad (3.30)$$

If the burnout height error is Δr_{bo}, then the downrange error at impact is

$$\Delta \Psi = \frac{4\mu}{v_{bo}^2 \cdot r_{bo}^2} \cdot \frac{\sin^2 \left(\frac{\Psi}{2} \right)}{\sin 2\phi_{bo}} \cdot \Delta r_{bo}. \qquad (3.31)$$

If the incorrect speed at burnout causing burnout speed error is Δv_{bo}, then the downrange error at impact is

$$\Delta \Psi = \frac{8\mu}{v_{bo}^3 \cdot r_{bo}} \cdot \frac{\sin^2 \left(\frac{\Psi}{2} \right)}{\sin(2\phi_{bo})} \cdot \Delta v_{bo}. \qquad (3.32)$$

3.2.2 Effect of the Earth's Rotation

In free-flight portion, the orbit is inertial, which is fixed in $\hat{I}\hat{J}\hat{K}$. Launch point and target are in motion. Launch site motion will add into the missile velocity at burnout. Target motion will compensate by "leading" the target. To analyze Earth rotation effect, three quantities at burnout are needed: speed v_e, elevation angle El, and azimuth angle Az. See Figure 1.8. The point on the equator's eastward speed is 1524 ft/sec = 464.5 m/sec. Let L_0 is the latitude of a launch site; its eastward speed is $v_0 = 0.4645 \cdot \cos L_0$ km/ sec.

Missile velocity in $\hat{S}\hat{E}\hat{Z}$ system (see Figure 2.19):

$$v_S = -v_e \cos El \cos Az \qquad (3.33)$$

$$v_E = v_e \cos El \sin Az + v_0 \qquad (3.34)$$

$$v_Z = v_e \sin El. \qquad (3.35)$$

True speed is $v = \sqrt{v_S^2 + v_E^2 + v_Z^2}$. Flight-path angle is $\sin El = v_Z / v$. Azimuth is $\tan Az = -v_E / v_S$.

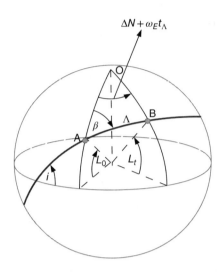

Figure 3.6   The actual range

3.2.2.1   Target Motion

If the launch point (A)'s latitude and longitude are (L_0, N_0) and the target (B)'s is at (L_t, N_t) with total range angle Λ (see Figure 3.6):

$$\cos \Lambda = \sin L_0 \cdot \sin L_t + \cos L_0 \cdot \cos L_t \cdot \cos(\Delta N + \omega_E t_\Lambda). \tag{3.36}$$

Launch azimuth β:

$$\sin L_t = \sin L_0 \cdot \cos \Lambda + \cos L_0 \cdot \sin \Lambda \cdot \cos \beta \tag{3.37}$$

$$\Rightarrow \cos \beta = \frac{\sin L_t - \sin L_0 \cdot \cos \Lambda}{\cos L_0 \cdot \sin \Lambda}. \tag{3.38}$$

To calculate Λ, we need to know time-of-flight t_Λ. To calculate t_Λ, total range Λ is needed. For the solution approach we need to iterate that. ω_E is the Earth's rotation rate.

Example 1

Missile launch at 29°N, 79.3°W, burns out at 30°N, 80°W, $v_e = 0.9$ DU/TU, $Al = 30°$, $Az = 330°$, $r_e = 1.1$ DU. Then the coordinates of reentry point are:

$$v_S = -v_e \cdot \cos El \cdot \cos Az = -0.6750$$

$$v_E = v_e \cdot \cos El \cdot \sin Az + v_0 = -0.3382$$

$$v_Z = v_e \cdot \sin El = 0.4500$$

$$v = 0.8789$$

$$\sin \phi_{bo} = \frac{v_Z}{v} \Rightarrow \phi_{bo} = 30.8$$

$$\tan \beta = -\frac{v_E}{v_S} \Rightarrow \beta = -26.7 \rightarrow 333.3$$

$$Q_{b0} = \frac{v^2 r}{\mu} = 0.85. \tag{3.39}$$

See Figure 6.2-9 in reference [1]. $Q_{b0} = 0.85$, $\phi_{bo} \approx 31 \Rightarrow \Psi = 90°$.
Reentry lat:

$$\sin L_{re} = \sin L_{bo} \cdot \cos \Psi + \cos L_0 \cdot \sin \Psi \cdot \cos(360 - \beta) \tag{3.40}$$

$$\Rightarrow L_{re} = 50.71° \tag{3.41}$$

$$\Delta N + \omega \cdot t_{ff}: \tag{3.42}$$

$$\cos(\Delta N + \omega \cdot t_{ff}) = \frac{\cos \Psi - \sin L_{re} \cdot \sin L_{bo}}{\cos L_{re} \cdot \cos L_{bo}} \Rightarrow \Delta N + \omega \cdot t_{ff} = 134.9°. \tag{3.43}$$

See also Figure 6.2-9 in reference [1].

$$Q_{b0} = 0.85, \ \phi_{bo} \approx 31, \ \Rightarrow \frac{t_{ff}}{TP_{cs}} \approx 0.46$$

$$TP_{cs} = 2\pi \sqrt{r_{bo}^3 / \mu} = 7.25 \Rightarrow t_{ff} = 3.334$$

$$\Delta N = 134.9 - \omega \cdot t_{ff} = 123.66$$

$$N_{re} = N_{bo} + \Delta N = -203.66 \rightarrow 156.34°E.$$

Example 2

Launch point: $L_0 = 60° $ N, $N_0 = 160°$ E
Target: $L_t = 0°$ N, $N_t = 115°$ W
Find: Distance from launch point to target.

As Λ is a function of t_{ff}, and longitude difference is also changed with the Earth's rotation, a numerical procedure is needed. To show the physical meaning, let us give a simple constraint of $\Lambda = \delta N + \omega_E t_{ff}$. At the initial stage, $\delta N = 180 - N_0 + 180 - N_t = 85°$,

$$\cos \Lambda = \sin(L_0) \cdot \sin(L_t) + \cos(L_0) \cdot \cos(L_t) \cdot \cos(\Delta N) \tag{3.44}$$

$$\Rightarrow \Lambda = 87.5023, \text{Target} : 115 - (87.5023 - 85) = 112.4976$$

$$\Rightarrow \Lambda = 88.7514, \text{Target} : 115 - (88.7514 - 85) = 111.2485$$

$$\Rightarrow \Lambda = 89.3757, \text{Target} : 115 - (89.3757 - 85) = 110.6242$$

$$\Rightarrow \Lambda = 89.6878, \text{Target} : 115 - (89.6878 - 85) = 110.3121$$

$$\Rightarrow \Lambda = 89.8439, \cdots$$

The algorithm converges. At final time $\Lambda = 90°$, the flight time is 20 min. At the same time the Earth rotates 5°. At launch time, the target is at 0°N, 110°W.

This analysis can be used to estimate missile type and launch azimuth when a launch's space-based sensors are quite good. Missile range and time parameters can be determined by missile type. Range is a 5th degree orthogonal polynomials which is a function of launch latitude and azimuth. Time is also a 5th degree orthogonal polynomials which is a function of launch latitude and range.

Launch azimuth is converted to azimuth plus error which creates an azimuth fan. The threat area can be created which includes minimum and maximum range, lower and upper azimuth. More detailed discussions on this section can be found in reference [1].

3.3   Lunar Missions

Before discussing lunar missions, it's best to be familiar with some constants regarding about the Moon. $m_{moon} = 7.348\,310^{22}$ kg, $R_{moon} = 1.738 \times 10^3$ km at its equator, $\mu_{moon} = Gm_{moon} = 4.903 \times 10^3$ km³/s², orbit eccentricity around the Earth $e_{moon} = 0.0549$, orbit inclination to ecliptic $i_{moon} = 5°09'$.

3.3.1   Possibility of Transfer

A convenient method is the **Simple Hohmann Trajectory** (recall Figure 3.1).

Assume S/C in circular parking orbit with $r = 6700$ km; the distance from the Earth to the Moon is $D = 384\,400$ km (assume circular orbit). The basic orbit maneuver is

$$a_t = \frac{6700 + 384\,400}{2} = 195\,550, \; \mathcal{E}_t = -\frac{\mu}{2a} = -1.0192 \tag{3.45}$$

$$v_1 = \sqrt{2\left(\frac{\mu}{r_1} + \mathcal{E}_t\right)} = 10.8142, \; v_{cs} = \sqrt{\frac{\mu}{r_1}} = 7.7132, \; \Delta v = 3.1010 \tag{3.46}$$

$$v_2 = \sqrt{2\left(\frac{\mu}{r_2} + \mathcal{E}_t\right)} = 0.1884, \; v_{cs} = \sqrt{\frac{\mu}{r_2}} = 1.0183, \; \Delta v = 0.8299. \tag{3.47}$$

Total $\Delta v = 3.9309$, $TOF = \pi\sqrt{a_t^3/\mu} \cong 119.5$ hrs.

TOF is excessive–recall that the Hohmann transfer is efficient (Δv), but time is long. In the case of manned flight, life support needs increase with mission duration and something different is required. For a more general case, r_1 is Earth parking orbit, r_2 is the Moon's orbit, and the transfer's orbit must satisfy $r_p \le r_1$ and $r_a \ge r_2$ for possible intersection. See Figure 3.7(a). The transfer orbit could be an ellipse, a parabola or a hyperbola. An integrated analysis is given in Figure 3.7(b). In this figure, the transfer orbit

$$r_p = \frac{p}{1+e} \le r_1 \implies e \ge \frac{p - r_1}{r_1} \tag{3.48}$$

$$r_a = \frac{p}{1-e} \ge r_2 \implies e \ge \frac{r_2 - p}{r_2} \tag{3.49}$$

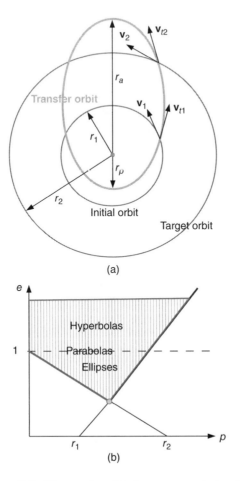

(a)

(b)

Figure 3.7   The transfer orbits between two circle orbits

of course p must be positive. Note that the circle point with minimum eccentricity in Figure 3.7(b) is the Hohmann transfer orbit. This possibility analysis can be extended to interplanetary travel, as for example: r_1 is the Earth orbit, r_2 is the Mars orbit is:

Numerical verification of the analysis is shown in Figure 3.8; the vertical axis is the velocity, Δv, which is the required for the transfer. It is a function of p and e. We can see the minimum fuel realized for the Hohmann transfer. Interested readers can verify this, in this example: $r_1 = 1$, $r_2 = 3$, $\mu = 1$; then the Hohmann orbit is $p = 1.5$, $e = 0.5$, the total $\Delta v = 0.3938$.

The actually Earth – Moon system is fairly complex. The system revolves about its common mass center. The orbital period of the Moon is slightly increasing and the Moon's orbital elements change with time. The mean distance between the Earth and Moon is 384 400 km. The Moon's orbit is inclined to the Earth's orbital plane; its eccentricity is very near 0.

Some simplifying assumptions are given here. The Moon's orbit is circular with $r = 384\,400$ km. The terminal attraction of the moon is neglected. Lunar trajectory is coplanar

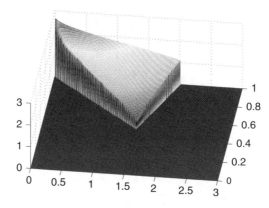

Figure 3.8 The numerical analysis of the transfer orbits between two circle orbits

with the Moon's orbit. The TOF can be calculated as a function of speed, again,

$$\mathcal{E} = \frac{v^2}{2} - \frac{\mu}{r}, \ h = rv \cos \phi_{fpa} \tag{3.50}$$

$$p = \frac{h^2}{\mu}, \ a = -\frac{\mu}{2\mathcal{E}}, \ e = \sqrt{1 - p/a}. \tag{3.51}$$

Finally,

$$\cos \theta = \frac{p - r}{e \cdot r}, \tag{3.52}$$

when r is the leaving orbit, θ_0 can be obtained; when r is the Moon's orbit, θ is obtained at arrival, then calculate E and time-of-flight (see Chapter 2). *TOF* is a function of injection speed (see more on in reference [1].

3.3.1.1 Trajectory Speeds

Assume transfer starts at perigee ($\phi_{fpa} = 0$); it is important to examine the trajectory vs. speed relationship.

When the S/C's speed is at infinity, the path to the Moon is a straight line, $TOF = 0$. Lower speed trajectory could be hyperbolic, parabolic or elliptical.

For Elliptical Hohmann transfer orbit is maximum in *TOF*, minimum in transfer (injection) speed, minimum in eccentricity, and slowest approach speed at the Moon.

3.3.1.2 Rendezvous Miss Distance

The miss distance is when the S/C is at the Moon's orbit where the instant Moon is at intercept point. Injection speed is approx. 11 km/sec; sweep angle is near 160 km/sec; and error effects can be canceled. The biggest contributor is the flight-path angle. Correction is needed when considering the lunar gravity.

Figure 3.9 Sphere of influence

3.3.1.3 Sphere of Influence

This analysis relies on construction of SOI – Csphere of influence. First assume S/C under gravitational influence of Earth's SOI, at some point S/C under gravitational influence of the Moon–Moon's SOI. This is an approximation analysis. More discussion is given in Section 3.4.1.

Laplace's work: let D to be the distance between the Earth and the Moon again.

$$R_{SOI} = D \cdot \left(\frac{m_M}{m_E}\right)^{2/5}$$

$$= (384,400)\left(\frac{7.3483 \times 10^{22}\text{kg}}{5.9737 \times 10^{24}\text{kg}}\right)^{2/5}$$

$$\approx 6.63 \times 10^4 \text{km}.$$

The Earth–Moon system is 5/6 of the way under Earth's SOI, 1/6 of the way under Moon's SOI, about 38 times the Moon's radius. See Figure 3.9.

3.3.1.4 Noncoplanar Lunar Trajectories

Assumeing that the lunar trajectory lay in the plane of the Moon's orbit, the Moon's orbit plane varies from 18.2° to 28.5° over 18.6 years. Cape Kennedy latitude is 28.5°. If the Moon is not at max inclination, a plane change must be performed. Alternatively a noncoplanar trajectory is needed, which is a function of launch latitude and azimuth, total sweep angle of trajectory and Moon's declination at intercept. This is a complex function of variables.

3.3.2 *More Practical Scenario*

To know the Earth departure conditions, four quantities need to be specified: $r_0, v_0, \phi_0, \gamma_0$, where γ_0 is the phase angle at departure between the S/C and the Moon. Detailed discussion can be found in reference [1], but a summary is given here. It requires *TOF* iteration.

Choose $r_0, v_0, \phi_0, \lambda_1$, where λ_1 is the point at which the geocentric trajectory crosses the lunar sphere of influence. See Figure 3.10.

$$\mathcal{E} = \frac{v_0^2}{2} - \frac{\mu}{r_0}. \text{ and } h = r_0 v_0 \cos\phi_0. \tag{3.53}$$

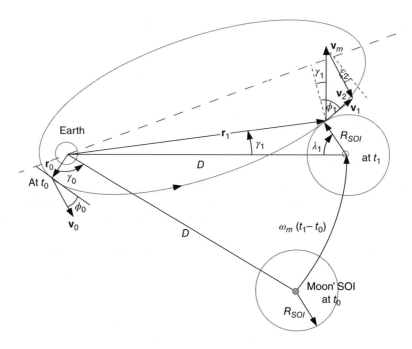

Figure 3.10 The Earth – Moon transfer adapted from reference [1]

Law of Cosines:

$$r_1 = \sqrt{D^2 + R_S^2 - 2 \cdot D \cdot R_{SOI} \cdot \cos \lambda_1} \tag{3.54}$$

$$v_1 = \sqrt{2\left(\mathcal{E} + \frac{\mu}{r_1}\right)} \text{ and } \cos \phi_1 = \frac{h}{r_1 \cdot v_1}.$$

Law of Sines:

$$\sin \gamma_1 = \frac{R_{SOI}}{r_1} \cdot \sin \lambda_1. \tag{3.55}$$

Geocentric Trajectory

$$p = \frac{h^2}{\mu}, \ a = -\frac{\mu}{2\mathcal{E}}, \ e = \sqrt{1 - p/a}.$$

$$\cos \theta_0 = \frac{p - r_0}{e \cdot r_0} \text{ and } \cos \theta_1 = \frac{p - r_1}{e \cdot r_1}. \tag{3.56}$$

Eccentric anomalies by Equation (2.155):

$$\cos(E_0) = \frac{e + \cos \theta_0}{1 + e \cdot \cos \theta_0} \text{ and } \cos(E_1) = \frac{e + \cos \theta_1}{1 + e \cdot \cos \theta_1}$$

$$t_1 - t_0 = \sqrt{a^3/\mu} \left[(E_1 - e \cdot \sin E_1) - (E_0 - e \cdot \sin E_0) \right].$$

Moon moves through angle $\omega_m \cdot (t_1 - t_0)$.

Phase angle at departure:

$$\gamma_0 = \theta_1 - \theta_0 - \gamma_1 - \omega_m \cdot (t_1 - t_0). \tag{3.57}$$

3.3.2.1 Patch Point

When S/C arrives at the Moon's SOI, it is also a two-body problem between S/C and the Moon. Variables with subscript 2 are relative to the center of the Moon and are needed here: $r_2 = R_{SOI}$, $\mathbf{v}_2 = \mathbf{v}_1 - \mathbf{v}_m$, where $\mathbf{v}_m$ is velocity of moon wrt center of the Earth. It equals $v_m = 1.018\,\text{km/sec}$. Then

$$v_2 = \sqrt{v_1^2 + v_m^2 - 2 \cdot v_1 \cdot v_m \cdot \cos(\phi_1 - \gamma_1)}. \tag{3.58}$$

Let ξ_2 be the angle between $\mathbf{v}_2$ and a line to the Moon's center (Figure 7.4-2 in reference [1]). As

$$v_2 \sin \xi_2 = v_m \cos \lambda_1 - v_1 \cos(\lambda_1 + \gamma_1 - \phi_1),$$

then ξ_2 can be solved.

3.3.2.2 Arrival Orbit

The mission is a lunar impact if $r_p < r_m = 1738\,\text{km}$. If we want the mission to be a lunar orbit, A Δv required for a circular lunar orbit can be calculated. Is the mission a circumlunar flight? Circumlunar conditions can be calculated and so do the conditions for exit of the Moon's SOI.

At perigee of lunar orbit:

$$\mathcal{E} = \frac{v_2^2}{2} - \frac{\mu_m}{r_2}, \quad \mu_m = 4.90\,287 \times 10^3 \text{km}^3 / \text{sec}^2$$

$$h = r_2 \cdot v_2 \sin \xi_2$$

$$p = \frac{h^2}{\mu_m} \quad \text{and} \quad e = \sqrt{1 + 2\mathcal{E} \cdot \frac{h^2}{\mu_m^2}}$$

$$r_p = \frac{p}{1+e} \quad \text{and} \quad v_p = \sqrt{2\left(\mathcal{E} + u_m/r_p\right)}, \tag{3.59}$$

the r_p and v_p can be checked if they have satisfied the circumlunar condition. If not, iterate with r_0, v_0, ϕ_0, or λ_1.

3.3.2.3 An Example

For a lunar probe mission to the Moon, its trajectory (injection) conditions are: $r_0 = 1.05\,\text{DU}, v_0 = 1.372\,\text{DU/TU}, \phi_0 = 0, \lambda_1 = 30°$, then

$\mathcal{E} = v_0^2/2 - \mu/r_0 = -0.0112\,(\text{DU/TU})^2$, $h = r_0 \cdot v_0 \cos \phi_0 = 1.4406$, as $D = 384400\,\text{km} = 60.2683\,\text{DU}$, $R_{SOI} = 66\,300\,\text{km} = 10.3949\,\text{DU}$, then

$$r_1 = \sqrt{D^2 + R_S^2 - 2D \cdot R_{SOI} \cos \lambda_1} = 51.5288\,\text{DU}$$

$$v_1 = \sqrt{2\,(\mathcal{E} + \mu/r_1)} = 0.1282\,\text{DU/TU}$$

$$\phi_1 = \cos^{-1}\left(\frac{h}{r_1 \cdot v_1}\right) = 77.40°$$

$$\gamma_1 = \sin^{-1}\left(\frac{R_S}{r_1} \cdot \sin \lambda_1\right) = 5.79°.$$

Conditions for geocentric transfer to the Moon's SOI

Need *TOF*:

$$p = \frac{h^2}{\mu} = 2.0753, \quad a = -\frac{\mu}{2\mathcal{E}} = 44.6869, \quad e = \sqrt{1 - p/a} = 0.9765$$

$\theta_0 = 0$ (probe burns out at perigee)

$$\theta_1 = \cos^{-1}\left(\frac{p - r_1}{e \cdot r_1}\right) = 169.36°(2.9560\,\text{rad})$$

$$E_0 = 0, \quad E_1 = \cos^{-1}\left(\frac{e + \cos \theta_1}{1 + e \cdot \cos \theta_1}\right) = 99.02°(1.7282\,\text{rad})$$

$$TOF = \sqrt{a^3/\mu}\left[(E_1 - e \cdot \sin E_1) - (E_0 - e \cdot \sin E_0)\right] = 228.17\,\text{TU} = 51.14\,\text{hrs}$$

Phase-angle:

$$\gamma_0 = \theta_1 - \theta_0 - \gamma_1 - \omega_m \cdot TOF = 135.43°$$

Now at the Moon's SOI – convert reference:
$v_1 = 0.1282\,\text{DU/TU} \rightarrow 1.0135\,\text{km/sec}$, $R_S = 10.3949\,\text{DU} \rightarrow 66\,300\,\text{km}$
$\mu_M = \mu_E/81.3 = 4902.84\,\text{km}^3/\text{sec}^2$, $r_M = 1738 - \text{km}$, then

$$v_2 = \sqrt{v_1^2 + v_m^2 - 2v_1 \cdot v_m \cdot \cos(\phi_1 - \gamma_1)} = 1.1885 - \text{km/sec}$$

$$\sin \xi_2 = \frac{v_m}{v_2} \cdot \cos \lambda_1 - \frac{v_1}{v_2} \cdot \cos(\lambda_1 + \gamma_1 - \phi_1) \Rightarrow \xi_2 = 5.9846°.$$

Closest approach of probe:

$$h = r_2 \cdot v_2 \cdot \sin \xi_2 = 8215.9\text{km}^2/\text{s}, \quad \mathcal{E} = \frac{v_2^2}{2} - \frac{\mu_m}{R_S} = 0.6324$$

$$p = \frac{h^2}{\mu_m} = 13768, \quad e = \sqrt{1 + 2\mathcal{E}(h/\mu_m)^2} = 2.1334$$

$$r_p = \frac{p}{1 + e} = 4393.8$$

$$\Rightarrow h_p = r_p - r_m = 2655.8\,\text{km}. \tag{3.60}$$

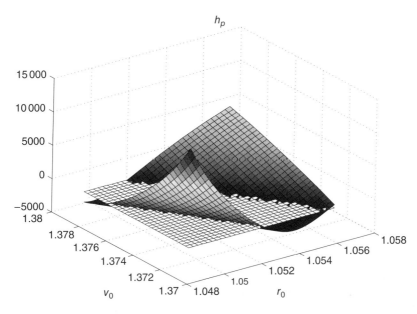

Figure 3.11 Relation between the minimum height h_p above the Moon and initial position r_0 and velocity v_0 about the Earth

The results are sensitive to initial conditions such as r_0, v_0, See Figure 3.11.

$$\mathcal{E} = \frac{v_0^2}{2} - \frac{\mu}{r_0} \text{ and } h = r_0 v_0 \cos \phi_0. \tag{3.61}$$

The value below the level plane is the collision orbit to the Moon since h_p is less then zero.

3.4 Interplanetary Travel

In an interplanetary mission, some knowledge about planets' orbital elements and their characteristics are required. Basically, they are also in the scope of the two-body problem.

Assumptions are that only two bodies involved. For spacecraft and the Earth, Earth's gravity is the only force on the spacecraft; Geocentric-Equatorial System (ECI, i.e. $\hat{I}\hat{J}\hat{K}$ system) is suitable for the Earth-based problems.

For an Interplanetary travel needs a different reference frame. Since the Sun's gravity becomes the dominant force, the Sun-centered system and heliocentric–ecliptic system are used.

In the heliocentric–ecliptic system, the Sun is centered at the origin. The fundamental plane is the plane of the Earth's orbit around the Sun, ecliptic plane. Principal direction $\hat{X}_e$ is the vernal equinox direction. $\hat{Y}_e$ in ecliptic plane, 90° from $\hat{X}_e$ in direction of the Earth's motion. $\hat{Z}_e$ normal to ecliptic plane. See Figure 1.4.

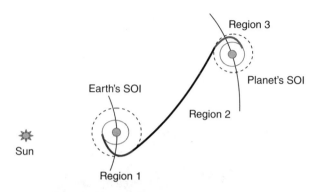

Figure 3.12   Patched conic approximation

Some simplifying assumptions are made. Though it is really a four-body problem, Spacecraft-Earth-Sun-Target Planet, the hard problem is broken into multiple easier problems, i.e. three two-body problems, Spacecraft-Sun, Spacecraft-Earth, Spacecraft-Target planet. The method to integrate (patch) these smaller problems into a single trajectory is the so-called Patched Conic Approximation (PCA). The approach is a very simple, convenient, and not altogether accurate method for performing calculations. Results will be accurate to a few percent, which is adequate at this level of design analysis. See Figure 3.12.

In this method, three analysis regions are defined:

Region 1: Earth departure – Earth's gravity dominates,
Region 2: Sun-centered transfer from Earth to target planet. Sun's gravity dominates.
Region 3: Arrival at target planet – planet's gravity dominates.

The boundary of these regions in PCA is the key.

3.4.1   Sphere of Influence (SOI)

By Newton's law, the gravity is inversely proportional to the distance, as the distance goes to infinity, the gravity approaches zero, that is:

$$F_g \propto 1/R^2 \quad R \to \infty \Rightarrow F_g \to 0.$$

F_g is never quite 0, but insignificant compared to other forces.

Laplace's work showed a practical view of sphere of influence (SOI): sphere of radius such that a body's gravitational force dominates; at distances greater than radius another body's gravity dominates.

Let us show the outline of the derivation. It is the same to the third body perturbation problem which is discussed in Chapter 5. See Equation (5.138). The vehicle (S/C) is attracted by the planet, perturbed by the Sun. See Figure 3.13.

$$\ddot{\mathbf{r}}_{pv} + \frac{G(m_p + m_v)}{r_{pv}^3}\mathbf{r}_{pv} = -Gm_S \left(\frac{\mathbf{r}_{Sv}}{r_{Sv}^3} - \frac{\mathbf{r}_{Sp}}{r_{Sp}^3} \right). \tag{3.62}$$

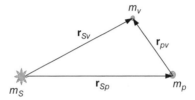

Figure 3.13   Vehicle (S/C) m_v attracted by a planet m_p perturbed by the Sun m_S

Simply, $\ddot{\mathbf{r}}_{pv} - \mathbf{A}_p = \mathbf{P}_S$. Similarly, a vehicle is attracted by the Sun, perturbed by a planet,

$$\ddot{\mathbf{r}}_{Sv} + \frac{G(m_S + m_v)}{r_{Sv}^3}\mathbf{r}_{Sv} = -Gm_p\left(\frac{\mathbf{r}_{pv}}{r_{pv}^3} + \frac{\mathbf{r}_{Sp}}{r_{Sp}^3}\right). \tag{3.63}$$

Simply, $\ddot{\mathbf{r}}_{Sv} - \mathbf{A}_S = \mathbf{P}_p$.

The definition of SOI is that it is the surface along which the magnitudes of the acceleration satisfy

$$\frac{P_p}{A_S} = \frac{P_S}{A_p}.$$

Then, the planet's SOI can be approximated, as solved in reference [3].

$$R_{SOI} = D_{\text{planet}} \cdot \left(\frac{m_p}{m_S}\right)^{2/5}, \tag{3.64}$$

where D_{planet} is the mean semi-major axis of the planets in the solar system. It can also be regarded as the mean distance from a planet to the Sun, where m_p is the mass of the planet, m_S is the mass of the Sun. And m_v is the mass of the vehicle, which is neglected.

SOIs of the planets in the solar system are listed in Table 3.1, where $m_S = 1.989 \times 10^{30}$ kg. Note that the Moon's SOI is for the Earth–Moon system.

3.4.2   Scenario

A space probe is being planned to map potential landing sites on Mars. The probe will leave Earth from a circular parking orbit of 6700 km and arrive at Mars for a circular parking orbit of 3580 km.

Determine initial estimates (approximations) of the total velocity change, transfer time, and planet phasing requirements for the mission.

3.4.2.1   Assumptions

Earth's orbit is circular with $R_E = 1.496 \times 10^8$ km, Mars' orbit is circular with $R_M = 2.278 \times 10^8$ km, the planets' orbits are coplanar, Patched-Conic Approximation is sufficiently accurate. The gravity parameters of μ_E, μ_S, μ_M can be found in Table 2.1.

Table 3.1 SOIs of the planets in the solar system. Comparison can be made with Table 2.1, Table 3.2 and Table 3.3

Planet	Distance from Sun (10^6 km)	Mass (10^{24} kg)	R_{SOI} (10^3 km)	Equatorial radius (10^3 km)	SOI radius (body radii)
Mercury	57.9	0.3302	112	2.436	46
Venus	108.2	4.8690	616	6.153	100
Earth	149.6	5.9742	925	6.378	145
Mars	228.0	0.6419	577	3.393	170
Jupiter	778.5	1898.8	48 216	71.491	677
Saturn	1433.4	568.5	54 802	60.527	905
Uranus	2872.3	86.625	51 740	27.167	1904
Neptune	4495.4	102.78	86 710	22.428	3886
Pluto	5870.2	0.015	3308	1.148	2881
Moon	0.3844	0.073 483	66.3	1.738	38

3.4.2.2 Region 2 – Sun-centered transfer

Hohmann transfer from Earth to Mars (see Figure 3.1):

$$a_t = \frac{R_E + R_M}{2} = 1.887 \times 10^8 \text{ km}$$

$$\mathcal{E}_t = -\frac{\mu_S}{2a_t} = -351.62 \text{ km}^2/\text{sec}^2, \; \mu_S = 1.327 \times 10^{11}$$

$$v_1 = \sqrt{2\left(\frac{\mu_S}{R_E} + \mathcal{E}_t\right)} = 32.72 \text{ km/sec}$$

$$v_2 = \sqrt{2\left(\frac{\mu_S}{R_M} + \mathcal{E}_t\right)} = 21.49 \text{ km/sec.}$$

For the transfer ellipse, the start transfer at Earth is $v = 32.72$. The end transfer at Mars: $v = 21.49$. $TOF = \pi \cdot \sqrt{a_t^3/\mu_S} = 22\,354\,876 \text{ sec} \approx 258.7 \text{ days} \approx 8.5 \text{ months} = 0.71 \text{ year.}$

3.4.2.3 Region 1 – Earth Departure

Spacecraft must leave Earth's SOI with a velocity for a hyperbolic trajectory to v_∞. v_∞ is hyperbolic excess velocity. This scenario is similar to the case in Figure 2.11. $\mathcal{E}_E = -\mu_S/(2R_E) = -443.52 \text{ km}^2/\text{sec}^2$. $v_E = \sqrt{2\left(\mu_S/R_E + \mathcal{E}_E\right)} = 29.78$ km/sec is velocity of the Earth around the Sun. Calculated $v_1 = 32.72$, so $v_\infty = |v_1 - v_E| = 2.294$ km/sec. $\mathcal{E}_\infty = v_\infty^2/2 = 4.32$ when $R_\infty = \infty$ and $\mathcal{E}_\infty > 0$.

For the circular parking orbit

$$v_p = \sqrt{\frac{\mu_E}{R_{\text{park}}}} = 7.71. \tag{3.65}$$

Table 3.2 The Hohmann transfer orbits to the planets in the solar system. Comparison can be made with Table 2.1, Table 3.1 and Table 3.3

Target planet	Launch velocity at Earth surface (km/s)	Flight duration (year)	Minimum lay-over (year)	Total 2-way duration (year)	Velocity at planet surface (km/s)	Orbital planet velocity (km/s)
Mercury	13.4	0.29	0.18	0.76	2.996	47.8725
Venus	11.3	0.40	1.28	2.08	7.256	35.0214
Mars	11.4	0.71	1.24	2.66	3.569	24.1309
Jupiter	14.2	2.73	0.59	6.05	42.15	13.0697
Saturn	15.2	6.05	0.94	13.04	25.07	9.6724
Uranus	15.9	16.04	0.93	33.01	15.73	6.8352
Neptune	16.2	30.62	0.77	62.01	17.58	5.4778
Pluto	16.3	45.47	0.06	91.00	0.49	4.749

Hyperbolic velocity wrt Earth is

$$v_h = \sqrt{2\left(\frac{\mu_E}{R_{\text{park}}} + \mathcal{E}_\infty\right)} = 11.30$$

$$\Delta v = \left|v_h - v_p\right| = 3.59 \text{ km/s}.$$

3.4.2.4   Region 3 – Mars Arrival

Spacecraft arrives at target planet on heliocentric transfer ellipse. When the target planet's orbit is outside the Earth's orbit (e.g. Mars), spacecraft to arrive at the apogee of the transfer ellipse must be ahead of and moving slower than the planet. The target planet overtakes the spacecraft and the spacecraft enters the planet's SOI in front of the planet.

When the target planet's orbit is inside the Earth's orbit (e.g. Venus), the spacecraft to arrive at perigee of transfer ellipse must be behind and moving faster than the planet. The spacecraft overtakes the planet and enters SOI from behind the planet.

The reference must be switched to a Mars-based equivalent of the Geocentric-Equatorial System. Spacecraft velocity wrt to Mars as it enters SOI is $v_{\infty M}$. Spacecraft velocity on heliocentric transfer ellipse is

$$v_{\text{transferatmars}} = \sqrt{2\left(\frac{\mu_S}{R_M} + \mathcal{E}_t\right)} = 21.49 \text{ km/ sec.} \tag{3.66}$$

Planet velocity is

$$\mathcal{E}_M = -\frac{\mu_S}{2R_M} = -291.26 \text{ km}^2/\text{sec}^2 \tag{3.67}$$

$$v_M = \sqrt{2\left(\frac{\mu_S}{R_M} + \mathcal{E}_M\right)} = 24.14 \text{ km/ sec} \tag{3.68}$$

$$v_{\infty M} = \left|v_{\text{transferatmars}} - v_M\right| = 2.65 \text{ km/ sec.} \tag{3.69}$$

Hyperbolic velocity wrt Mars is

$$\mathcal{E}_{\infty M} = \frac{v_{\infty M}^2}{2} = 3.51, \text{ as } R \to \infty, \tag{3.70}$$

$$v_h = \sqrt{2 \left(\frac{\mu_M}{R_{\text{parkatmars}}} + \mathcal{E}_{\infty M} \right)} = 5.57. \tag{3.71}$$

Velocity for parking orbit at Mars is

$$v_p = \sqrt{\frac{\mu_M}{R_{\text{parkatmars}}}} = 3.47 \tag{3.72}$$

$$\Delta v = \left| v_h - v_p \right| = 2.10 \text{km/sec}. \tag{3.73}$$

Total mission $\Delta v = \Delta v_E + \Delta v_M = 3.59 + 2.10 = 5.69$ km/sec, and the TOF ≈ 8.5 months.

3.4.2.5   Planet Phasing

Angular velocities:

$$\omega = \sqrt{\mu_S/R_{\text{orbit}}^3} \tag{3.74}$$

$$\omega_E = 1.9908 \times 10^{-7} \text{rad/sec} = 2\pi/\text{year} \tag{3.75}$$

$$\omega_M = 1.0595 \times 10^{-7} \text{rad/sec}. \tag{3.76}$$

Lead angle:

$$\alpha_{\text{lead}} = \omega_M \cdot TOF = 2.3685 \text{ rad} = 135.71°.$$

Phase angle:

$$\phi_{\text{phase}} = 180 - \alpha_{\text{lead}} = 44.29°.$$

See Figure 3.1. Suppose the Earth is 50° behind Mars. How long is "wait time" until the proper relative position? $\phi_{\text{initial}} = 50°(0.8727 \text{ radians})$, $\phi_{\text{phase}} = 44.29°(0.7730 \text{ radians})$. $\omega_E = 1.994 \times 10^{-7}$ rad/sec , $\omega_M = 1.060 \times 10^{-7}$ rad/sec.

$$\text{Wait time} = \frac{\phi_{\text{phase}} - \phi_{\text{initial}}}{\omega_M - \omega_E} = 1\,067\,451.8 \text{ sec} \approx 12.35 \text{ days}. \tag{3.77}$$

If the opportunity for launch at proper phasing is missed, how long will it be until the next opportunity? This is the synodic period problem.

$$SP = \frac{2\pi}{\omega_S - \omega_M} \approx 2.13 \text{ years}. \tag{3.78}$$

After arrival at Mars, the waiting time to be back by Hohmann transfer is minimum lay-over time, which is

$$t_{\text{lay-over}} = (1 \text{ year} - TOF) \times \omega_E \times 2/(\omega_E - \omega_M) = 1.24 \text{ years}. \tag{3.79}$$

Table 3.3 The planets' synodic periods in the solar system with some data [4]. Comparison can be made with Table 2.1, Table 3.1 and Table 3.2

Planet	Mean distance from Sun (AU)	e (-)	Sidereal period (trop. yr)	Synodic period (trop. yr)	Inclination orbital to ecliptic (°)	Mean velocity (km/s)
Mercury	0.387 099	0.2056	0.24 085	0.317	7.00	47.89
Venus	0.723 332	0.0068	0.61 521	1.599	3.39	35.03
Earth	1.000 000	0.0167	1.00 004	–	0.00	29.78
Mars	1.523 688	0.0934	1.88 089	2.135	1.85	24.13
Jupiter	5.202 561	0.0485	11.8623	1.092	1.30	13.06
Saturn	9.554 747	0.0556	29.4577	1.035	2.49	9.64
Uranus	19.218 114	0.0472	84.0139	1.012	0.77	6.81
Neptune	30.109 57	0.0086	164.79	1.006	1.77	5.43
Pluto	39.44	0.248	248.5	1.004	17.17	4.74

The total two-way duration is $2 \times TOF + t_{\text{lay-over}} = 2 \times 0.712 + 1.24 = 2.66$ year.

Table 3.2 shows these data of Hohmann transfer between the Earth and the planets.

Table 3.3 shows the synodic periods and others properties of the planets in the solar system.

3.4.2.6 Non coplanar Trajectories

It is assumed that all planetary orbits lie in the ecliptic plane, but this is not quite true as only the Earth is in this plane, while others vary. So a simple plane change during the mid-course is used.

3.4.3 Gravity Assist

When a S/C enters a planet's SOI, it is pulled in the direction of the planet's motion (ΔV applied to SC). Hence its velocity can be changed wrt the Sun. When S/C passes behind the planet, normally its velocity increases; S/C passes in front of the planet, its velocity decreases. This is not always the case. See Figure 3.15 and Figure 3.14. In the Galileo mission to Jupiter, the probe performed three gravity-assist maneuvers, $\Delta V_1 = 2.2$ km/s by Venus, $\Delta V_2 = 5.7$ km/s by Venus again, and $\Delta V_3 = 3.7$ km/s from the Earth.

A gravity-assist maneuver is also called a planetary flyby trajectory , or a swingby trajectory. It has been used by interplanetary spacecrafts to obtain "free" velocity change, such as Voyager 1 and 2, Galileo etc. This is because angular momentum is transfered from the flyby planet to the satellite in the heliocentric system, though the inbound and outbound velocities of the satellite in the planet system are the same.

Here we introduce this maneuver by an example of Jupiter flyby in detail.

Assume that the Earth and Jupiter are in co-planar circular orbits with the mean radial distance from the Sun to Jupiter at 5.20 AU. We are to analyze the Hohmann transfer using the patched-conic approximation. Begin in a geocentric parking orbit at 300 nmi (555.6 km) and find the velocity change needed to escape Earth's gravity along a (heliocentric) Hohmann orbit that will encounter Jupiter. The required geometry between Earth–Sun and Jupiter at departure is similar to the discussions for an Earth–Mars mission in the previous section.

We first consider the heliocentric orbit from a radius of 1 AU (Earth) to a radius of 5.2 AU (Jupiter). We consider the Hohmann case; again see Figure 3.1:

$$\mathcal{E} = \frac{v_p^2}{2} - \frac{\mu_\odot}{r_p} = \frac{\mu_\odot}{2a_t}. \tag{3.80}$$

For the Hohmann case $2a_t = r_1 + r_2$ and this leads to:

$$v_p = \sqrt{\frac{2\mu_\odot r_2}{r_1(r_1 + r_2)}} = 1.2953 \, \text{SU}_\odot \tag{3.81}$$

Since circular speed at the Earth's mean solar distance is $1\text{SU}_\odot$ we get

$$\Delta V_1 = v_p - v_{c1} = 0.2953 \, \text{SU}_\odot = 8.791 \, \text{km/s}. \tag{3.82}$$

From the Patched Conic approximation we interpret this Δv_1 as the hyperbolic escape speed in a geocentric hyperbola. Again, an energy analysis is applied

$$\mathcal{E} = \frac{v_p^2}{2} - \frac{\mu_\oplus}{r_p} = \frac{v_\infty^2}{2} \Rightarrow \left[v_\infty^2 + \frac{2\mu_\oplus}{r_p} \right]^{1/2}. \tag{3.83}$$

The numerical values are $r_p = 1.0871$ DU (300 nmi orbit) and $v_\infty = 8.791$ km/s $= 1.1124$ $\text{SU}_\oplus$, leading to $v_p = 1.7542 \, \text{SU}_\oplus$. Since we begin in a circular parking orbit $\Delta v_p = v_p - v_c = 0.7951 \, \text{SU}_\oplus = 6.283$ km/s.

Finally we consider the time-of-flight and required phase angle. Since the Hohmann transfer uses precisely one-half the ellipse we have $TOF = \pi\sqrt{a_t^3} = 17.15 \, \text{TU}_\odot$.

Jupiter's angular rate is $0.0843/\text{TU}_\odot$, so that during the TOF Jupiter will traverse a Sun central angle of 1.446 radians, while the spacecraft will traverse π. Thus, at insertion into the heliocentric orbit Jupiter must lead by $\phi_{\text{phase}} = 1.695 \approx 97.1°$.

A further analysis can be developed to the above Earth–Jupiter problem.

The heliocentric orbit should be Hohmann-like at Earth departure but will be a (nominal) free-return orbit that returns to Earth at the first opportunity.

We have previously analyzed the initial velocity impulse, as well as the required geometry between Earth–Sun and Jupiter at departure; now, however, we are not in the purely Hohmann case.

Now proceed to consider the encounter at Jupiter. Find the phase angle adjustment so that we will pass in front of Jupiter with a peri-Jove radius of 5 planetary radii (the mean radius of Jupiter is about 11.2 times that of Earth, while its mass is about 318 times that of Earth).

We will compute the velocity impulse (in km/s) required at peri-Jove to circularize at that altitude.

Suppose the rocket motor does not work (at all) and we continue through peri-Jove along the (Jovian) hyperbolic orbit. Then we describe the resulting heliocentric orbit after this encounter to see if we could return to the Earth.

Even further, we can briefly describe how to modify the calculations to produce a path that would return to Earth should the rocket motor fail before we reached Jupiter.

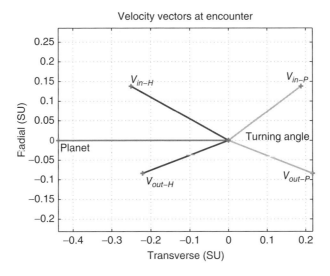

Figure 3.14 Gravity decelerate example.

In the example, the ellipse orbit entry to the Jupiter in heliocentric frame is $a = 3.3019$ AU, period: 6 year, eccentricity $e = 0.6971$; the exit orbit in heliocentric orbit is $a = 3.0412$ AU, $e = 0.7523$, radius at periapsis $r_p = 0.7532$ AU.

The planet velocity $v_p = -0.4385$ $SU_\odot$.

The entry velocity of probe in heliocentric frame: $\mathbf{v}_{in-H} = [-0.2505\ 0.1378]$ $SU_\odot$; the entry velocity of probe in planetary frame: $\mathbf{v}_{in-P} = [0.1880\ 0.1378]SU_\odot$.

Here the hyperbolic periapsis radius at Jupiter is $r_{pJ} = 40$ DU.

Probe-planetary (exit) $\mathbf{v}_{out-P} = [0.2176\ -0.0836]$ $SU_\odot$; probe-heliocentric (exit); $\mathbf{v}_{out-H} = [-0.2209\ -0.0836]$ $SU_\odot$.

Turning angle $\delta = 0.9995$ rad by Equation (2.85).

Note that

$$\mathbf{v}_{in-H} = \mathbf{v}_P + \mathbf{v}_{in-P}, \quad \mathbf{v}_{out-H} = \mathbf{v}_P + \mathbf{v}_{out-P},$$

This is a typical flyby trajectory which is ahead of the planet. Normally the orbit is a decelerate flyby, but this is not always true. In the following we give another example, this time of a gravity-assist orbital though it is still a flyby ahead of the planet. See Figure 3.15. In the example, the ellipse orbit entry to the Jupiter in helicentric frame is $a = 3.3019$ AU, period: 6 year, eccentricity $e = 0.6971$; the orbit exit orbit in helicentric orbit is $a = 6.7483$ AU, $e = 0.5195$, radius at periapsis $r_p = 3.2428$ AU.

The planet velocity is still $v_p = -0.4385$ $SU_\odot$.

The entry velocity of probe in heliocentric frame: $v_{in-H} = [-0.2505\ 0.1378]$ $SU_\odot$; the entry velocity of probe in planetary frame $v_{in-P} = [0.1880\ 0.1378]$ $SU_\odot$.

This time the periapsis radius of the hyperbola at Jupiter is $r_{pJ} = 5$ DU.

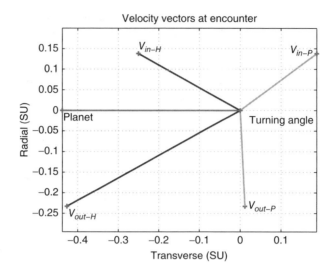

Figure 3.15 Gravity assist example.

For probe-planetary (exit), $\mathbf{v}_{out-P} = [0.0117 \; -0.2328]$ SU$_\odot$; probe-heliocentric (exit), $\mathbf{v}_{out-H} = [-0.4269 \; -0.2328]$ SU$_\odot$.

This time the turning angle $\delta = 2.1534$ rad.

3.5 Launch Issues, Starting the Mission

A spacecraft orbit is determined by its mission requirements. Hence these requirements dictate launch location, launch time and launch direction. Launch window is the period of time to launch a S/C directly into a specified orbit from a given launch location. As a desired orbital plane is fixed in inertial space and the Earth rotates underneath, a launch window occurs when the launch site and orbital plane intersect.

3.5.1 Launch Time

The Earth's rotation under S/C intended orbit is periodic. Knowing the period can determine the launch window which can determine the launch time.

To determine the time, some information previously given in Chapter 1 is analyzed here.

International time reference is Greenwich Mean Time (GMT), UK. The Sun's successive passage over a given point is called the Apparent Local Solar Day. Watches, clocks, etc. keep time which is a reference to a mean solar day. 24 mean solar hours means 1 mean solar day. The Earth's $e = 0.017$ around the Sun (almost circular), so apparent day varies a little; its average is the mean solar day.

The coordinate system (Geocentric-Equatorial) is Earth centered. Sun (hence time reference) appears to move with respect to the coordinate system. The solar day is the successive passages of the Sun over a given longitude, whereas the sidereal day is the successive passages of the vernal equinox over a given longitude. Local Sidereal Time (LST) is the time since the vernal equinox last passed over the local longitude, which is generally measured as an angle in degrees, $Time(degrees) = Time(hours) \times \omega_E$, $\omega_E = 15°$/hour.

In a sidereal day, Earth rotates $360°$ because the vernal equinox direction is fixed. But in a solar day, Earth rotates more than $360°$ because of its revolution about the Sun, (see Figure 1.9). One mean solar day equals 1.0027 sidereal days, whereas a sidereal day equals 24 sidereal hours, which means 23 hr 56 min 04 s mean solar time. Fairly elaborate time propagation needs computer routines, but here just understand the concepts.

3.5.2 When and Where to Launch

In this section we discuss how to launch directly to a given orbit in the Northern Hemisphere. Note that launch window calculations always depend on geometry.

Let us exam the launch opportunities' constraints. The S/C should be directly launched into a given orbital plane; launch site rotates underneath the orbital plane; proper angles for launch should be determined; the critical relationship is Orbit inclination i, which should be greater than launch site latitude L_0. So normally there is only one launch opportunity per day. Consider an example first.

A scenario: At Kennedy Space Center (KSC), its latitude $L_0 = 28.5°$. As a geostationary orbit ($i = 0$) is often the final designed orbit, it cannot be launched directly from KSC. Expensive ΔV maneuvers are required for the final orbit.

This is an opportunity for the Boeing Sea Launch. Boeing is Norwegian Ship Co, whose port is at Long Beach, CA. They moved out to the Pacific and have launched the satellite at the equator by Russian booster.

Launch Window Sidereal Time (LWST) is LST of launch site when the site is under the orbital plane – measured from vernal equinox direction ($\hat{I}$). When launch site LST equals LWST, the correct geometry exists for the launch.

In case one, one day only, one launch opportunity when $i = L_0$. Let Ω is the right ascension of ascending node, then LWST $= \Omega + 90°$ and LST of launch site is equal to LWST.

In case two, when inclination is greater than local latitude, there are two launch opportunities. One is near ascending node opportunity, the other is near descending node opportunity.

Let's define some notations first. Inclination auxiliary angle, α($\alpha = i$ for direct orbit); launch-direction auxiliary angle, γ; launch-site latitude, L_0; launch window location angle, δ; launch azimuth, β. See Figure 3.16; see also Figure 1.8.

It is easy to derive some relations using spherical geometry in Appendix A. For example using angle law of cosines:

$$\cos \alpha = -\cos(\pi/2) \cdot \cos \gamma + \sin(\pi/2) \cdot \sin \gamma \cdot \cos L_0 \tag{3.84}$$

$$\cos \gamma = -\cos(\pi/2) \cdot \cos \alpha + \sin(\pi/2) \cdot \sin \alpha \cdot \cos \delta, \tag{3.85}$$

then we have

$$\sin \gamma = \frac{\cos \alpha}{\cos L_0}, \quad \cos \delta = \frac{\cos \gamma}{\sin \alpha}. \tag{3.86}$$

LWST $= \Omega + \delta$ when the launch site is near ascending node; and LWST $= \Omega + 180 - \delta$ when it is near descending node. The launch time is LST of launch site LWST.

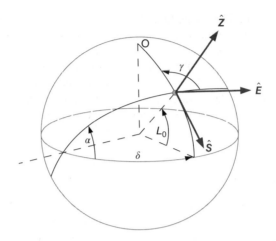

Figure 3.16   Launch window calculation.

The needed launch azimuth angle β is defined as $\beta = 0°$, which means the launch direction is the North direction, $90° = $ East direction, $180° = $ South, and $270° = $ West.

For only one launch opportunity, $\beta = \gamma = 90°$. For the two launch opportunities, if the launch site is near the ascending node, $\beta = \gamma$; if it is near the descending nodes $\beta = 180° - \gamma$.

3.5.3   Launch Velocity

We already know when to launch by a lunch window sidereal time (LWST) and what direction by azimuth angle β. Now we should determine how much velocity is needed to a desired orbit.

There are four phases to launch a spacecraft (S/C) to an orbit. First is vertical assent phase where the S/C gains altitude and may possibly roll; the second phase is pitch over to gain down-range velocity; the third phase is gravity pull to increase horizontal velocity; the fourth phase is vacuum flight to continue acceleration to achieve burnout velocity. The burnout conditions are mainly referred to as the burnout velocity.

To analyze the orbital velocity two components are required to be found: the launch vehicle burnout velocity and the launch site tangential velocity.

Tangential velocity, $V = R_E \omega_E$, where V is tangential velocity (m/s), R_E is the radius from center of rotation, and ω_E is angular velocity (rad/s) of the Earth.

Earth's rotation eastward, at equator, $V = R_E \omega_E = (6378.137 \text{ km})$ $(15.04\,107°/\text{hr}) = 0.4651$ km/s, at a launch site: $V = R_E \cos(L_0)\omega_E = 0.4561 \cos(L_0)$ km/s.

Launch site is centered at Topocentric-Horizon Frame; its origin is at the launch site, and its fundamental frame is the frame tangent to the Earth's surface at the launch site. It is a right-handed system ($\hat{S}\hat{E}\hat{Z}$). Due south from launch site is $\hat{S}$ direction, the straight-up (zenith) is the direction of $\hat{Z}$; the east vector is $\hat{E}$. See Figure 3.16.

Velocity components in the Simplified Model are:

1. $\mathbf{V}_{\text{lossgravity}} = $ velocity needed to overcome gravity and reach correct altitude;
2. $\mathbf{V}_{\text{burnout}} = $ velocity needed at burnout to be in desired orbit;

3. $\mathbf{V}_{launchsite}$ = velocity of launch site;
4. $\Delta\mathbf{V}_{needed}$ = total velocity change needed to meet requirements.

3.5.3.1 Calculation

$$\mathbf{V}_{launchsite} = 0.4651 \cdot \cos(L_0)\hat{E}. \tag{3.87}$$

At desired altitude h, $V = 0$

$$\frac{V^2_{lossgravity}}{2} - \frac{\mu}{R_S} = \frac{0^2}{2} - \frac{\mu}{R_E + h} \tag{3.88}$$

$$\Rightarrow \mathbf{V}_{lossgravity} = \sqrt{\frac{2\mu h}{R_E \cdot (R_E + h)}}\hat{Z} \tag{3.89}$$

$$\mathbf{V}_{burnout} = -V_{burnout} \cdot \cos(\phi)\cos(\beta)\hat{S} + V_{burnout} \cdot \cos(\phi)\sin(\beta)\hat{E}$$
$$+ V_{burnout} \cdot \sin(\phi)\hat{Z}. \tag{3.90}$$

where ϕ = flight path angle at burnout (0° for circular orbit), β = launch azimuth angle.

$$\Delta\mathbf{V}_{needed} = \mathbf{V}_{lossgravity} + \mathbf{V}_{burnout} - \mathbf{V}_{launchsite}. \tag{3.91}$$

Then compute magnitude, ΔV, of the design parameter for launch vehicle:

$$\Delta V_{design} = \Delta V_{needed} + \Delta V_{losses}, \tag{3.92}$$

where ΔV_{losses} is the combination losses due to other factors – drag; pressure, winds, …
Rule-of-thumb (ROT): $\Delta V_{losses} \approx 1.5$ km/s.

3.5.3.2 Example ΔV_{design}

Orbit altitude $h = 600$ km; orbit inclination $i = 35°$; assume launch on ascending node opportunity and $\Delta V_{losses} = 1.5$ km/s.
  Calculate vector components of $\Delta\mathbf{V}_{needed}$.

$$\mathbf{V}_{lossgravity} = \sqrt{\frac{2\mu h}{R_E(R_E + h)}}\hat{Z} = \sqrt{\frac{(2)(3.986 \times 10^5)(600)}{(6378)(6978)}}\hat{Z} = 3.278\hat{Z} \tag{3.93}$$

and

$$V_{burnout} = V_{circularorbit} = \sqrt{\frac{\mu_S}{R_{circular}}} = \sqrt{\frac{3.986 \times 10^5}{6978}} = 7.558.$$

$\phi = EL = 0$ for circular orbit, $\beta = \gamma = Az$ for ascending node opportunity. Then

$$\sin\gamma = \frac{\cos(i)}{\cos(L_0)} = \frac{\cos(35)}{\cos(28.5)} = 0.9321 \Rightarrow \beta = \gamma = 68.77. \tag{3.94}$$

So

$$\mathbf{V}_{burnout} = -V_{burnout}\cos(El)\cos(Az)\hat{S} + V_{burnout}\cos(El)\sin(Az)\hat{E} \quad (3.95)$$
$$+ V_{burnout}\sin(El)\hat{Z} = -2.737\hat{S} + 7.045\hat{E}$$
$$\mathbf{V}_{launchsite} = (0.4651)\cos(28.5)\hat{E} = 0.4087\hat{E},$$

hence,

$$\mathbf{\Delta V}_{needed} = \mathbf{V}_{lossgravity} + \mathbf{V}_{burnout} - \mathbf{V}_{launchsite}$$
$$= 3.278\hat{Z} + [-2.737\hat{S} + 7.045\hat{E}] - 0.4087\hat{E}$$
$$= -2.737\hat{S} + 6.636\hat{E} + 3.278\hat{Z}$$
$$\Delta V_{needed} = |\mathbf{\Delta V}_{needed}| = \sqrt{(-2.737)^2 + (6.636)^2 + (3.2278)^2} = 7.891 \text{ km/s}$$
$$\Delta V_{design} = \Delta V_{needed} + \Delta V_{losses}$$
$$= 7.891 + 1.5 = 9.391 \text{ km/s}.$$

SUMMARY

Know ΔV for design of launch, and know how to calculate ΔV for orbit maneuvers. We need to estimate: ΔV for station keeping and ΔV for reentry/disposal. Obtain velocity requirements for the entire mission.

3.5.4 Rockets and Launch Vehicles

The conservation of momentum leads to the so-called rocket equation, which trade off exhaust velocity with payload fraction. It is based on the assumption of short impulses with coast phase between them and applies to chemical and nuclear-thermal rockets. The rocket equation was first derived by Tsiolkovsky in 1895 for straight-line rocket motion with constant exhaust velocity. It is also valid for elliptical trajectory with only initial and final impulses.

From Newton's Second Law

$$F = \frac{dp}{dt} = 0 \quad (3.96)$$

$$\Rightarrow \frac{d(mv)}{dt} = m\frac{dv}{dt} + v\frac{dm}{dt} = 0 \quad (3.97)$$

$$\Rightarrow \frac{v_{ex}}{m}\frac{dm}{dt} = -\frac{dv}{dt} \quad (3.98)$$

$$\Rightarrow \frac{m_{final}}{m_{initial}} = \exp(-\frac{\Delta v}{v_{ex}}). \quad (3.99)$$

Then we can have the **ideal rocket equation**:

$$\Delta v = v_{ex} \cdot \ln\left(\frac{m_{initial}}{m_{final}}\right), \quad (3.100)$$

where Δv = velocity change (m/s); v_{ex} = effective exhaust velocity (m/s); $m_{initial}$ = initial mass (kg); m_{final} = final mass (kg). Thrust is produced by high speed exhaust-velocity, of a vehicle in the opposite direction. $v_{ex} = I_{sp} \cdot g_0$, g_0 = gravitational acceleration = 9.81 m/s^2. I_{sp} = specific impulse(s), which is a measure of merit of effectiveness of rocket thrust. Larger I_{sp} is better. Then the rocket equation becomes

$$\Delta v = I_{sp} \cdot g_0 \cdot \ln\left(\frac{m_{initial}}{m_{final}}\right). \tag{3.101}$$

Typical I_{sp}s for different engines are 180 – 480 for thermodynamic; $\approx$ 800 for solar-thermal; $\approx$ 800 for thermoelectric; $\approx$ 1000 for nuclear-thermal.

To understand more, relevant courses are: thermodynamics, fluid mechanics, heat and mass transfer or rocket propulsion.

To select a rocket, some factors should be considered: thrust performance – I_{sp}, time performance – Cimpulse burns, safety, logistics, integration and technical risk.

Launch vehicle is a key subsystem of a space mission. Its propulsion system includes thrust-to-weight ratio > 1.0 (chemical rockets), throttling and thrust vector control, dynamic pressure control, G-load control, nozzle design.

The vehicle also includes guidance, navigation and control (GNC), maintain desired profile, communication and data handling, nominal operations. structure and mechanisms.

Approximate 80% of vehicle's lift-off mass is propellant, Structures [tanks, engines,...] should be disposed when fuel used up. Actually, split propellant into multiple tanks and release when empty. This is called **staging**. Multiple stage launch vehicle is most efficient with today's technology. Apply rocket equation to each stage to determine Δv required.

SUMMARY

Have another piece of the Δv puzzle, keep track of individual pieces, then add together to get mission total.

3.5.5   Reentry

The requirements of a reentry mission are mainly in three aspects, namely: deceleration, heating, and accuracy.

3.5.5.1   Trajectory and Deceleration

Let β = atmospheric scale height, γ = flight-path angle, BC = vehicle ballistic coefficient [$BC = m/(C_D \cdot A)$], ρ_0 = sea-level atmospheric density. The vehicle maximum deceleration is

$$a_{max} = \frac{V_{reentry}^2 \cdot \beta \cdot \sin\gamma}{2e}, \tag{3.102}$$

which occurs at altitude:

$$Alt_{a\,max} = \frac{1}{\beta} \cdot \ln\left[\frac{\rho_0}{BC \cdot \beta \cdot \sin\gamma}\right]. \tag{3.103}$$

For **trajectory and heating**, let $\dot{q}$ = heating rate.

$$\dot{q} \approx 1.83 \times 10^{-4} \cdot V^3 \cdot \sqrt{\frac{\rho}{r_{\text{nose}}}}, \tag{3.104}$$

where ρ is air density, r_{nose} is vehicle nose radius.

Maximum heating rate at the altitude

$$Alt_{h\,\text{max}} = \frac{1}{\beta} \ln \left[\frac{\rho_0}{3BC \cdot \beta \cdot \sin \gamma} \right], \tag{3.105}$$

where the velocity $\approx 0.846 \cdot V_{\text{reentry}}$. Steep reentry angles cause high maximum heating rates, but for ashort time; Shallow reentry angles cause low maximum heating rates, but for a long time.

3.6   Summary and Keywords

This chapter introduces orbital maneuver related problems, based on the basic two body in-plane and out-of-plane transfer, which include Hohmann transfer, Earth departure, ballistic missile, lunar and interplanetary exploration. Many detailed practical examples of these missions are given.

Keywords: Hohmann transfer, ICBM, lunar mission, sphere of influence, interplanetary travel, patched conic approximation, gravity assist, launching window, reentry, specific impulse, etc.

Problems

3.1   Calculate the period and radius of a geosynchronous orbit. Starting from a circular parking orbit with a 6578 km radius and 28.5 deg inclination; calculate a 2 maneuver transfer to reach geosynchronous, equatorial orbit. Assume the entire inclination change is performed as part of the apoapsis maneuver.

3.2   A satellite leaves a circular parking orbit at inclination i and executes a two-maneuver transfer to a geosynchronous, equatorial orbit. Part of the required inclination change Δi_1 is performed during the first maneuver and the remainder $\Delta i_2 = (i - \Delta i_1)$ is performed during the second maneuver. Derive an expression for the total V as a function of the velocity of the circular parking orbit, geosynchronous orbit velocity, transfer ellipse periapsis and apoapsis velocities and Δi_1 and i. For a circular parking orbit radius of 6578 km, make a plot of the total ΔV required as a function of Δi_1, the amount of inclination change performed during the first maneuver. What does this tell you about the inclination change performed in Problem 1?

3.3   A lunar module (LM) lifts off from a spherical lunar surface and flies a powered ascent to its 30-km altitude burnout condition, at which time its velocity vector is parallel to the lunar surface. The LM coasts half-way around the Moon where it must rendezvous with a command module (CM) in a 250-km altitude circular orbit. The mass of the Moon is 0.0123 times the mass of the Earth and the radius of the Moon is 1740 km.

    (a) What is the 30-km altitude burnout speed of the LM?

    (b) What is the magnitude of the maneuver required to rendezvous with the CM?

    (c) What is the LM coast time prior to rendezvous?

    (d) For the LM and CM to arrive at the rendezvous point together (or at least close), where must the CM be in relation to the LM at burnout? Give your answer as either an angular or time differential.

3.4   The Space Shuttle does not directly inject into its prescribed mission orbit. Rather, the main engine shutdown occurs at the apogee of an elliptical orbit with 100-km altitude and the external tank is discarded. Perigee of this orbit is at 30-km altitude. After the tank is released, the Shuttle uses its orbital maneuvering system (OMS) to maneuver into an elliptical orbit with perigee at 100-km altitude and apogee at 250-km altitude. This is called the OMS-1 burn. At the new apogee, the Shuttle performs the OMS-2 burn to circularize into the 250-km altitude mission orbit. Calculate:

    (a) the burnout speed at apogee in the external tank disposal orbit;

    (b) the ΔV required for the OMS-1 maneuver;

    (c) the ΔV required for the OMS-2 maneuver;

    (d) if launching from the Kennedy Space Center, without a plane change maneuver, what is the range of inclinations possible for the Shuttle mission orbit? Why?

3.5   A Mars S/C approaches the planet on a hyperbola with eccentricity of 3 and a hyperbolic excess velocity of 3.3 km/sec. How much time is required to travel from the Mars sphere-of-influence to periapsis?

3.6   The Pioneer X spacecraft flew by Jupiter and on to solar system escape. During the planning phase of this mission, the question of using a direct Hohmann transfer to Jupiter was considered. No ΔV was to be added after leaving LEO. Determine whether or not the S/C could pick up enough energy via a Hohmann transfer and Jupiter flyby to escape the solar system.

3.7   Compare the ΔV and time-of-flight required for a Hohmann transfer from Earth to Mars with that required for a bi-elliptic transfer from Earth to Mars, where the bi-elliptic transfer reaches an intermediate apoapsis equal to that of Jupiter's orbit. Assume circular, co-planar orbits for Earth, Mars and Jupiter. What intermediate apoapsis radius is required for the bi-elliptic transfer to require less ΔV than the Hohman transfer? What is the time-of-flight ratio for this transfer relative to that for the Hohmann?

3.8   A Mercury flyby spacecraft is in orbit about the Sun with $a = 0.5$ AU, $e = 0.5$ and $i = 7$ deg. The spacecraft's longitude of periapsis is 150 deg greater than that of Mercury. The mission plan includes a science observation and gravity assist of Mercury when the spacecraft is outbound from its perihelion.

    (a) Find the spacecraft's r, v, and δ at the time of the Mercury encounter.

    (b) The closest approach to Mercury is 200-km altitude. The S/C passes Mercury such that its orbital energy will increase. Will the S/C pass ahead or behind the planet?

    (c) Find r, v, γ and $\Delta\omega$ of the spacecraft with respect to the Sun immediately after the Mercury encounter.

    (d) What flyby ΔV was achieved? Where did this energy come from?

    (e) If instead of performing a flyby (as in parts b–d), a circular orbit about Mercury is desired, determine the circular orbit altitude which minimizes the required circularization ΔV.

3.9   For a Hohmann transfer to Mars, how large an error in Earth departure speed from LEO will cause an approach error at Mars of 100 km? Modern propulsion systems are capable of keeping V errors below 1 m/s. If this is larger than can be tolerated with a 100-km approach window, how can the trajectory accuracy be brought within tolerable limits?

3.10   A spacecraft is on a Hohmann transfer from Earth to Mars with an arrival offset or aiming distance (Δ_a) of 0 km. As the spacecraft passes the interplanetary transfer minor axis a small trajectory correction maneuver of +10 m/s is propulsively applied in a direction collinear to the spacecraft velocity vector to modify its flight-path. At apoapsis of the modified interplanetary transfer, find the spacecraft's miss distance to Mars and V_∞ magnitude for Mars. What does this example illustrate regarding the sensitivity of Δ and V_∞ with magnitude through small trajectory correction maneuvers?

3.11   It is desired that a lunar spacecraft be placed in a posi-grade circular orbit with an altitude of 262 km. At the lunar sphere of influence, $V_{\text{inf}} = 0.5$ km/sec. Calculate the value of V_{inf} which satisfies these conditions. Determine the lunar orbit insertion ΔV (magnitude and direction) necessary to place this probe into the required circular orbit.

3.12   A spacecraft departs a 200-km circular Earth orbit on a 3-day posigrade transfer to the Moon. What are the values of the Earth phase angle at lunar arrival (λ_1) and resulting V_{inf} that yield a retrograde lunar periapsis altitude of 50 km? Upon lunar arrival, the spacecraft is to be inserted into a circular orbit of radius equal to its periapsis. What is the lunar orbit insertion V (magnitude and direction)?

3.13   On its way to the Moon, Apollo 17 altered its initial free-return trajectory in order to provide an opportunity for a more precise landing on the lunar surface. It arrived at the Moon following a direct outbound transfer of 3.4 days with an Earth-relative radial velocity component of +0.5 km/s. The lunar periapsis radius was 1848 km as Apollo 17 past in front of the Moon.

    (a) Determine the eccentricity of the hyperbolic approach trajectory and the offset distance (Δ).

    (b) Assuming the initial circular Earth parking orbit had an altitude of 200 km, determine the ΔV (magnitude and direction) to reach the Moon.

    (c) If the S/C had entered lunar orbit with a periapsis altitude 111 km and an apoapsis altitude of 314 km, determine the ΔV (magnitude and direction) at the lunar orbit insertion point. Assume lunar orbit insertion is performed at periapsis.

    (d) Determine the ΔV required (magnitude and direction) to depart lunar orbit on a 3-day direct transfer to Earth from a circular lunar orbit of 111 km. Approximate the Earth atmospheric entry velocity.

3.14 Apollo 13 had the same translunar trajectory as Apollo 17 (see previous problem), but a major failure was experienced en route to the Moon after leaving the free return trajectory.

(a) If all of Apollo 13's rockets were inoperable, determine its final orbit and the fate of the Apollo 13 astronauts.

(b) As it turned out the descent engine on the Apollo 13 Lunar Module was operable and was used 2.5 days into the lunar trajectory. The spacecraft approached the Moon with $V_\theta = 0$ and $V_r = +0.244$ km/sec. Determine the required ΔV and r_p (with respect to the Moon) such that a direct 3-day Earth return results with no additional maneuvers.

3.15 NASA's Crew Exploration Vehicle (CEV) is designed to arrive at the Moon with an Earth-relative velocity magnitude of 0.38 km/s and an Earth-relative flight-path angle of $+50°$.

(a) Find the lunar flyby periapsis altitude that results in a free-return trajectory.

(b) Sketch the lunar flyby geometry.

(c) Find the altitude of the circular Earth parking orbit from which this Earth–Moon transfer was initiated, the ΔV at departure and the roundtrip transfer time.

3.16 Is it possible for a spacecraft on a Hohmann translunar trajectory to escape the Earth's gravitational field? Quantify your answer and justify your rationale. If it is possible, what is the maximum value for the lunar periapsis? The Michelson chart may be of qualitative assistance in this problem.

3.17 A total ΔV of 4.29 km/s is required to transfer from low Earth orbit (28.5 deg inclination) to geostationary orbit. A space tug with a restartable hydrogen-oxygen engine has an I_{sp} of 453 s, 16 000 kg of propellant and a structural mass of 1300 kg.

(a) How much payload can the tug deliver from this LEO to geostationary orbit?

(b) Can the tug perform a round-trip transfer without payload? If so, how much payload could it deliver to geostationary orbit and then return to LEO (without payload)?

3.18 The solid rocket boosters of the space shuttle have an I_{sp} of 290 s, a dry mass of 81 900 kg and a propellant-loaded mass of 584 600 kg each. The shuttle external tank has a dry mass of 35 000 kg and contains 703 000 kg of propellant; whereas the space shuttle main engines (SSMEs) have an I_{sp} of 455 s. Consider a shuttle-derived launch system consisting of four solid rocket boosters, burning together as the first-stage, and an external tank with two SSMEs as the second stage.

(a) Approximately how much payload can this vehicle deliver to low Earth orbit (assume a total V requirement including losses of 9.448 km/s and neglect the mass of the SSMEs)?

(b) How does this payload mass compare to that delivered by the Space Shuttle and Saturn V launch systems?

3.19 You are the chief engineer for an emerging launch vehicle company. Your goal is to produce a launch system with the lowest overall structural mass (m_s) that will launch a 10 000 lb m payload into LEO. Your design subteams have produced three candidate designs: A, B, and C.

Design (A) is a single-stage-to-orbit system using a LOX/LH2 RBCC airbreathing engine with a relatively high average I_{sp}; however, the lower bulk density of the propellants requires larger tanks (high λ). In addition, ascent losses are higher due to the lower acceleration altitude requirements of this airbreathing system. Designs B and C are both rocket vehicles.

Design (B) is a single-stage-to-orbit system. It uses high performance LOX/LH2 cryogenic engines and advanced structures technology to produce a low λ.

Design (C) is a series two-stage system with a lower performance LOX/RP1 first stage and a high performance LOX/LH2 second stage. Due to the high propellant density of the first stage, a relatively low λ_1 is possible with near-term technology.

(a) Which of the three design options would you select for the lowest m_s (for vehicle C, use the combined structural mass of the two stages)?

(b) Which would you pick if your goal was to have the highest overall payload fraction?

3.20 A spherical stony meteorite with density of 3000 kg/m^3 and radius of 5 m is on an Earth impact trajectory with entry flight-path angle of $-45°$ and entry velocity of 15 km/s. Determine altitude and velocity of peak deceleration as well as the peak deceleration magnitude.

3.21 A GPS replacement satellite needs to be launched into a 55° inclination with a longitude of the ascending node of 150°. If the launch location is Cape Canaveral (28.5° N, 80.5° W) and the launch date is December 10, 2006, what are the launch times that will work for a direct injection into this orbital plane. Give the launch azimuth for both of these times.

References

[1] Bate, R, Mueller, D, and White, J. (1971) *Fundamentals of Astrodynamics*. New York: Dover Publications.

[2] Chobotov, V. (2002) *Orbital Mechanics*, 3rd edn. AIAA Edu. Series, Reston, VA: AIAA.

[3] Battin R.H. and Vaughan, R.J. (1987) *An Introduction to the Mathematics and Methods of Astrodynamics*. AIAA, New York.

[4] Vallado, D. (1997) *Fundamentals of Astrodynamics and Applications*. El Segundo, CA: Microcosm Press.

4

Special Topics

This chapter introduces a few orbit problems which are relatively complicated for beginners. But all these subjects are important for practical applications. The linear orbital theory (CW equation) is important for problems of spacecraft rendezvous; orbit determination is crucial for autonomous navigation; the Lambert problem is the key for S/C targeting and intercept; the three-body problem has more and more applications in interplanetary missions; optimal control is practical as continuous low thrust engines are becoming available. All these methods will be discussed in this chapter with special simulations.

4.1 Relative Motion – CW Equation

There are many scenarios of space mission in which the subject of interest is only the relative positions of two spacecraft of two spacecraft and also the two spacecraft are near to each other. For these cases a simpler, approximate mathematical model can be used. The model is a linearized equation that approximately describes the motion of a spacecraft (chaser S/C) relative to a neighboring reference orbit (target S/C). The advantage of linear equations of motion is their solutions' superposition. The linear analysis can be used for mid-course guidance, close-range interception and rendezvous, formation flying, etc. The linearized equation of relative motion of two bodies is called Clohessy-Wiltshire (CW) equations, also known as Hill's equation.

In this method, a radius-transverse-normal $(\hat{R}\hat{S}\hat{W})$ coordinate system (different from NASA) is used. This is also called the $\hat{x}$-up $(\hat{R})$, $\hat{y}$-forward $(\hat{S})$, $\hat{z}$-left $(\hat{W})$ frame system, See Figure 4.1 and 4.3 also. Note again, that the assumptions for above equation are: (1) the two satellites are only a few km apart; (2) the target is in a circular orbit; (3) there are no external forces (drag, etc.).

4.1.1 Equations of Motion

See Figure 4.1. The target satellite has two-body motion:

$$\ddot{\mathbf{r}}_t = -\frac{\mu}{r_t^3}\mathbf{r}_t. \tag{4.1}$$

Fundamental Spacecraft Dynamics and Control, First Edition. Weiduo Hu.
© 2015 John Wiley & Sons (Asia) Pte Ltd. Published 2015 by John Wiley & Sons (Asia) Pte Ltd.

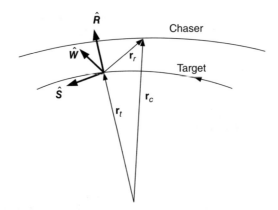

Figure 4.1 Relative motion

The chaser is allowed to have thrusting

$$\ddot{\mathbf{r}}_c = -\frac{\mu}{r_c^3}\mathbf{r}_c + \mathbf{F}. \tag{4.2}$$

Then relative motion $\mathbf{r}_r = \mathbf{r}_c - \mathbf{r}_t, \Rightarrow \ddot{\mathbf{r}}_r = \ddot{\mathbf{r}}_c - \ddot{\mathbf{r}}_t$, So,

$$\ddot{\mathbf{r}}_r = -\frac{\mu}{r_c^3}\mathbf{r}_c + \mathbf{F} + \frac{\mu}{r_t^3}\mathbf{r}_t. \tag{4.3}$$

Now,

$$r_c^2 = r_t^2 + 2r_t r_r \cos\alpha + r_r^2 \tag{4.4}$$

$$\frac{\mathbf{r}_c}{r_c^3} = \frac{\mathbf{r}_t + \mathbf{r}_r}{\left(r_t^2 + 2\mathbf{r}_t \cdot \mathbf{r}_r + r_r^2\right)^{3/2}}. \tag{4.5}$$

α is the angle between $\mathbf{r}_t$ and $\mathbf{r}_r$. If r_r^2 is small relative to r_t^2 then

$$\frac{\mathbf{r}_c}{r_c^3} \approx \frac{\mathbf{r}_t + \mathbf{r}_r}{r_t^3} \frac{1}{\left(1 + 2\mathbf{r}_t \cdot \mathbf{r}_r/r_t^2\right)^{3/2}}. \tag{4.6}$$

Simplify using binomial series $(1 + x)^n = 1 + nx + n(n - 1)x^2/2! + \cdots$ when the absolute of x is much smaller than one.

$$\frac{\mathbf{r}_c}{r_c^3} = \frac{\mathbf{r}_t + \mathbf{r}_r}{r_t^3}\left[1 - \frac{3}{2}\left(\frac{2\mathbf{r}_t \cdot \mathbf{r}_r}{r_t^2}\right) + \cdots\right]. \tag{4.7}$$

Substituting in:

$$\ddot{\mathbf{r}}_r = -\mu\left\{\frac{\mathbf{r}_t + \mathbf{r}_r}{r_t^3}\left[1 - \frac{3}{2}\left(\frac{2\mathbf{r}_t \cdot \mathbf{r}_r}{r_t^2}\right) + \cdots\right]\right\} + \mathbf{F} + \frac{\mu\mathbf{r}_t}{r_t^3} \tag{4.8}$$

$$= -\frac{\mu}{r_t^3} \left\{ -\frac{3\mathbf{r}_t}{2} \left(\frac{2\mathbf{r}_t \cdot \mathbf{r}_r}{r_t^2} \right) + \mathbf{r}_r - \frac{3\mathbf{r}_r}{2} \left(\frac{2\mathbf{r}_t \cdot \mathbf{r}_r}{r_t^2} \right) \right\} + \mathbf{F} \tag{4.9}$$

$$= -\frac{\mu}{r_t^3} \left\{ -\frac{3\mathbf{r}_t}{2r_t} \left(\frac{2\mathbf{r}_t \cdot \mathbf{r}_r}{r_t} \right) + \mathbf{r}_r \right\} + \mathbf{F}. \tag{4.10}$$

For application interest, the equation should be transformed from inertial frame to target's moving relative frame, and $\mathbf{r}_t / r_t = \hat{R}$, $\hat{R} \cdot \mathbf{r}_r = x$,

$$\ddot{\mathbf{r}}_{rI} = -\frac{\mu}{r_t^3} \left\{ -3x\hat{R} + \mathbf{r}_r \right\} + \mathbf{F}. \tag{4.11}$$

Recall the relation of acceleration between inertial and rotational frame; see Equation (2.115):

$$\ddot{\mathbf{r}}_I = \ddot{\mathbf{r}}_R + \dot{\boldsymbol{\omega}}_R \times \mathbf{r}_R + 2\boldsymbol{\omega}_R \times \dot{\mathbf{r}}_R + \boldsymbol{\omega}_R \times (\boldsymbol{\omega}_R \times \mathbf{r}_R) \Rightarrow \tag{4.12}$$

$$\ddot{\mathbf{r}}_R = \ddot{\mathbf{r}}_I - \dot{\boldsymbol{\omega}}_R \times \mathbf{r}_R - 2\boldsymbol{\omega}_R \times \dot{\mathbf{r}}_R - \boldsymbol{\omega}_R \times (\boldsymbol{\omega}_R \times \mathbf{r}_R). \tag{4.13}$$

So, if $\mathbf{r}_{rR}$ is shortened as $\mathbf{r}_r = [x, y, z]^T = x\hat{R} + y\hat{S} + z\hat{W}$ in rotational frame:

$$\ddot{\mathbf{r}}_r = \ddot{\mathbf{r}}_{rI} - \dot{\boldsymbol{\omega}}_R \times \mathbf{r}_r - 2\boldsymbol{\omega}_R \times \dot{\mathbf{r}}_r - \boldsymbol{\omega}_R \times (\boldsymbol{\omega}_R \times \mathbf{r}_r) \tag{4.14}$$

$$\boldsymbol{\omega}_R \times \mathbf{r}_r = \begin{bmatrix} \hat{R} & \hat{S} & \hat{W} \\ 0 & 0 & \dot{\omega} \\ x & y & z \end{bmatrix} = -\dot{\omega} y \hat{R} + \dot{\omega} x \hat{S} \tag{4.15}$$

$$\boldsymbol{\omega}_R \times \dot{\mathbf{r}}_r = -\omega \dot{y} \hat{R} + \omega \dot{x} \hat{S} \tag{4.16}$$

$$\boldsymbol{\omega}_R \times (\boldsymbol{\omega}_R \times \mathbf{r}_r) = -\omega^2 x \hat{R} - \omega^2 y \hat{S} \tag{4.17}$$

$$\Rightarrow \ddot{\mathbf{r}}_r = -\frac{\mu}{r_t^3} \left\{ -3x\hat{R} + \mathbf{r}_r \right\} + \mathbf{F} + \dot{\omega} y \hat{R} - \dot{\omega} x \hat{S} \tag{4.18}$$

$$-2(-\omega \dot{y} \hat{R} + \omega \dot{x} \hat{S}) + \omega^2 x \hat{R} + \omega^2 y \hat{S}. \tag{4.19}$$

If we assume circular motion, $\mu / r_t^3 = \omega^2$, $\dot{\omega} = 0$,

$$\ddot{\mathbf{r}}_r = -\omega^2 \left\{ -3x\hat{R} + x\hat{R} + y\hat{S} + z\hat{W} \right\} + \mathbf{F} \tag{4.20}$$

$$+2\omega \dot{y} \hat{R} - 2\omega \dot{x} \hat{S} + \omega^2 x \hat{R} + \omega^2 y \hat{S}. \tag{4.21}$$

Thus, we have the CW or Hill's equation:

$$\ddot{x} - 2\omega \dot{y} - 3\omega^2 x = f_x \tag{4.22}$$

$$\ddot{y} + 2\omega \dot{x} = f_y \tag{4.23}$$

$$\ddot{z} + \omega^2 z = f_z. \tag{4.24}$$

For unthrusted motion, $\mathbf{F} = [f_x, f_y, f_z]^T = 0$, in plane motion is coupled, see Equation (4.22) and Equation (4.23), but the out of plane motion, Equation (4.24), is decoupled from the in plane motion, whose solution is undamped linear oscillator, which can be solved first. And Equation (4.23) involves only derivatives, so that it can be integrated directly. Hence the above equations can be solved (see reference [5] for details) without difficulty:

$$x(t) = \frac{\dot{x}_0}{\omega} \sin(\omega t) - \left(3x_0 + \frac{2\dot{y}_0}{\omega}\right) \cos(\omega t) + \left(4x_0 + \frac{2\dot{y}_0}{\omega}\right) \tag{4.25}$$

$$y(t) = \left(6x_0 + \frac{4\dot{y}_0}{\omega}\right) \sin(\omega t) + \frac{2\dot{x}_0}{\omega} \cos(\omega t) - (6\omega x_0 + 3\dot{y}_0)t + \left(y_0 - \frac{2\dot{x}_0}{\omega}\right) \tag{4.26}$$

$$z(t) = z_0 \cos(\omega t) + \frac{\dot{z}_0}{\omega} \sin(\omega t) \tag{4.27}$$

$$\dot{x}(t) = \dot{x}_0 \cos(\omega t) + (3\omega x_0 + 2\dot{y}_0) \sin(\omega t) \tag{4.28}$$

$$\dot{y}(t) = (6\omega x_0 + 4\dot{y}_0) \cos(\omega t) - 2\dot{x}_0 \sin(\omega t) - (6\omega x_0 + 3\dot{y}_0) \tag{4.29}$$

$$\dot{z}(t) = -z_0 \omega \sin(\omega t) + \dot{z}_0 \cos(\omega t). \tag{4.30}$$

So, given $x_0, y_0, z_0, \dot{x}_0, \dot{y}_0, \dot{z}_0$ of chaser, we can compute $x, y, z, \dot{x}, \dot{y}, \dot{z}$ of interceptor at a future time.

We can also determine ΔV needed for rendezvous, which means $x(t) = y(t) = z(t) = 0$. Given x_0, y_0, z_0, we want to determine $\dot{x}_0, \dot{y}_0, \dot{z}_0$.

$$\dot{y}_0 = \frac{(6x_0(\omega t - \sin(\omega t)) - y_0)\omega \sin(\omega t) - 2\omega x_0(4 - 3\cos(\omega t))(1 - \cos(\omega t))}{(4\sin(\omega t) - 3\omega t)\sin(\omega t) + 4(1 - \cos(\omega t))^2} \tag{4.31}$$

$$\dot{x}_0 = -\frac{\omega x_0(4 - 3\cos(\omega t)) + 2(1 - \cos(\omega t))\dot{y}_0}{\sin(\omega t)} \tag{4.32}$$

$$\dot{z}_0 = -z_0 \omega \coth(\omega t). \tag{4.33}$$

4.1.2 Examples

Example 1. Scenario of the Tiangong 1 Space Station.
Given: Assume a satellite is about to be released from the Tiangong 1, which is in a circular orbit at 353.5 km altitude. The relative velocity (from the Tiangong 1) of the ejection probe is 0.12 m/s up, 0.05 m/s backwards, 0.03 m/s to the right.
Find: Position and velocity of the probe after 3 and 10 minutes.

First, convert the given data to the correct sign convention for the relative motion definitions. The initial velocities are then

$$\dot{x}_0 = 0.12, \ \dot{y} = -0.05, \ \dot{z}_0 = -0.03 \text{ m/s}.$$

Next, find the angular rate of the target (Tiangong). In a 353.5 km altitude orbit:

$$\omega = \sqrt{\frac{\mu}{r_c^3}} = \sqrt{\frac{398\,600.5}{6731.5^3}} = 0.00\,114\,310\,955\,415 \text{ rad/s}. \tag{4.34}$$

The initial position of 0.0 simplifies the equation in algorithm to

$$x(t) = \frac{\dot{x}_0}{\omega} \sin(\omega t) - \frac{2\dot{y}_0}{\omega} \cos(\omega t) + \frac{2\dot{y}_0}{\omega}$$

$$y(t) = \frac{4\dot{y}_0}{\omega} \sin(\omega t) + \frac{2\dot{x}_0}{\omega} \cos(\omega t) - 3\dot{y}_0 t - \frac{2\dot{x}_0}{\omega}$$

$$z(t) = \frac{\dot{z}_0}{\omega} \sin(\omega t)$$

$$\dot{x}(t) = \dot{x}_0 \cos(\omega t) + 2\dot{y}_0 \sin(\omega t)$$

$$\dot{y}(t) = 4\dot{y}_0 \cos(\omega t) - 2\dot{x}_0 \sin(\omega t) - 3\dot{y}_0$$

$$\dot{z}(t) = \dot{z}_0 \cos(\omega t).$$

Substituting the time of 3 minutes (180s) yields

$$x(3) = 19.6\,025\,956\,\text{m}, \qquad \dot{x}(3) = 0.0\,970\,376\,\text{m/s}$$

$$y(3) = -13.1\,752\,666\,\text{m}, \qquad \dot{y}(3) = -0.0\,948\,158\,\text{m/s}$$

$$z(3) = -5.3\,619\,772\,\text{m}, \qquad \dot{z}(3) = -0.0\,293\,672\,\text{m/s}$$

and, for the case of 10 minutes (600s)

$$x(10) = 46.7\,044\,376\,\text{m}, \qquad \dot{x}(10) = 0.0\,295\,302\,\text{m/s}$$

$$y(10) = -68.2\,871\,866\,\text{m}, \qquad \dot{y}(10) = -0.1\,567\,766\,\text{m/s}$$

$$z(10) = -16.6\,215\,883\,\text{m}, \qquad \dot{z}(10) = -0.0\,232\,161\,\text{m/s}.$$

Notice how both the position and velocity change dramatically.

Example 2. Solving for Velocity in Hill's Equation
Given: A satellite in circular orbit at 353.5 km altitude. Suppose we continue Example 1; at 10 minutes, we decide to retrieve the probe.
Find: Δv to rendezvous with the space station in 6 and 20 minutes.
Examining Equation (4.34) reveals quantities similar to those in Example 1.

$$\omega = 0.00\,114\,310\,955\,415\,\text{rad/s.}$$

Now, solve for the required velocity to rendezvous Equations (4.32)–(4.33) in 6 and 20 minutes (360 s and 1200 s), but be sure to calculate $\dot{y}_0$ first because you need it to calculate $\dot{x}_0$. At t = 6 minutes (360 seconds) and at t = 20 minutes (1200 seconds),

$$\dot{x}_0 = -0.2\,185\,857\,\text{m/s}, \qquad \dot{x}_0 = -0.1\,221\,180\,\text{m/s}$$

$$\dot{y}_0 = 0.1\,238\,232\,\text{m/s}, \qquad \dot{y}_0 = -0.0\,387\,497\,\text{m/s}$$

$$\dot{z}_0 = 0.0\,435\,348\,\text{m/s}, \qquad \dot{z}_0 = 0.0\,038\,331\,\text{m/s}.$$

Notice how much the required velocity drops as the allowable time to rendezvous increases.

This problem has a practical application:rescuing astronauts who become detached from the space station. Willians and Baughman (1994) show how these calculations are used for designing self-rescue modules for astronauts.

Finally, we give two example of CW Equations. (4.22)–(4.24) and their comparisons with the nonlinear full model of relative motion Equation (4.3). Still using the Tiangong 1 as the reference orbit, we assume a large deviation $x_0 = 50$ km, $y_0 = 0$, $z_0 = 0$, but choose appropriate $\dot{x}_0 = 0$, $\dot{y}_0 = -0.11\,410\,367\,469\,885$ km/s $\dot{z}_0 = 0$. This orbit has the same period with reference orbit. This is a typical example for a formation flying and fly-around mission.

The simulation time is 5 hrs and is also an example when $x_0 = 20$ km. See Figure 4.2 and Figure 4.3. We find when the initial relative distance is larger, the error of the linearized CW equation grows larger with time.

As another example, $\dot{x}_0 = 0.0$ km/s, $\dot{y}_0 = -0.01$ km/s, $\dot{z}_0 = 0.0$ km/s, and another orbit $\dot{y}_0 = -0.004$ km/s with the same other initial values and the same simulation time (5 hrs). We can see the difference between a nonlinear model and linear model in Figure 4.4 and Figure 4.5. One practical problem of this case can be the rendezvous problem, see Figure 3.4.

4.1.3 J_2 Perturbed No-circular Target Orbit

Similarly, we can get a more general form of eccentric reference CW-equation with J_2 perturbation (see Chapter 5 concerning J_2):

$$\ddot{\mathbf{r}}_r + \dot{\boldsymbol{\omega}} \times \mathbf{r}_r + \boldsymbol{\omega} \times (\boldsymbol{\omega} \times \mathbf{r}_r) + 2\boldsymbol{\omega} \times \dot{\mathbf{r}}_r + \ddot{\mathbf{r}}_t = -\left(\frac{\mu}{r_c^3} \mathbf{r}_c - \frac{\mu}{r_t} \mathbf{r}_t \right). \tag{4.35}$$

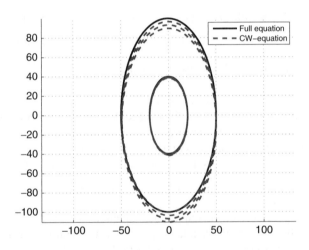

Figure 4.2 Examples of comparison when $x_0 = 50$ km and $x_0 = 20$ km

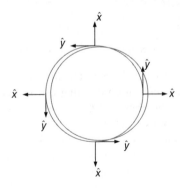

Figure 4.3 The orbits and coordinates in inertial space for Figure 4.2 (not to scale)

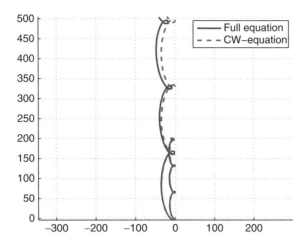

Figure 4.4 Examples of comparison when $\dot{y}_0 = -10$ m/s and $\dot{y}_0 = -4$ m/s

then the scalar form is

$$\ddot{x} - 2\omega\dot{y} - \dot{\omega}y - \omega^2 x = -\frac{\mu(r_t + x)}{\left[(r_t + x)^2 + y^2 + z^2\right]^{3/2}} + \frac{\mu}{r_t^2} + J_2 f_x(\mathbf{e}, \mathbf{r}_r, \dot{\mathbf{r}}_r) \qquad (4.36)$$

$$\ddot{y} + 2\omega\dot{x} + \dot{\omega}y - \omega^2 y = -\frac{\mu y}{\left[(r_t + x)^2 + y^2 + z^2\right]^{3/2}} + J_2 f_y(\mathbf{e}, \mathbf{r}_r, \dot{\mathbf{r}}_r) \qquad (4.37)$$

$$\ddot{z} = -\frac{\mu z}{\left[(r_t + x)^2 + y^2 + z^2\right]^{3/2}} + J_2 f_z(\mathbf{e}, \mathbf{r}_r, \dot{\mathbf{r}}_r) \qquad (4.38)$$

$$\ddot{r}_t = r_t \omega^2 - \frac{\mu}{r_t^2} \left[1 + J_2 f_r(\mathbf{e})\right] \qquad (4.39)$$

$$\dot{\omega} = -\frac{2\dot{r}_t \omega}{r_t} \left[1 + J_2 f_\theta(\mathbf{e})\right], \ \omega = \dot{\theta}_t, \qquad (4.40)$$

where θ_t is the true anomaly of the target orbit, $\mathbf{e}$ is an element set such as $[a, i, q_1 = e\cos\omega, q_2 = e\sin\omega, \Omega, \theta_t]^T$. As $r_t = (h^2/\mu)/(1 + e\cos\theta_t)$, part of above equation can be transformed

$$\frac{\mu}{\left[(r_t + x)^2 + y^2 + z^2\right]^{3/2}} = \frac{(1 + e\cos\theta_t)^3 \mu^4}{h^6}. \qquad (4.41)$$

More discussion about J_2 perturbation to CW equation can be found in reference [7].

4.2   Lambert Problem

A important problem in orbital mechanics is the transfer of a spacecraft from one position to another within a given specified time. For example, a spacecraft is often required to target a planet or another spacecraft in a known trajectory. This targeting could be an interception problem in which only, the position needs to be matched or a rendezvous application in which

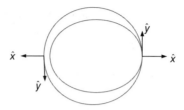

Figure 4.5 The orbits and coordinates in inertial space for Figure 4.4 (not to scale)

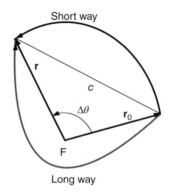

Figure 4.6 The geometry of Lambert problem

both position and velocity should be matched. The underlying theory of these applications is: given two positions and the time-of-flight between them, determine the orbit between the two positions. See Figure 4.6. Generally speaking, this is both an orbital maneuver problem and also an orbital determination problem.

4.2.1 Lambert's Theorem

The orbital transfer time depends only upon the semi-major axis, the sum of the distances of the initial and final points of the arc from the center of force, and length of the chord joining these points.

$$\Delta t = f(a, r_0 + r, c). \tag{4.42}$$

This is also known as a Gaussian problem in reference [2], or a universal problem. Two transfer methods exist: short way ($\Delta\theta < 180°$) and long way ($\Delta\theta > 180°$). See Figure 4.8.

Find the chord, c, using the cosine law and $\cos \Delta\theta = \mathbf{r}_0 \cdot \mathbf{r}/r_0 r$:

$$c = \sqrt{r_0^2 + r^2 - 2r_0 r \cos(\Delta\theta)}. \tag{4.43}$$

Define the semi-perimeter, s, as half the sum of the sides of the triangle created by the position vectors and the cord:

$$s = (r_0 + r + c)/2. \tag{4.44}$$

We know the sum of the distances from the foci to any point on the ellipse equals twice the semi-major axis, thus

$$2a = r + (2a - r).$$ (4.45)

We will look at the minimum-energy solution, where the chord length equals the sum of the two radii from the vacant focus (a single secondary focus), i.e. the vacant foci is on the chord.

$$2a - r + 2a - r_0 = c.$$ (4.46)

Thus,

$$a_{\min} = \frac{s}{2} = \frac{r_0 + r + c}{4}.$$ (4.47)

Using the inset of the geometry (see Figure 4.7) and the Pythagorean theorem, the square of the two foci distance ($FF^* = 2ae$) can be written:

$$4a_{\min}^2 e_{\min}^2 = [(s - r)\sin\alpha]^2 + [r - (s - r)\cos\alpha]^2$$ (4.48)

$$= (s - r)^2(1 - \cos^2\alpha) + r^2 + (s - r)^2\cos^2\alpha - 2r(s - r)\cos\alpha$$ (4.49)

$$= s^2 - 2r(s - r)(1 + \cos\alpha).$$ (4.50)

The cosine law gives:

$$r_0^2 = r^2 + c^2 - 2rc\cos\alpha.$$ (4.51)

Thus,

$$\cos\alpha = \frac{-r_0^2 + r^2 + c^2}{2rc}.$$ (4.52)

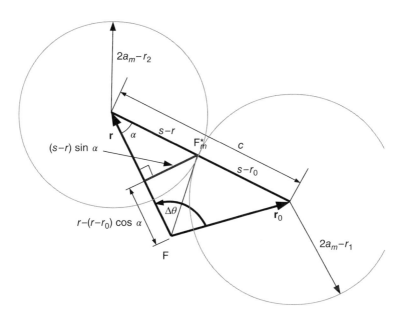

Figure 4.7 Geometry of minimum energy ellipse adapted from reference [3]

Using $s = (r_0 + r + c)/2$ can show:

$$2s(s - r_0) - rc = (-r_0^2 + r^2 + c^2)/2. \tag{4.53}$$

Thus,

$$\cos \alpha = \frac{2s(s - r_0)}{rc} - 1 \tag{4.54}$$

and

$$4a_{min}^2 e_{min}^2 = s^2 - \frac{4s}{c}(s - r_0)(s - r) \tag{4.55}$$

$$4a^2 e^2 = 4a(ae^2) = 4a(a - p) \tag{4.56}$$

$$4a_{min}^2 e_{min}^2 = s^2 - 2sp_{min}, \qquad \text{since } a_{min} = s/2. \tag{4.57}$$

Thus (for more, see reference [3]),

$$s^2 - 2sp_{min} = s^2 - \frac{4s}{c}(s - r_0)(s - r) \tag{4.58}$$

$$p_{min} = \frac{2}{c}(s - r_0)(s - r) = \frac{r_0}{r}(1 + \cos \Delta\theta) \tag{4.59}$$

$$e_{min} = \sqrt{1 - \frac{2p_{min}}{s}}. \tag{4.60}$$

If $\Delta\theta > 180°$, then $\beta_e = -\beta_e$; if $t > t_m$, then $\alpha_e = 2\pi - \alpha_e$, where $t = t_m$ when $a = a_m = s/2$, $\alpha_e = \pi$

$$\sin\left(\frac{\alpha_e}{2}\right) = \sqrt{\frac{r_0 + r + c}{4a}} = \sqrt{\frac{s}{2a}} \tag{4.61}$$

$$\sin\left(\frac{\beta_e}{2}\right) = \sqrt{\frac{r_0 + r - c}{4a}} = \sqrt{\frac{s - c}{2a}} \tag{4.62}$$

$$\Rightarrow \quad t - t_0 = \sqrt{\frac{a^3}{\mu}} \left[\alpha_e - \sin \alpha_e - (\beta_e - \sin \beta_e)\right]. \tag{4.63}$$

For parabola transfer:

$$t - t_0 = \frac{1}{3}\sqrt{\frac{2}{\mu}} \left(s^{3/2} \pm (s - c)^{3/2}\right). \tag{4.64}$$

For hyperbola:

$$\sinh\left(\frac{\alpha_h}{2}\right) = \sqrt{\frac{r_0 + r + c}{-4a}} = \sqrt{\frac{s}{-2a}} \tag{4.65}$$

$$\sinh\left(\frac{\beta_h}{2}\right) = \sqrt{\frac{r_0 + r - c}{-4a}} = \sqrt{\frac{s - c}{-2a}} \tag{4.66}$$

$$\Rightarrow \quad t - t_0 = \sqrt{\frac{a^3}{\mu}} \left[-\alpha_h + \sinh \alpha_h + (\beta_h - \sinh \beta_h)\right]. \tag{4.67}$$

For minimum energy, it is an elliptic orbit $\alpha_e = \pi$

$$\sin\left(\frac{\beta_e}{2}\right) = \sqrt{\frac{s-c}{2}} \tag{4.68}$$

$$t_{min} = \sqrt{\frac{a^3_{min}}{\mu}}(\pi - \beta_e + \sin\beta_e) \tag{4.69}$$

$$\mathbf{v}_0 = \frac{\sqrt{\mu p_{min}}}{r_0 r \sin(\Delta\theta)}\left\{\mathbf{r} - \left[1 - \frac{r}{p_{min}}(1 - \cos\Delta\theta)\right]\mathbf{r}_0\right\}. \tag{4.70}$$

4.2.2   Culp's Proof of Lambert's Theorem

A brief introduction is given here. More can be found in reference [2]. Let E_0 and E be the eccentric anomalies corresponding to $\mathbf{r}_0$ and $\mathbf{r}$ and define:

$$2G = E + E_0 \tag{4.71}$$

$$2g = E - E_0. \tag{4.72}$$

Using $r = a(1 - e\cos E)$:

$$r_0 + r = 2a\left[1 - \frac{e}{2}(\cos E_0 - \cos E)\right]. \tag{4.73}$$

Using the trig identity

$$\cos x + \cos y = 2\cos\frac{x+y}{2}\cos\frac{x-y}{2} \tag{4.74}$$

and $x = a\cos E$, $y = b\sin E = a(1 - e^2)^{1/2}\sin E$. See Figure 2.21

$$r_0 + r = 2a(1 - e\cos G\cos g) \tag{4.75}$$

$$c^2 = (x - x_0)^2 + (y - y_0)^2 \tag{4.76}$$

$$= a^2(\cos E - \cos E_0)^2 + a^2(1 - e^2)(\sin E - \sin E_0)^2 \tag{4.77}$$

$$= a^2(2\sin G\sin g)^2 + a^2(1 - e^2)(2\sin g\cos G)^2. \tag{4.78}$$

since

$$\cos x - \cos y = 2\sin\frac{x+y}{2}\sin\frac{x-y}{2} \tag{4.79}$$

$$\sin x - \sin y = 2\sin\frac{x-y}{2}\cos\frac{x+y}{2}. \tag{4.80}$$

As for elliptical motion $e \le 1$, can define angle h as

$$\cos h = e\cos G \tag{4.81}$$

$$c^2 = 4a^2\sin^2 g(\sin^2 G + \cos^2 G - e^2\cos^2 G) \tag{4.82}$$

$$= 4a^2\sin^2 g\sin^2 h. \tag{4.83}$$

Hence, $r_0 + r$ can be rewritten as

$$r_0 + r = 2a(1 - e \cos G \cos g) = 2a(1 - \cos g \cos h). \tag{4.84}$$

Define $\alpha_e = h + g$, $\beta_e = h - g$, then $\alpha_e - \beta_e = 2g = E - E_0$ and

$$\cos\left(\frac{\alpha_e + \beta_e}{2}\right) = \cos h = e \cos G = e \cos\left(\frac{E + E_0}{2}\right). \tag{4.85}$$

So

$$r_0 + r + c = 2a(1 - \cos g \cos h) + 2a \sin g \sin h \tag{4.86}$$

$$= 2a(1 - \cos(h + g)) = 2a(1 - \cos \alpha_e). \tag{4.87}$$

Using $1 - \cos \alpha_e = 2 \sin^2(\alpha_e/2)$, $r_0 + r + c = 4a \sin^2(\alpha_e/2)$. Similarly,

$$r_0 + r - c = 2a(1 - \cos g \cos h) - 2a \sin g \sin h \tag{4.88}$$

$$= 2a(1 - \cos(h - g)) \tag{4.89}$$

$$= 2a(1 - \cos \beta_e) = 4a \sin^2\left(\frac{\beta_e}{2}\right). \tag{4.90}$$

With $E > E_0$ and $\mu = n^2 a^3$, the time interval can be written using Kepler's equation:

$$n(t - t_0) = E - E_0 - e(\sin E - \sin E_0) \tag{4.91}$$

$$= \alpha_e - \beta_e - e\left(2 \sin\left(\frac{E - E_0}{2}\right) \cos\left(\frac{E + E_0}{2}\right)\right) \tag{4.92}$$

$$= \alpha_e - \beta_e - 2 \sin\left(\frac{E - E_0}{2}\right) \cos\left(\frac{\alpha_e + \beta_e}{2}\right) \tag{4.93}$$

$$= \alpha_e - \beta_e - 2 \sin\left(\frac{\alpha_e - \beta_e}{2}\right) \cos\left(\frac{\alpha_e + \beta_e}{2}\right). \tag{4.94}$$

Finally,

$$n(t - t_0) = \alpha_e - \beta_e - (\sin \alpha_e - \sin \beta_e). \tag{4.95}$$

We have α_e and β_e as function of $r_1 + r_2$, c and a, so we have proved Lambert's Theorem.

Determining the orbit

$$\hat{h} = \frac{\mathbf{r}_0 \times \mathbf{r}}{|\mathbf{r}_0 \times \mathbf{r}|}. \tag{4.96}$$

Get Ω and i.

Compute cord c as:

$$c^2 = r_0^2 + r^2 - 2\mathbf{r}_0 \cdot \mathbf{r} = r_0^2 + r^2 - 2r_0 r \cos(\Delta\theta). \tag{4.97}$$

We also have

$$r_0 + r + c = 4a \sin^2\left(\frac{\alpha_e}{2}\right) \tag{4.98}$$

$$r_0 + r - c = 4a \sin^2 \left(\frac{\beta_e}{2} \right) \tag{4.99}$$

$$(t - t_0)(\frac{\mu}{a^3})^{1/2} = \alpha_e - \beta_e - (\sin \alpha_e - \sin \beta_e). \tag{4.100}$$

Three equations, three unknowns (a, α_e, β_e), can be estimated. As $\Delta E = \alpha_e - \beta_e$, then $f, g, \dot{f}, \dot{g}$ can be obtained by Equations (4.116)–(4.119). Then $\mathbf{r}, \mathbf{v}$ can be calculated by Equations (4.103) and (4.104). Finally, e, ω can be solved by normal orbital element equations.

4.2.3 f, g Function Algorithm

As we know from Chapter 2, in perifocal coordinate:

$$\mathbf{r} = x\hat{P} + y\hat{Q} \tag{4.101}$$

$$\mathbf{v} = \dot{x}\hat{P} + \dot{y}\hat{Q}. \tag{4.102}$$

The f, g expansion is

$$\mathbf{r} = f\mathbf{r}_0 + g\mathbf{v}_0 \tag{4.103}$$

$$\mathbf{v} = \dot{f}\mathbf{r}_0 + \dot{g}\mathbf{v}_0. \tag{4.104}$$

Start by crossing the position vector into the initial velocity vector:

$$\mathbf{r} \times \mathbf{v}_0 = (f\mathbf{r}_0 + g\mathbf{v}_0) \times \mathbf{v}_0 \tag{4.105}$$

$$= f(\mathbf{r}_0 \times \mathbf{v}_0) + g(\mathbf{v}_0 \times \mathbf{v}_0). \tag{4.106}$$

The second term is zero, and the other terms are normal to the plane:

$$x\dot{y}_0 - \dot{x}_0 y = fh \Rightarrow f = \frac{x\dot{y}_0 - \dot{x}_0 y}{h}. \tag{4.107}$$

Differentiating this last equation:

$$\dot{f} = \frac{\dot{x}\dot{y}_0 - \dot{x}_0 \dot{y}}{h}. \tag{4.108}$$

Now cross the initial position vector into the position vector:

$$\mathbf{r}_0 \times \mathbf{r} = \mathbf{r}_0 \times (f\mathbf{r}_0 + g\mathbf{v}_0) \tag{4.109}$$

$$= f(\mathbf{r}_0 \times \mathbf{r}_0) + g(\mathbf{r}_0 \times \mathbf{v}_0) = f\mathbf{h}. \tag{4.110}$$

The first term is zero, and the other terms are normal to the plane:

$$x_0 y - x y_0 = gh \Rightarrow g = \frac{x_0 y - x y_0}{h}. \tag{4.111}$$

Differentiating this last equation:

$$\dot{g} = \frac{\dot{x}_0 \dot{y} - \dot{x}\dot{y}_0}{h}. \tag{4.112}$$

Look at the cross-product:

$$\mathbf{h} = \mathbf{r} \times \mathbf{v} = (f\mathbf{r}_0 + g\mathbf{v}_0) \times (\dot{f}\mathbf{r}_0 + \dot{g}\mathbf{v}_0) \tag{4.113}$$

$$= f\dot{f}(\mathbf{r}_0 \times \mathbf{r}_0) + f\dot{g}(\mathbf{r}_0 \times \mathbf{v}_0) + \dot{f}g(\mathbf{v}_0 \times \mathbf{r}_0) + g\dot{g}(\mathbf{v}_0 \times \mathbf{v}_0) \tag{4.114}$$

$$= f\dot{g}\mathbf{h} - \dot{f}g\mathbf{h}. \tag{4.115}$$

which can only be true if $1 = f\dot{g} - \dot{f}g$, i.e. Equation (2.202).

As $x = r\cos\theta = a\cos E - ae$, $y = r\sin\theta = a\sqrt{1 - e^2}\sin E$, $\dot{x} = \sqrt{\mu/p}\sin\theta$, $\dot{y} = \sqrt{\mu/p}$
$(e + \cos\theta)$, which gives:

$$f = 1 - \frac{r}{p}(1 - \cos\Delta\theta) = 1 - \frac{a}{r_0}(1 - \cos\Delta E) \tag{4.116}$$

$$g = \frac{rr_0\sin\Delta\theta}{\sqrt{\mu p}} = t - \sqrt{\frac{a^3}{\mu}}(\Delta E - \sin\Delta E) \tag{4.117}$$

$$\dot{f} = \sqrt{\frac{\mu}{p}}\tan\left(\frac{\Delta\theta}{2}\right)\left(\frac{1 - \cos\Delta\theta}{p} - \frac{1}{r} - \frac{1}{r_0}\right) = -\frac{\sqrt{\mu a}}{r_0 r}\sin\Delta E \tag{4.118}$$

$$\dot{g} = 1 - \frac{r_0}{p}(1 - \cos\Delta\theta) = 1 - \frac{a}{r}(1 - \cos\Delta E). \tag{4.119}$$

So, to summarize, given an initial position $\mathbf{r}_0$ and final position $\mathbf{r}$, we can calculate the initial and final velocity given the change in the true anomaly $\Delta\theta$. The above four equations (4.116)–(4.119), are not independent as the constraint Equation (2.202). So, we have three unknown $p, a, \Delta E$, three independent equations. Some numerical algorithms are discussed in reference [4], such as p-iteration, z-iteration methods.

If the initial position and velocity are given, we can also calculate a future position and velocity, given the change in the true anomaly $\Delta\theta$, as $\mathbf{r} \cdot \mathbf{r}_0 = rr_0\cos\Delta\theta$.

f and g series has close relation with State Transition Matrix (STM).

We can reexpress our f and g series representation in terms of a state-variable relationship:

$$\mathbf{X} = \begin{bmatrix} \mathbf{r} \\ \mathbf{v} \end{bmatrix}, \quad \mathbf{X}_0 = \begin{bmatrix} \mathbf{r}_0 \\ \mathbf{v}_0, \end{bmatrix} \tag{4.120}$$

then,

$$\mathbf{X} = \mathbf{\Phi}\mathbf{X}_0 = \begin{bmatrix} F & G \\ \dot{F} & \dot{G} \end{bmatrix}\mathbf{X}_0, \quad \text{where } F = \begin{bmatrix} f & 0 & 0 \\ 0 & f & 0 \\ 0 & 0 & f \end{bmatrix} \tag{4.121}$$

and $\mathbf{\Phi}$ is the STM.

4.2.4 Examples

An Earth satellite is observed to have the following positions:

$$\mathbf{r}_1 = [1, 0, 0]^T \, \text{DU} \tag{4.122}$$

$$\mathbf{r}_2 = [-0.0767, 1.5217, 0]^T \, \text{DU} \tag{4.123}$$

at two times, separated by ($TOF = 1, 2, 5, 10, 15, 20$ TU).

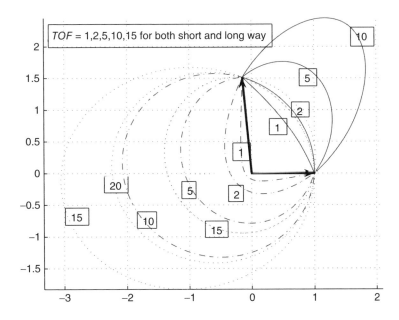

Figure 4.8   Examples of Lambert transfer with different *TOF*

It is easy to obtain $\Delta\theta = 1.6212$ rad $= 92.8855°$ for short way orbit. $\Delta\theta = 4.6620$ rad for long way orbit, $r_1 = 1$, $r_2 = 1.5236$. See Figure 4.8. The two thick arrows are initial and final positions. All the orbits in the figure are in the same orbital plane, i.e. their $i = 0°$ for short way; $i = 180°$, for long way. For *TOF* = 15, which we show here, are two multiple revolution Lambert orbits. The small *TOF* = 15 one is a long-way two revolutions. And the large *TOF* = 15 one is a short-way two-revolution orbit. *TOF* = 20 is also a two-revolution long-way orbit. Only the two orbit of *TOF* = 1 are hyperbolic orbit; all others are elliptic. See the data in Table 4.1.

Interested reader can verify that the $a_{min} = 1.07588$; at this time its $p_{min} = 0.81293$, $e_{min} = 0.56594$, $t_{min} = 3.37918$.

Table 4.1   The data of the orbits in Figure 4.8

TOF	type	a	e	i	$\Omega + \omega$	θ_2
1	Short way	−0.602	2.5136	0	28.9	64.0201
1	Long way	−0.3303	1.2393	180	−131.6	135.4975
2	Short way	1.5648	0.3666	0	−14.7	107.6260
2	Long way	1.9791	0.8665	180	−125.8	141.3
5	Short way	1.1609	0.6268	0	−119.09	210.98
5	Long way	1.1488	0.3266	180	−85.39	181.72
10	Short way	1.5556	0.8057	0	−124.3	217.2
10	Long way	1.5408	0.3580	180	−16.4	250.7
20	short way	2.0319	0.8917	0	−126.4	291.2
20	long way	1.4564	0.3277	180	−23.72	243.4

The minimum eccentricity ellipse is sometimes called the fundamental ellipse [2], whose value in this example is

$$e_s = \frac{r_2 - r_1}{c} = 0.2942. \tag{4.124}$$

Note that in the figure solid lines represent short-way orbits, dash-dot lines represent long-way orbits, dot lines represent multi-orbits. The key to finding multiple Lambert orbit is to set the initial z for the universal variable, see Figure 2.24.

4.2.5 Comparison of CW and Nonlinear Lambert Analysis

For the linear analysis of the Lambert problem, first rewrite the CW-equation as

$$\begin{bmatrix} \mathbf{r}_2 \\ \mathbf{v}_2 \end{bmatrix} = \begin{bmatrix} \phi_{11} & \phi_{12} \\ \phi_{21} & \phi_{22} \end{bmatrix} \begin{bmatrix} \mathbf{r}_1 \\ \mathbf{v}_1 \end{bmatrix}. \tag{4.125}$$

Given: $TOF = t_2 - t_1, \mathbf{r}_1, \mathbf{r}_2$, we can find $\mathbf{v}_1, \mathbf{v}_2$. As

$$\mathbf{v}_1 = \phi_{12}^{-1} \left(\mathbf{r}_2 - \phi_{11}\mathbf{r}_1 \right). \tag{4.126}$$

then

$$\mathbf{v}_2 = \phi_{21}\mathbf{r}_1 - \phi_{22}\mathbf{v}_1 \tag{4.127}$$

$$= \phi_{21}\mathbf{r}_1 - \phi_{22}\phi_{12}^{-1} \left(\mathbf{r}_2 - \phi_{11}\mathbf{r}_1 \right). \tag{4.128}$$

The Space Shuttle, in a 250 km circular orbit, is to rendezvous with the Mir space station, which is in a 360 km circular orbit. The maneuver begins with the Shuttle trailing Mir by 15° central angle. We want to rendezvous in the time Mir moves through a central angle of $\pi/2$. Use the (nonlinear) Lambert analysis to find the required velocity impulse. Repeat the calculation using the (linearized) Clohessy-Wiltshire (CW) analysis. Briefly compare and discuss the results. See Figure 4.9.

When $\theta = 10°$, the required impulse from a (nonlinear) Lambert analysis is: $\Delta\mathbf{v}_{lam} = [-0.4531, -0.5958, 0]$;
The required impulse from a linear analysis is: $\Delta\mathbf{v}_{cw} = [-0.5480, -0.5394, 0]$;
$\Delta v_{error} = 0.0204$

When $\theta = 15°$, the required impulse from a (nonlinear) Lambert analysis is: $\Delta\mathbf{v}_{lam} = [-0.7396, -0.9114, 0]$;
The required impulse from a linear analysis is: $\Delta\mathbf{v}_{cw} = [-0.9464, -0.7737, 0]$;
$\Delta v_{error} = 0.0487$

When $\theta = 30°$, the required impulse from a (nonlinear) Lambert analysis is: $\Delta\mathbf{v}_{lam} = [-1.7541, -1.5849, 0]$;
The required impulse from a linear analysis is: $\Delta\mathbf{v}_{cw} = [-2.5721, -1.0098, 0]$;
$\Delta v_{error} = 0.3991$.

This example show when the trailing angle is larger than 20°, the linear CW method can have relatively large errors.

Some other examples can also show that fewer errors appear for the R-bar approach than for the V-bar approach between these two methods.

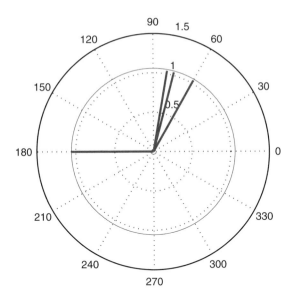

Figure 4.9   Comparison of CW and nonlinear Lambert results for $\theta = 10°$, $15°$ and $30°$

This analysis is compatible with the discussion in examples in Figure 4.2 and 4.4 as CW-equations involve a linear approximation of the relative motion.

4.3   Orbit Determination

Orbital determination is classified into two groups, initial orbital determination (IOD), which uses a few observations to determinate a orbit, whereas differential correction (DC) of Equation (A.25) in the Appendix of this book uses more observations to improve its accuracy. In differential correction, there are two basic types, batch (least squares estimate) and sequential (Kalman filter). The batch method processes the data in a group after a period of measurement and the sequential method processes measurement data simultaneously with each updated measurement.

The Gauss and Laplace methods uses three angular position observations only; they are called angle-only method. The difference between the Gauss and Laplace methods is that Gauss used the solution two-body problem, whereas Laplace applyied the relative motion equation of two-bodies directly. These two method are more theoretically than practically important nowadays additional information is usually available using modern measurement such as radar data. So we will not give a detailed introduction of these two methods in this book.

We have already discussed some methods of orbit determinations in previous chapters. If we have a position vector and its velocity vector of a satellite at a given time from a radar site, we can calculate its orbital elements easily. And also if we know three position vectors at three different times by using a GPS receiver we can determine the orbit using Gibber's method. And if we know two positions and the time of flight between these two position vectors, it has been referred to as Lambert's problem of orbital transfer in this book, and can also be regarded as an orbital determination problem. In these two cases, the principle is the same.

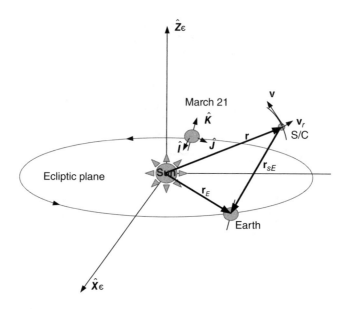

Figure 4.10   The spacecraft and the Earth position in HECS

In this section we introduce extended Kalman filter application to orbital determination with an interplanetary flight example in the Sun-centered heliocentric frame. See Figure 4.10. where the spacecraft position vector is $\mathbf{r} = [x, y, z]^T$, the Earth position vector is $\mathbf{r}_E = [x_E, y_E, z_E]^T$, the position vector from the satellite to the Earth is $\mathbf{r}_{sE} = [x_E - x, y_E - y, z_E - z]^T$.

The method was proposed by reference [6] realized and verified by reference [8] etc.

Rewrite the relative equation of two bodies in state equation format:

$$\dot{\mathbf{x}} = \mathbf{f}(\mathbf{x}, t) + \mathbf{w} = \left[\dot{x}, \dot{y}, \dot{z}, -\frac{\mu_s x}{r^3}, -\frac{\mu_s y}{r^3}, -\frac{\mu_s z}{r^3}\right]^T + \mathbf{w}, \tag{4.129}$$

where state vector $\mathbf{x} = [x, y, z, \dot{x}, \dot{y}, \dot{z}]^T$, radius $r = (x^2 + y^2 + z^2)^{1/2}$,
$\mathbf{w} = [w_x, w_y, w_z, w_{vx}, w_{wv}, w_{vz}]^T$ is a white Gaussian process noise term and μ_s is the gravity mass constant of the Sun. The measurement equation is

$$\mathbf{z} = \mathbf{h}(\mathbf{x}, t) + \mathbf{v}_n. \tag{4.130}$$

$\mathbf{h} = [h_1, h_2, h_3]^T$ is a measurement vector, and $\mathbf{v}_n$ is a measurement noise which is assumed to be a white Gaussian noise process. The radial Doppler velocity measurement to the Sun is

$$h_1 = v_r = \dot{r} = \frac{\mathbf{r} \cdot \dot{\mathbf{r}}}{r} = \frac{\mathbf{r} \cdot \mathbf{v}}{r} = \frac{x\dot{x} + y\dot{y} + z\dot{z}}{r}. \tag{4.131}$$

For the azimuth angle (celestial longitude angle α) and elevation angle (or celestial latitude δ) about the line of sight (LOS) vector to the Sun, see Figure 4.11.

$$\mathbf{h}_2 = [\alpha_S, \delta_S]^T = [\arctan(y/x), \arcsin(z/r)]^T. \tag{4.132}$$

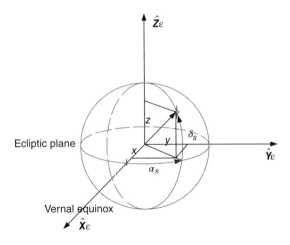

Figure 4.11   The measurement angles

The azimuth angle and elevation of the Earth viewed from an interplanetary spacecraft to the Earth is represented in the inertial reference frame

$$
\mathbf{h}_3 = [\alpha_{sE}, \delta_{sE}]^T = \left[\arctan\left(\frac{y_E - y}{x_E - x} \right), \arcsin\left(\frac{z_E - z}{|\mathbf{r}_E - \mathbf{r}|} \right) \right]^T.
\tag{4.133}
$$

where $[x_E, y_E, z_E]$ is the Earth's position in the heliocentric frame.

The error covariance matrix can be expressed as

$$
\mathbf{P}(t) = \mathbf{E}\left[[\hat{\mathbf{x}}(t) - \mathbf{x}(t)][\hat{\mathbf{x}}(t) - \mathbf{x}(t)]^T \right].
\tag{4.134}
$$

In order to use the Kalman filter, the linearization about the current best estimate of the nonlinear trajectory should be performed.

$$
\Delta\dot{\mathbf{x}} = \mathbf{F}\Delta\mathbf{x} + \mathbf{G}\mathbf{w}
\tag{4.135}
$$

$$
\Delta\mathbf{z} = \mathbf{H}\Delta\mathbf{x} + \mathbf{v},
\tag{4.136}
$$

where the Jacobians are

$$
\mathbf{F} \equiv \frac{\partial \mathbf{f}}{\partial \mathbf{x}} = \left[\mathbf{f}_r, \mathbf{f}_v \right]
\tag{4.137}
$$

$$
\mathbf{f}_r =
\begin{bmatrix}
0 & 0 & 0 \\
0 & 0 & 0 \\
0 & 0 & 0 \\
\mu_s(3x^2 - r^2)/r^5 & 3\mu_s xy/r^5 & 3\mu_s xz/r^5 \\
3\mu_s xy/r^5 & \mu_s(3y^2 - r^2)/r^5 & 3\mu_s yz/r^5 \\
3\mu_s xz/r^5 & 3\mu_s yz/r^5 & \mu_s(3z^2 - r^2)/r^5
\end{bmatrix}
\tag{4.138}
$$

$$\mathbf{f_v} = \begin{bmatrix} 1 & 0 & 0 \\ 0 & 1 & 0 \\ 0 & 0 & 1 \\ 0 & 0 & 0 \\ 0 & 0 & 0 \\ 0 & 0 & 0 \end{bmatrix} \tag{4.139}$$

$$\mathbf{G} \equiv \frac{\partial \mathbf{f}}{\partial \mathbf{w}} \tag{4.140}$$

$$\mathbf{H} \equiv \frac{\partial \mathbf{f}}{\partial \mathbf{x}} = \left[\mathbf{H_r}, \mathbf{H_v}\right] \tag{4.141}$$

$$\mathbf{H_r} = \left[\mathbf{H}_x, \mathbf{H}_y, \mathbf{H}_z\right] \tag{4.142}$$

$$\mathbf{H}_x = \begin{bmatrix} \dfrac{\dot{x}r^2 - x(x\dot{x}+y\dot{y}+z\dot{z})}{r^3} \\[2ex] -\dfrac{y}{x^2+y^2} \\[2ex] \dfrac{y_E-y}{(x_E-x)^2+(y_E-y)^2} \\[2ex] \dfrac{(x_E-x)(z_E-z)}{\sqrt{((x_E-x)^2+(y_E-y)^2+(z_E-z)^2)}((x_E-x)^2+(y_E-y)^2)} \end{bmatrix} \tag{4.143}$$

$$\mathbf{H}_y = \begin{bmatrix} \dfrac{\dot{y}r^2 - y(x\dot{x}+y\dot{y}+z\dot{z})}{r^3} \\[2ex] \dfrac{x}{x^2+y^2} \\[2ex] -\dfrac{yz}{r^2\sqrt{x^2+y^2}} \\[2ex] \dfrac{x_E-x}{(x_E-x)^2+(y_E-y)^2} \\[2ex] \dfrac{(y_E-y)(z_E-z)}{\sqrt{((x_E-x)^2+(y_E-y)^2+(z_E-z)^2)}((x_E-x)^2+(y_E-y)^2)} \end{bmatrix} \tag{4.144}$$

$$\mathbf{H}_x = \begin{bmatrix} \dfrac{\dot{z}r^2 - z(\mathbf{r}\cdot\mathbf{v})}{r^3} \\[2ex] 0 \\[2ex] -\dfrac{\sqrt{r^2-z^2}}{r^2} \\[2ex] 0 \\[2ex] \dfrac{\sqrt{(x_E-x)^2+(y_E-y)^2}}{((x_E-x)^2+(y_E-y)^2+(z_E-z)^2)} \end{bmatrix} \tag{4.145}$$

$$\mathbf{H_v} = \begin{bmatrix} x/r & y/r & z/r \\ 0 & 0 & 0 \\ 0 & 0 & 0 \\ 0 & 0 & 0 \\ 0 & 0 & 0 \end{bmatrix} \tag{4.146}$$

and we assume a zero mean white noise process and the measurement noise with known covariance matrices:

$$\mathbf{E}\left\{\mathbf{w}\mathbf{w}^T\right\} = \mathbf{Q}, \quad \mathbf{E}\left\{\mathbf{v}\mathbf{v}^T\right\} = \mathbf{R}, \tag{4.147}$$

where $\mathbf{w}$ and $\mathbf{v}$ are assumed to be uncorrelated with each other. $\mathbf{Q}$ and $\mathbf{R}$ are the process noise covariance matrix and the measurement noise covariance matrix, respectively. For discrete time

$$\Delta \mathbf{x}_k = \mathbf{\Phi}_{k,k-1}\Delta \mathbf{x}_{k-1} + \mathbf{w}_{k-1}. \tag{4.148}$$

where $\mathbf{\Phi}$ is STM by Taylor first order expansion

$$\mathbf{\Phi}_{k,k-1} \approx \mathbf{I} + \mathbf{F}(t_{k-1}) \cdot T = \mathbf{I} + \frac{\partial \mathbf{f}(\mathbf{x}(t_{k-1}), t_{k-1})}{\partial \mathbf{x}(t_{k-1})} \cdot T \tag{4.149}$$

and T is the small sampling period.

The overall precess of the extended Kalman filter is

1. one step state prediction:

$$\hat{\mathbf{x}}_{k/k-1} = \hat{\mathbf{x}}_{k-1} + \mathbf{f}(\hat{\mathbf{x}}_{k-1}, t_{k-1}) \cdot T; \tag{4.150}$$

2. one step covariance prediction:

$$\mathbf{P}_{k/k-1} = \mathbf{\Phi}_{k,k-1}\mathbf{P}_{k-1}\mathbf{\Phi}_{k,k-1}^T + \mathbf{Q}_{k-1}; \tag{4.151}$$

3. the filter gain:

$$\mathbf{K}_k = \mathbf{P}_{k/k-1}\mathbf{H}_k^T[\mathbf{H}_k\mathbf{P}_{k/k-1}\mathbf{H}_k^T + \mathbf{R}_k]^{-1}; \tag{4.152}$$

4. state error estimate:

$$\Delta \hat{\mathbf{x}} = \mathbf{K}_k[\mathbf{z}_k - \mathbf{h}(\hat{\mathbf{x}}_{k/k-1}, t_k)]; \tag{4.153}$$

5. estimation error covariance matrix:

$$\mathbf{P}_k = (\mathbf{I} - \mathbf{K}_k\mathbf{H}_k)\mathbf{P}_{k/k-1}(\mathbf{I} - \mathbf{K}_k\mathbf{H}_k)^T + \mathbf{K}_k\mathbf{R}_k\mathbf{K}_k^T; \tag{4.154}$$

6. state estimation:

$$\hat{\mathbf{x}}_k = \hat{\mathbf{x}}_{k/k-1} + \mathbf{K}_k[\mathbf{z}_k - \mathbf{h}(\hat{\mathbf{x}}_{k/k-1}, t_k)], \tag{4.155}$$

where $\mathbf{I}$ is the identity matrix, $\hat{\mathbf{x}}_{k/k-1}$ is $\hat{\mathbf{x}}(t_k)$ before processing the measurement from $t = t_k$, $\hat{\mathbf{x}}_k$ is $\hat{\mathbf{x}}(t_k)$ after processing the measurement from $t = t_k$, $\mathbf{P}_{k/k-1}$, is $\mathbf{P}(t_k)$ before processing the measurement from $t = t_k$, $\hat{\mathbf{P}}_k$ is $\hat{\mathbf{P}}(t_k)$ after processing the measurement from $t = t_k$.

To show the solution format, we intentionally give a simulation result with big initial errors to see the difference between true trajectory, $\mathbf{x}$, and its estimate, $\hat{\mathbf{x}}$; see Figure 4.12. This is because we just use two measurements h_1, h_2 and choose a large sampling period with large initial estimation errors. The orbit elements of the trajectory are: $a = 154\,316\,072\,\text{km}$, $e = 0.03\,285\,574$, $i = 1.9\,216\,765°$, $\Omega = 0°$, $\omega = 398.4°$. With three measurement h_1, h_2, h_3 the filter performance becomes much better. See example in Figure 4.13 and Figure 4.14. The spacecraft orbit elements are still $a = 154\,316\,071.71\,781$, $e = 0.0\,328\,557\,377\,521\,196$, $i = 1.92\,167\,651\,332\,882°$, $\Omega = 359.997\,243\,840\,277°$, $\omega = 298.397\,427\,664\,513°$.

4.4   Optimal Control

In the previous chapter the Hohmann transfer was discussed. Actually it is the optimal control for two-impulse maneuver, as we show in Section 3.3.1. As the electrical propulsion engines

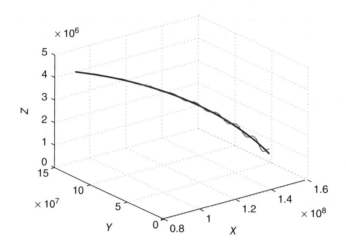

Figure 4.12 The true orbit trajectory (smooth thick line) and its estimate (thin oscillated line) by extended Kalman filter with two measurements $h_1, \mathbf{h}_2$

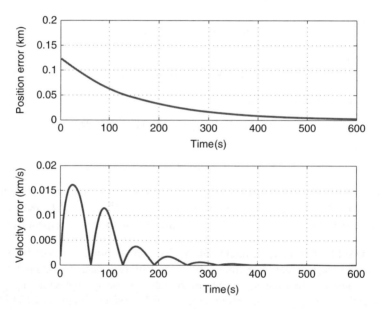

Figure 4.13 The estimate errors of position and velocity by the extended Kalman filter with three measurements $h_1, \mathbf{h}_2, \mathbf{h}_3$

become available, continuous optimal control (Bryson and Ho) of low thrust trajectories is now practical. For a chemical engine, its specific impulse is about 3 km/s, whereas the exhaust velocity of electrical propulsion is of the order 30 km/s or even higher. The solar or nuclear energy can be the source of electrical energy for electric propulsion.

   In this section an optimal control example for orbital transfers is presented by using the classical calculus of variation. The goal of this example is to determine the maximum radius

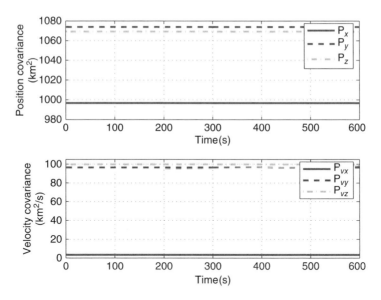

Figure 4.14 The covariances of position and velocity by the extended Kalman filter with three measurement h_1, h_2, h_3

orbit transfer in a given time t_f assuming a constant thrust engine whose thrust is T. Its initial and final orbits are circular, similar to Hohmann transfer. See Figure 4.15.

The control variable during the transfer is the thrust direction angle $\phi(t)$. See Figure 4.15. The nomenclatures are: r is the radial distance from the attracting center, with gravitational constant μ, v_r and $v_\theta = r\dot{\theta}$ are radial and tangential velocity components, m is the mass of S/C, and $\dot{m}$ is the fuel consumption rate, which is constant here, θ is the true anomaly of the orbit, and $\mathbf{e}$ is at the periapsis direction. The problem is to find $\phi(t)$ to maximize $r(t_f)$. The dynamics of the system are:

$$\ddot{r} = -\frac{\mu}{r^2} + r\dot{\theta}^2 + \frac{T \sin \phi}{m}, \tag{4.156}$$

$$\ddot{\theta} = \frac{1}{r}(-2\dot{r}\dot{\theta} + \frac{T \cos \phi}{m}) \tag{4.157}$$

$$\dot{m} = -c, \tag{4.158}$$

where $c = \dot{m}$ is a constant here. Define the three state variable as $x_1 = r, x_2 = \dot{r} = v_r, x_3 = r\dot{\theta} = v_\theta$, re-write dynamics:

$$\dot{x}_1 = x_2 \tag{4.159}$$

$$\dot{x}_2 = -\frac{\mu}{x_1^2} + \frac{x_3^2}{x_1} + \frac{T \sin \phi}{m_0 - |\dot{m}|t} \tag{4.160}$$

$$\dot{x}_3 = -\frac{x_2 x_3}{x_1} + \frac{T \cos \phi}{m_0 - |\dot{m}|t}. \tag{4.161}$$

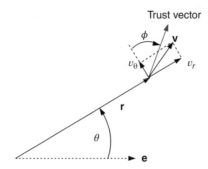

Figure 4.15 Definition of the continuous low thrust control

The initial conditions are:

$$x_1(0) = r(0) = r_0, \quad x_2(0) = v_r(0) = 0, \quad x_3(0) = v_\theta(0) - \sqrt{\frac{\mu}{r_0}} \tag{4.162}$$

and final conditions:

$$x_2(t_f) = v_r(t_f) = 0, \quad x_3(t_f) - \sqrt{\mu/x_1(t_f)} = 0. \tag{4.163}$$

The Hamiltonian is:

$$H = \lambda_1 x_2 + \lambda_2 \left(-\frac{\mu}{x_1^2} + \frac{x_3^2}{x_1} + \frac{T \sin \phi}{m_0 - |\dot{m}|t} \right) + \lambda_3 \left(-\frac{x_2 x_3}{x_1} + \frac{T \cos \phi}{m_0 - |\dot{m}|t} \right). \tag{4.164}$$

The costate equations $\dot{\lambda}_i = -\partial H / \partial x_i$ are:

$$\dot{\lambda}_1 = -2\lambda_2 \mu / x_1^3 + \lambda_2 x_3^2 / x_1^2 - \lambda_3 x_2 x_3 / x_1^2 \tag{4.165}$$

$$\dot{\lambda}_2 = -\lambda_1 + \lambda_3 x_3 / x_1 \tag{4.166}$$

$$\dot{\lambda}_3 = -2\lambda_2 x_3 / x_1 + \lambda_3 x_2 / x_1. \tag{4.167}$$

By the necessary condition for optimal control $\partial H / \partial \phi = 0$

$$\lambda_2 \left(\frac{T \cos \phi}{m_0 - |\dot{m}|t} \right) + \lambda_3 \left(-\frac{T \sin \phi}{m_0 - |\dot{m}|t} \right), \tag{4.168}$$

which gives that:

$$\tan \phi = \frac{\lambda_2(t)}{\lambda_3(t)}, \tag{4.169}$$

which can be solved for the control input given costate, that is

$$\sin \phi = \frac{\lambda_2}{\sqrt{\lambda_2^2 + \lambda_3^2}}, \quad \cos \phi = \frac{\lambda_3}{\sqrt{\lambda_2^2 + \lambda_3^2}}. \tag{4.170}$$

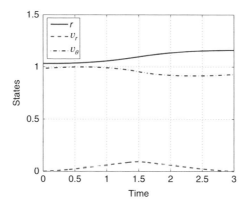

Figure 4.16   Optimal control: state

The control variable ϕ can then be replaced by costate in the dynamics equations. The physical meanings of λ_2 and λ_3 can be understood this way: λ_2 is the thrust in radius direction v_r, whereas λ_3 is the thrust in the tangential direction v_θ.

The terminal constraints are adjoined to the function to be maximized (the final radius) by means of the constant Lagrangian multipliers (v_1, v_2) :

$$w = -x_1(t_f) + v_1 x_2(t_f) + v_2(x_3(t_f) - \sqrt{\mu/x_1(t_f)}). \tag{4.171}$$

Since the first state x_1 is not specified at the final time, then

$$\lambda_1(t_f) = \frac{\partial w}{\partial x_1(t_f)} = -1 + \frac{v_2}{2}\sqrt{\frac{\mu}{r^3(t_f)}}. \tag{4.172}$$

Also that

$$\lambda_3(t_f) = \frac{\partial w}{\partial x_3(t_f)} = v_2, \tag{4.173}$$

which gives v_2 in terms of the costate.

When a very low Earth orbit is not a special working requirement, such as disaster monitoring conditions at some special time, it's better to raise its orbit to a higher orbit to reduce air resistance, and to expand its lifetime. The specific final height is not required, but it should be as high as possible for a given time. This is an application example of the above optimal control problem. See Figures 4.16–4.19 for a numerical solution. In this example, $\mu = 1\text{DU}^3/\text{TU}^2$, $t_f = 3\text{TU}$, $m_0 = 1$, $\dot{m} = -0.007\,485$, $T = 0.04\,683$, $r_0 = 1.0314\,\text{DU}$, $\dot{r}_0 = 0$, $\dot{r}(t_f) = 0$. Generally speaking, this is a two-point boundary value problem which can not be solved analytically since the dynamics are nonlinear. Numerous solution techniques are needed, such as shooting methods and collocation. There are some standard package which can convert the problem to a series of nonlinear algebraic equations in the unknowns.

This example is a fixed final time optimization problem. There are other cases whose final time is free, but the final state is fixed, such as the scenario of the rendezvous problem, that is,

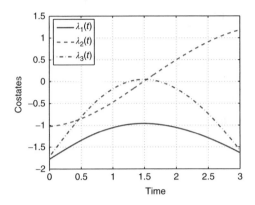

Figure 4.17 Optimal control: costate

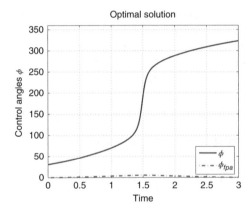

Figure 4.18 Optimal control: control variable ϕ and flight path angle ϕ_{fpa}

fixed $r(t_f)$, free t_f. The method in this section can be extended to solve this kind of problem. By expressing the final times differently, interested readers can find the approximate relationship between r_f and t_f. Then you can find the correspondent t_f, for given r_f.

4.5 Three-Body Problem – CRTBP

Newton told us that two masses attract each other under the law that gives us the nonlinear system of second-order differential equations:

$$m_i \ddot{\mathbf{r}}_i = -\sum_{j \neq i}^{n} \frac{m_i m_j}{r_{ij}^3}(\mathbf{r}_i - \mathbf{r}_j). \tag{4.174}$$

where $i = 1, 2, \cdots, n$, $\mathbf{r}_i$ is the position vector of m_i in an inertial frame.

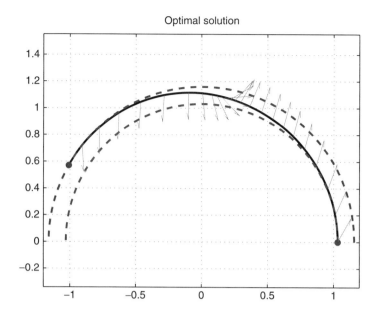

Figure 4.19 Optimal control: trajectory, the inner dash line is the initial circular orbit, the outer dash line is the final circular orbit, in-between solid line is the optimal trajectory. The arrows represent the thrust directions

The two-body problem was analyzed by Johannes Kepler in 1609 and solved by Isaac Newton in 1687. There are many systems we would like to calculate. For instance, a flight of a spacecraft from the Earth to Moon, or flight path of a meteorite. So we need to solve the few bodies problem of interactions.

In the mid-1890s Henri Poincaré showed that there could be no such quantities analyzed in positions, velocities and mass ratios for $N > 2$. In 1912 Karl Sundman found an infinite series that could in principle be summed to give the solution – but which converged exceptionally slowly. Henri Poincaré identified very sensitive dependence on initial conditions and developed a topology to provide a simpler overall description.

In this section we discuss a special kind of three-bodies problem, that is, the Circular Restricted Three-Body Problem(CRTBP).

Assume: (1) the primary and secondary bodies move in circular orbits about the center of mass, which lies between the two objects; (2) the mass of the third body (satellite) is negligible compared to that of the major bodies; (3) we will use a synodic coordinate frame whose origin coincides with the barycenter and is rotating with an angular velocity ω_s.

Geometry for (CRTBP). See Figure 4.20. The origin is at the mass center of the system. Notice the $\hat{x}_s$–axis is in the direction of the secondary object (Moon), and the $\hat{y}_s$–axis lies in the plane of the Earth–Moon. Remember, this synodic coordinate system is rotating with respect to the fixed barycentric frame $(\hat{x}_b, \hat{y}_b, \hat{z}_b)$. The $\hat{z}_b$ and $\hat{z}_s$ axes are aligned. The distances from the barycenter to each object (r_{b1}, r_{b2}) are used to develop the equation of motion. The mass ratio μ^* allows us to normalize the problem.

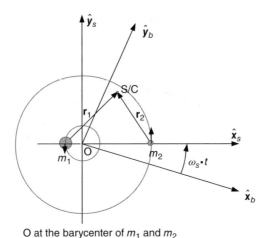

O at the barycenter of m_1 and m_2

Figure 4.20   The fixed barycentric and the rotating synodic frame

4.5.1   Equations of Motion

S/C's inertial acceleration relative to the barycenter (see Figure 4.22) is:

$$\ddot{\mathbf{r}}_{bsat} = -Gm_1 \frac{\mathbf{r}_{1sat}}{r_{1sat}^3} - Gm_2 \frac{\mathbf{r}_{2sat}}{r_{2sat}^3} \tag{4.175}$$

$$= \nabla \left(\frac{\mu_1}{r_1} + \frac{\mu_2}{r_2} \right). \tag{4.176}$$

Note: $\nabla \Phi = \partial R/\partial x \hat{x}_s + \partial R/\partial y \hat{y}_s + \partial R/\partial z \hat{z}_s$ is a gradient of potential function Φ.

$$\ddot{\mathbf{r}}_{bsat} = \mathbf{a}_s + \dot{\boldsymbol{\omega}}_s \times \mathbf{r}_s + \boldsymbol{\omega}_s \times (\boldsymbol{\omega}_s \times \mathbf{r}_s) + 2\boldsymbol{\omega}_s \times \mathbf{v}_s + \mathbf{a}_{org} \tag{4.177}$$

$$= \ddot{r}_s - \omega_s^2 (x \hat{x}_s + y \hat{y}_s) - 2\omega_s (\dot{y} \hat{x}_s - \dot{x} \hat{y}_s). \tag{4.178}$$

where $\boldsymbol{\omega} = [0, 0, \omega_s]^T$ is a constant vector, the acceleration of origin $\mathbf{a}_{org} = 0$. Then the scalar form is:

$$\ddot{x} - 2\omega_s \dot{y} - \omega_s^2 x = \frac{\partial}{\partial x} \left(\frac{\mu_1}{r_1} + \frac{\mu_2}{r_2} \right) = -\frac{\mu_1(x + r_{b1})}{r_1^3} - \frac{\mu_2(x - r_{b2})}{r_2^3} \tag{4.179}$$

$$\ddot{y} + 2\omega_s \dot{x} - \omega_s^2 y = \frac{\partial}{\partial y} \left(\frac{\mu_1}{r_1} + \frac{\mu_2}{r_2} \right) = -\frac{\mu_1 y}{r_1^3} - \frac{\mu_2 y}{r_2^3} \tag{4.180}$$

$$\ddot{z} = \frac{\partial}{\partial z} \left(\frac{\mu_1}{r_1} + \frac{\mu_2}{r_2} \right) = -\frac{\mu_1 z}{r_1^3} - \frac{\mu_2 z}{r_2^3}. \tag{4.181}$$

where

$$r_1 = \sqrt{(x + r_{b1})^2 + y^2 + z^2} \tag{4.182}$$

$$r_2 = \sqrt{(x - r_{b2})^2 + y^2 + z^2}. \tag{4.183}$$

Multiply each equation by $\dot{x}$, $\dot{y}$, and $\dot{z}$ respectively and sum them:

$$2\dot{x}\ddot{x} + 2\dot{y}\ddot{y} - 2\omega_s^2(x\dot{x} + y\dot{y}) \tag{4.184}$$

$$= 2\dot{x}\frac{\partial}{\partial x}\left(\frac{\mu_1}{r_1} + \frac{\mu_2}{r_2}\right) + 2\dot{y}\frac{\partial}{\partial y}\left(\frac{\mu_1}{r_1} + \frac{\mu_2}{r_2}\right) + 2\dot{z}\frac{\partial}{\partial z}\left(\frac{\mu_1}{r_1} + \frac{\mu_2}{r_2}\right). \tag{4.185}$$

Integrating this equation results in the Jacobian Integral:

$$\dot{x}^2 + \dot{y}^2 + \dot{z}^2 - \omega_s^2(x^2 + y^2) = \frac{2\mu_1}{r_1} + \frac{2\mu_2}{r_2} - C, \tag{4.186}$$

where C is known as Jacobi's Constant.

Choose the unit of mass so that the sum is unity:

$$m_1 = 1 - \mu^*, \quad m_2 = \mu^* \tag{4.187}$$

$$\mu^* = \frac{M_2}{M_1 + M_2}(= 0.01\,215 \text{ for Earth and Moon System}). \tag{4.188}$$

μ^* is called the mass ratio and is always less than 0.5.

If we use unity for the distance between the two masses, the larger and smaller masses are at a distance from the origin of $\mu^*, 1 - \mu^*$ respectively. We can now write:

$$r_1 = \sqrt{(x + \mu^*)^2 + y^2 + z^2} \tag{4.189}$$

$$r_2 = \sqrt{(x + \mu^* - 1)^2 + y^2 + z^2}, \tag{4.190}$$

since at this time, $\omega_s = 1$ with these new units, then the Jacobian Integral becomes:

$$v^2 = x^2 + y^2 + \frac{2(1 - \mu^*)}{r_1} + \frac{2\mu^*}{r_2} - C + \mu^*(1 - \mu^*) \tag{4.191}$$

When $v = 0$, Equation (4.191) can be used to calculate zero velocity curves. See Figure 4.22 and Figure 4.23.

The rotating nondimensional equations are:

$$\ddot{x} - 2\dot{y} - x = -\frac{(1 - \mu^*)(x + \mu^*)}{r_1^3} - \frac{\mu^*(x - 1 + \mu^*)}{r_2^3} \tag{4.192}$$

$$\ddot{y} + 2\dot{x} - y = -\frac{(1 - \mu^*)y}{r_1^3} - \frac{\mu^* y}{r_2^3} \tag{4.193}$$

$$\ddot{z} = -\frac{(1 - \mu^*)z}{r_1^3} - \frac{\mu^* z}{r_2^3}. \tag{4.194}$$

4.5.2 Lagrange Points

Set velocity and acceleration to be zero and solve for x, y for the equilibrium positions when $z = 0$. See Figure 4.21.

$$x - \frac{(1 - \mu^*)(x + \mu^*)}{r_1^3} - \frac{\mu^*(x - 1 + \mu^*)}{r_2^3} = 0 \tag{4.195}$$

$$y - \frac{(1 - \mu^*)y}{r_1^3} - \frac{\mu^* y}{r_2^3} = 0, \tag{4.196}$$

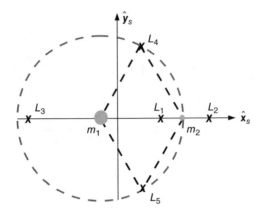

Figure 4.21   The five equilibrium points

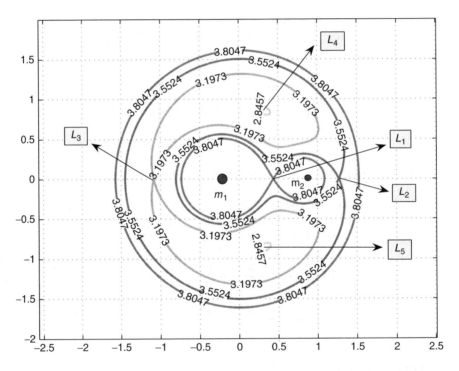

Figure 4.22   Contour of zero velocity curve when $\mu^* = 0.2$

which gives the five "Lagrange Points," denoted as L_1, L_2, L_3, L_4, L_5, in which L_1, L_2, L_3 are all on the x-axis, L_1 is between m_1 and m_2, L_2 beyond m_2, L_3 is beyond m_1, L_4, L_5 are leading and trailing equilateral libration points, respectively.

L_1, L_2, L_3 are always unstable, L_4, L_5 are stable when $\mu^* \leq 0.0385$, and are when $\mu^* > 0.0385$; more discussion can be found in reference [1]. "Trojan Asteroids" are found

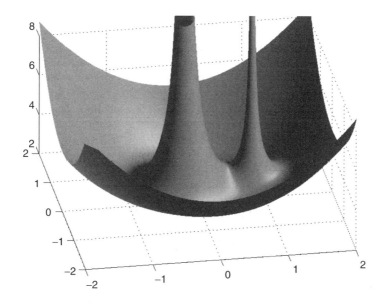

Figure 4.23 Zero velocity curve in 3 dimensions corresponds to Figure 4.22

at stable L_4, L_5 of the Sun–Jupiter system. These Lagrange points exist only in the rotating coordinate system.

See also an example of contour plot with different Jacobi constant in Figure 4.22. In this figure, the origin is at the barycenter (the mass center of m_1 and m_2), m_1 is at $(-\mu^*, 0) = (-0.2, 0)$, m_2 is at $(1 - \mu^*, 0) = (0.8, 0)$. It can be seen that the Lagrange points L_1 at $[0.4381,0,0]^T$ are on the curve of $C_{L1} = 3.80\,465$; and L_2 is at $[1.2710,0,0]^T$, when $C_{L2} = 3.55\,239$; L_3 is at $[-1.0828, 0, 0]^T$ when $C_{L3} = 3.19\,732$; L_4, L_5 are at $[0.3000, 0.866, 0]^T$ and $[0.3000, -0.866, 0]^T$ which are circumscribed by the contour $C \approx 2.8456$, actually $C_{L4} = C_{L5} = 2.840$. Potential energy wells appear in the vicinity of both m_1 and m_2; peaks are at equilateral triangular points L_4 and at L_5, and saddle points L_1, L_2 and L_3; see also Figure 4.23. Note that a S/C may have relative motion only in regions corresponding to values of C higher than its own; for example, if C is very large, than only when the S/C is very far away or when S/C is very near to m_1 or m_2.

4.5.3 Examples of CRTBP

If the Earth–Moon system is simplified to a circular restricted three body problem. Some related data can be calculated. The mass ratio is $\mu^* = 0.01\,215\,060\,379\,322$. The five Lagrange points are at

$$L_1 = [0.83\,691\,503\,629\,958, 0, 0]^T, \tag{4.197}$$

$$L_2 = [1.15\,568\,223\,538\,058, 0, 0]^T, \tag{4.198}$$

$$L_3 = [-1.00\,506\,265\,338\,634, 0, 0]^T, \tag{4.199}$$

$$L_4 = [0.48\,784\,939\,620\,678, 0.86\,602\,540\,378\,444, 0]^T, \tag{4.200}$$

$$L_5 = [0.48\,784\,939\,620\,678, -0.86\,602\,540\,378\,444, 0]^T. \tag{4.201}$$

Sometimes, L_1 is called a cislunar libration point which is between the Earth and Moon. L_2 is called a translunar libration point which is beyond the Moon's orbit. L_4, L_5 are leading and trailing equilateral libration points, respectively. Their corresponding Jacobi constants are:

$$C_{L1} = 3.18\,834\,128\,542\,812, \tag{4.202}$$

$$C_{L2} = 3.17\,216\,060\,448\,591, \tag{4.203}$$

$$C_{L3} = 3.01\,214\,716\,885\,328, \tag{4.204}$$

$$C_{L4} = C_{L5} = 2.98\,799\,703\,337\,932. \tag{4.205}$$

Since the Moon can be regarded as an Earth-oriented satellite, we can not see its reverse side to the Earth directly. In an effort to open up the far side of the moon to exploration, studies have been conducted to determine the feasibility of operating a communication satellite around the translunar equilibrium point L_2 in the Earth–Sun–Moon system. The desired satellite orbit is known as a halo orbit. The objective of the controller is to keep the satellite on a halo orbit trajectory that can be seen from the Earth so that the lines of communication are accessible at all times. The communication link is from the Earth to the satellite and then to the far side of the Moon.

As the μ^* of the Earth–Moon system is smaller than 0.0385, apparently the L_4 and L_5 points are stable, but with the perturbation from the Sun, they actually become unstable.

One method to analyze the system is a Poincaré map in which a series of simulations are performed. Hence it gives a global picture of the system, compared with the linear analysis of equilibrium points. In Figure 4.24, the data of x and $\dot{x}$ are recorded when the trajectory is across is the x-axis and when $y = 0$. Rich features of complicated dynamics can be revealed from these kinds of figures. For example, the point near $(x, \dot{x}) = (0.5, 0)$ is a periodic orbit. The rich structure including tori and periodic orbits, chaotic motions for an energy level between C_{L1} and C_{L2} can be seen. As $C_{L1} > C > C_{L2}$, the S/C can move in regions close both to the m_1 and m_2, and transports from each other, or at a great distance from both. Interested readers can read reference [1] for more detailed discussion.

For the Sun–Earth CRTBP, its $\mu^* = 3.040\,705\,167\,685\,162\,e - 006$, 1 AU = 1.496×10^8 km, its five Lagrange points are at

$$L_1 = [0.98\,998\,567\,386\,758, 0, 0]^T, \tag{4.206}$$

$$L_2 = [1.01\,007\,551\,208\,916, 0, 0]^T, \tag{4.207}$$

$$L_3 = [-1.00\,000\,126\,696\,049, 0, 0]^T, \tag{4.208}$$

$$L_4 = [0.49\,999\,695\,929\,483, 0.86\,602\,540\,378\,444, 0]^T, \tag{4.209}$$

$$L_5 = [0.49\,999\,695\,929\,483, -0.86\,602\,540\,378\,444, 0]^T. \tag{4.210}$$

and their corresponding Jacobi constants are

$$C_{L1} = 3.00\,089\,799\,664\,718, \tag{4.211}$$

$$C_{L2} = 3.00\,089\,394\,233\,228, \tag{4.212}$$

$$C_{L3} = 3.00\,000\,304\,070\,498, \tag{4.213}$$

$$C_{L4} = C_{L5} = 2.99\,999\,695\,930\,408. \tag{4.214}$$

In Figure 4.25, the inner isolated four twisted orbits are similar to "figure 8" orbits of geosynchronous orbits, which have already been used in navigation. The outer four near circular orbits are periodic orbits in three dimensions, which are often called Halo orbits. These have been used for deep space observations, such as the "Solar and Heliospheric Observatory" (SOHO) mission and Genesis mission. In the viewer on the Earth, the Halo orbits seem to be circling around the Sun, making them good places for communication with the Earth. All these orbits are symmetrical about the oxz plane. To find these orbits needs some numerical skills, which is beyond the scope of this book. The small circular "o"s in the middle of this figure are the L_1, L_2 point in the Sun–Earth system. The symbol star "*" is the Earth position in this system.

4.6 Summary and Keywords

This chapter introduces some relatively advanced topics in orbital mechanics, such as relative motion of two bodies the Lambert problem, orbital determination, continuous impulse optimal control, and three-body problems. Special simulation examples are given for all these problems with figures and comparisons.

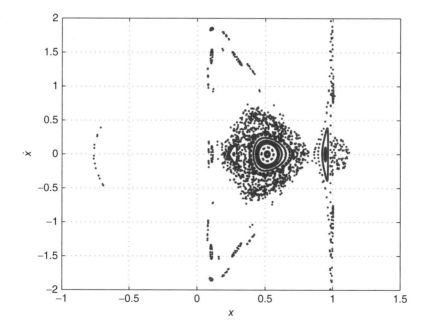

Figure 4.24 A Poincaré map of the Earth–Moon system

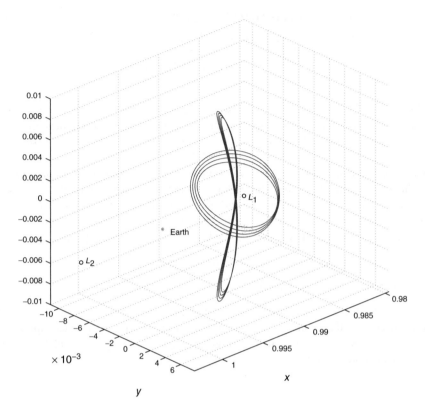

Figure 4.25   A closer look at periodic orbits near L_1 of the Sun–Earth system in three dimensions

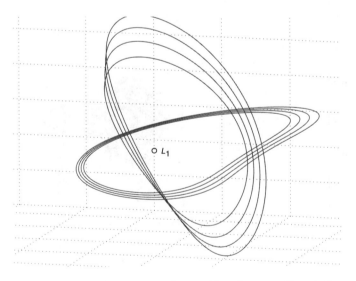

Figure 4.26   Halo orbits (3D) and Lyapunov (planar orbits whose $z = 0$) periodic orbits around L_1 points. A periodic orbit is a close-loop smooth trajectory whose derivative is continuous

Keywords: CW equation, Lambert theory, f,g function algorithm, continuous impulse, optimal control, circular restricted three body problem, Jacobian integral, zero velocity curve, Lagrange points, Poincaré map, Halo orbits, Lyapunov orbits, etc.

Problems

4.1 The Mir space station is located in a 300 km circular orbit. The Progress supply vessel is located a short distance away: 200 meters higher, 300 meters forward along Mir's velocity vector, and 50 meters to the left of Mir's orbit plane in the direction of the angular momentum vector. Use the Clohessey-Wiltshire (aka Hill's) equations to determine delta-V needed to rendezvouz Progress with Mir in 5 minutes. What is the delta-V needed if the transfer time is 30 minutes?

4.2 You are to design an approach maneuver between the shuttle and space station. Assume the space station is in a 6700 km circular initial orbit. The initial orbit of the shuttle is a 6698 km circular initial orbit and it is initially lagging the ISS by 10 km.

(a) What is the initial velocity of the space shuttle relative to the ISS with respect to the rotating reference frame (i.e. Δv)? Neglect the small curvature in the orbit over the 10 km separation (analogous to working in "curvilinear coordinates")

(b) What is the required ΔV to move the shuttle to a point 1 km behind the ISS (in the same 6700 km orbit) in 30 minutes?

4.3 In a CW equation if $x_0 = y_0 = z_0 = 0$, $\dot{y}_0 = 0$, $\dot{x}_0 = \dot{z}_0/3$, verify that the orbit of the chaser is an ellipsoid around the target.

4.4 Under what conditions is the solution of CW equation stationary, i.e. so that $\dot{x}, \dot{y}, \dot{z}, \ddot{x}, \ddot{y}, \ddot{z}$ are all zero? What is its physical meaning in inertial frame?

4.5 Prove that in a CW equation, if $\dot{y}_0 + 2\omega x_0$, then the chaser and target have the same period.

4.6 Prove that for chaser orbit, $e = \sqrt{4\dot{y} + \dot{x}/(\omega a)}$ where $a \approx r_t$.

4.7 If $r_1 = 1$, $r_2 = 2$, $\mu = 1$, $c = 2.5$, what is a_{min}, t_{min}, e_{min} and e_s in Equation (4.124) for possible transfer orbit?

4.8 An orbit with parameter $\mu = 1$, $p = 2$ DU connects two position vectors: $r_1 = 3\hat{I}$ DU, $r_2 = 4\hat{J}$ DU. Assuming a "long-way" trajectory, how long does a spacecraft take to get from r_1 to r_2? $t = ?$ [TU]

4.9 An Earth satellite is observed to have the following positions:

$$\mathbf{r}_1 = [0.5, 0.6, 0.7]\text{DU}, \mathbf{r}_2 = [0.0, -1.0, 0.0]\text{DU}$$

at two times, separated by 20 TU. Use the p-iteration technique to determine the orbit elements, $\mathbf{r}_1$ and $\mathbf{r}_2$. Solve for both the long-way and short-way transfers.

4.10 For the following Earth satellite data sets determine the orbit elements for the transfer from r_1 to r_2:

(a) $r_1 = [1.2, 0, 0]$ DU, $r_2 = [0.0, 2.0, 0.0]$ DU, $TOF = 10$ TU,
(b) $r_1 = [4, 0, 0]$ DU, $r_2 = [-2, 0, 0]$ DU, $TOF = 10$ TU,
(c) $r_1 = [2.0, 0, 0]$ DU, $r_2 = [-2.0, -0.2, 0.0]$ DU, $TOF = 20$ TU,

Solve for both the short-way and long-way transfers.

4.11 Show Equation (4.64), beginning with Equation (4.63) as $a \to \infty$.

4.12 An interplanetary probe departs the Earth's heliocentric orbit enroute to Mars on a transfer with $\Delta\theta = 170$ deg and TOF of 250 Earth days. Assume a circular orbit for the Earth and a Mars orbit eccentricity of 0.0934, inclination of 1.85 degrees, semi-major axis of 1.524 AU, and longitude of periapsis of 0 degrees. For communication purposes, we wish to arrive in Mars' heliocentric orbit when Mars is at its periapsis.

(a) Determine the orbital elements, v_1 and v_2 of this heliocentric transfer.
(b) Determine the Earth-Mars phase angle required for Mars intercept/rendezvous.
(c) Determine the synodic period for this geometric alignment.

4.13 Launched on a single launch vehicle, an orbiter and lander are designed for a 210-day interplanetary transfer from Earth to Mars. The heliocentric departure and arrival positions as specified in the heliocentric ecliptic and equinox Cartesian coordinate system are:

$$r_1 = [-0.707, -0.707, 0.0]\text{AU}, r_2 = [0., 1.6655, 0.0]\text{AU}.$$

The spacecraft depart a 200-km circular LEO and travel to Mars as a single unit. Once at the Mars SOI, the lander and orbiter separate and target the appropriate miss distances for atmospheric entry and orbit insertion, respectively. The lander must enter the Mars atmosphere (interface defined at a Mars radius of 3522.2 km) with a flight-path angle between -11 and -12 deg. The orbiter is designed to propulsively enter a circular science orbit with altitude between 396 and 404 km. Assume the planets move in a counterclockwise direction about the Sun.

(a) Determine the heliocentric velocity vectors at Earth departure and Mars arrival.
(b) Determine the Earth departure ΔV (magnitude and direction) and arrival Vinf vector (in the heliocentric frame).
(c) For the lander, determine the Mars entry velocity and required miss distance precision.
(d) For the orbiter, determine the Mars orbit insertion V (magnitude and direction) and required miss distance precision.

4.14 A human Mars exploration mission is planned based on use of the outbound transfer defined in problem 3. If a return transfer to the Earth with $\Delta\theta = 170$ deg and TOF of 250 Earth days is planned, how long must the astronauts stay at Mars before the proper planetary alignment occurs? If both transfers could be reduced to 100 Earth days with a $\Delta\theta = 80$ deg, how long must the astronauts stay at Mars before the proper planetary alignment occurs?

4.15 Three bodies of identical, spherically-distributed mass are acted on only by their mutual gravitational attraction. They are located at vertices of an equilateral triangle with sides of length d. Consider the motion of any one of the bodies about the system center of mass and determine the angular velocity required for d to remain constant.

4.16 The value of the Jacobi integral at the Earth-Moon L_2 point is 3.17 216. Ignore small terms in the expression for Jacobi's integral and

(a) Approximate the ΔV required to depart a 200-km circular low Earth orbit and arrive in the vicinity of the Earth-Moon L_2 point.
(b) Use the geometry of the L_4 point to determine the value of the Jacobi integral at this Earth-Moon location.
(c) Approximate the ΔV needed to depart the same LEO orbit and arrive in the vicinity of the Earth-Moon L_4 point.

4.17 The Deep Impact spacecraft was launched on January 12, 2005 at 18:08:29 UTC on a trajectory targeted to impact comet Tempel-1 as this comet reached its perihelion. Based on Earth-based observations made during the Tempel-1 approach to perihelion in 1999, the following heliocentric orbital elements were predicted for the 2005 perihelion approach:

Comet Tempel-1 estimated orbital elements:
Semi-Major Axis: 4.66 974 119 E08 km, Eccentricity: 0.51 751,
Inclination: 10.5301 deg, Ascending Node 68.9373 deg,
Argument of Perihelion: 178.8389 deg, Mean Anomaly 0.000 deg,
Epoch: July 5, 2005, 07:34:11 UTC.
Coordinate System here is the Sun-centered Ecliptic and Equinox of J2000.

(a) Calculate the orbit period of Tempel-1.
(b) Calculate the heliocentric Cartesian state of Tempel-1 at the 2005 perihelion passage.
(c) Given the following heliocentric position and velocity of Earth at the time of Deep Impact launch, calculate the heliocentric velocity vector of the spacecraft at departure from Earth.

Earth state at launch:
X: 5.6 508 552 E07 km, Y: 1.358 526 E08 km, Z: 1.954 E03 km,
V_x: −27.997 701 km/s, V_y: −11.562 501 km/s, V_z: −0.000 641 km/s.
Epoch: January 12, 2005, 18:08:29 UTC.
Coordinate System Sun-centered Ecliptic and Equinox of J2000

(d) Calculate the Earth-relative V_{inf} of Deep Impact following launch.
(e) Calculate the heliocentric arrival velocity of Deep Impact at Tempel-1. What is the comet relative impact velocity? For an impactor mass of 370 kg, approximate the kinetic energy of the impact event in Joules. If the energy release associated with exploding a ton of TNT is approximately 4.184 E09 J, what is the TNT equivalent of the impact event.

4.18 Tempel-1 has been observed from Earth since 1967, but its orbital elements are known to change over time due to Jupiter's influence and the outgassing of volatile near perihelion.

(a) Calculate the magnitude of the velocity change (ΔV) for a perihelion outgassing event that increases the Tempel-1 orbit period by 5 hours.

(b) If the ΔV outgassing event at the 1999 perihelion was not detected by Earth-based observations on its outbound trajectory and the Deep Impact spacecraft was targeted based upon the Tempel-1 perihelion approach phase orbital elements given in the previous problem, what would the resulting miss distance be at the planned time of impact?

(c) The Tempel-1 comet nucleus is 5–10 km in diameter. What does this exercise tell us about our ability to target a spacecraft onto a comet-intercept trajectory in the presence of orbit perturbations? How was the Deep Impact spacecraft successfully targeted for its impact with the comet Tempel-1?

4.19 For the Earth–Moon system, assuming the approximation that their paths both describe circular orbits about the common center of mass, we can make use of circular restricted three-body problem to compute the straight-line libration points. For the system the mass ration μ is given by $\mu = 1/(81.3 + 1)$.

(a) Solving equation numerically, L_1: $x - (1 - \mu)/(\mu + x)^2 + \mu/(x - 1 + \mu)^2 = 0$

(Hint: using MATLAB® function "roots" or "solve").

(b) Give the similar equations for L_2, L_3 in (a).

4.20 The linearized (and normalized) equations of Equation (4.192)–(4.194) around the equilibrium points are:

$$\dot{\mathbf{x}} = \begin{bmatrix} 0 & 0 & 0 & 1 & 0 & 0 \\ 0 & 0 & 0 & 0 & 1 & 0 \\ 0 & 0 & 0 & 0 & 0 & 1 \\ U_{xx} & U_{xy} & 0 & 0 & 2 & 0 \\ U_{xy} & U_{yy} & 0 & -2 & 0 & 0 \\ 0 & 0 & U_{zz} & 0 & 0 & 0 \end{bmatrix} + \begin{bmatrix} 0 \\ 0 \\ 0 \\ 1 \\ 0 \\ 0 \end{bmatrix} u_1 + \begin{bmatrix} 0 \\ 0 \\ 0 \\ 0 \\ 1 \\ 0 \end{bmatrix} u_2 + \begin{bmatrix} 0 \\ 0 \\ 0 \\ 0 \\ 0 \\ 1 \end{bmatrix} u_3.$$

The state vector, $\mathbf{x}$, is the satellite position and velocity, and the inputs u_i, $i = 1, 2, 3$, are the engine thrust accelerations in the $\hat{x}_s, \hat{y}_s$ and $\hat{z}_s$ direction of synodic coordinate in Figure 4.21, respectively. Prove:

(a) For L_1, $U_{xx} = 11.2952$, $U_{yy} = -4.1476$, $U_{zz} = -5.1476$, $U_{xy} = 0$.

(b) For L_2, $U_{xx} = 7.3809$, $U_{xy} = 0$, $U_{yy} = -2.1904$, $U_{zz} = -3.1904$.

(c) For L_3, $U_{xx} = 3.0214$, $U_{yy} = -0.0107$, $U_{zz} = -1.0107$, $U_{xy} = 0$.

(d) Proof $U_{yy} = U_{zz} + 1$ for L_1, L_2 and L_3.

(e) The positions for L_4, L_5 are $[1/2 - \mu^*, \pm\sqrt{3}/2, 0]^T$ by Equations (4.195)–(4.196) and $U_{xx} = 0.75$, $U_{yy} = 2.25$, $U_{xy} = -1.2675$, $U_{zz} = -1$.

4.21 For the previous problem, about the cislunar L_1 point, see Figure 4.26.

(a) Is the equilibrium point a stable location? What are their in-plane and out-of-plane non dimensional frequencies for their oscillatory mode? Compare them with the Moon's orbital period.

(b) Is the system controllable from u_1 alone?

(c) Repeat part (b) for u_2.

(d) Repeat part (b) for u_3.

(e) Suppose that we can observe the position in the y direction. Determine the transfer function from u_2 to y. (Hint: let $\mathbf{y}=[0\ 1\ 0\ 0\ 0\ 0]\ \mathbf{x}$.)

(f) Compute a state-space representation of the transfer function in part(e) using $\mathtt{ss}$ function. Verify that the system is controllable.

(g) Using the state feedback $u_2 = \mathbf{Kx}$, design a controller (i.e., find $\mathbf{K}$) for the system in part (f) such that the close-loop system poles are at and $s_{1,2} = -1 \pm j$ and $s_{3,4} = -10$.

References

[1] Szebehely, V. (1967) *Theory of Orbits*. New York: Academic Press Inc., 1967.

[2] Battin, R.H. and Vaughan, R.J., (1987) *An Introduction to the Mathematics and Methods of Astrodynamics*. New York: AIAA.

[3] Kaplan M, (1976) *Modern Spacecraft Dynamics and Control*. New York: John Wiley & Sons Inc.

[4] Bate, R., Mueller, D., and White, J. (1971) *Fundamentals of Astrodynamics*. New York: Dover Publications, Inc.

[5] Schaub, H. Junkins, and J. (2003) *Analytical Mechanics of Space Systems*. AIAA Education Series. Reston, VA. 2003: AIAA.

[6] Yim, J.R., Crassidis, J.L., and Junkins, J.L. (2000) Autonomous orbit navigation for interplanetary spacecraft, AIAA Guidance, Navigation and Control Conference, Denver, CO, Aug.2000:53–61.

[7] Alfriend, K., Vadali, S., Gurfil, P., *et al. Spacecraft Formation Flying*. Oxford: Elsevier Astrodynamics Series.

[8] Fufei Yang and Weiduo Hu (2006) *A Method of Interplanetary Autonomous Orbital Determination* (in Chinese). Beijing: Graduate Design of Beihang University.

[9] Bryson, A. and Ho, Y. (1975) *Applied Optimal Control*. Levittown, New York: Taylor & Francis.

5

Perturbed Orbital Motions

For the two-body problem, it is assumed that gravity is the only force; the Earth's mass is much greater than a spacecraft's mass; the Earth is spherically symmetric with uniform density and a spacecraft's mass is constant. These assumptions led to the restricted two-body problem. Its solution yields: a-semi-major axis, e-eccentricity, i-inclination, Ω-right ascension of ascending node, ω-argument of perigee, θ-true anomaly. Only θ varies with time – others are constant for a given orbit.

If these assumptions change, other elements may change. In this chapter we consider the cases where perturbations exist but the central body's attraction force is dominated, that is much larger than perturbation forces.

5.1 Special Perturbation

When considering perturbation, there are two general categories of techniques: (1) special perturbations techniques, which deal with numerical integration of equation of motion, for particular problem; (2) general (absolute) perturbations techniques, which deal with analytic integration of series expansion of perturbing accelerations for general problems.

5.1.1 General Concept of Perturbation

Gravity is no longer the only force. For example, consider atmospheric drag. Earth's atmosphere has an effect as high as 600 km. Drag is a non conservative force - it takes energy away from the orbit in terms of friction, which results not only in the semi-major axis geting smaller, but in eccentricity decreasing. Drag is difficult to model. To keep a spacecraft in orbit, fire thrusters are used, but now they have small mass change.

There are some other perturbations, such as solar radiation pressure, third body gravitational effects, and unexpected thrusting (outgassing).

Here we examine three special perturbation techniques: Cowell's Method, Encke's Method, Variation of Elements.

Fundamental Spacecraft Dynamics and Control, First Edition. Weiduo Hu.
© 2015 John Wiley & Sons (Asia) Pte Ltd. Published 2015 by John Wiley & Sons (Asia) Pte Ltd.

5.1.2   Cowell's Method

Look at analytic formulation, the two-body perturbed equation:

$$\ddot{\mathbf{r}} + \frac{\mu}{r^3}\mathbf{r} = \mathbf{a}_p. \tag{5.1}$$

Reduce to first order differential equation (DE):

$$\dot{\mathbf{r}} = \mathbf{v} \quad \text{and} \quad \dot{\mathbf{v}} = \mathbf{a}_p - \frac{\mu}{r^3}\mathbf{r} \tag{5.2}$$

Then

$$\dot{x} = v_x, \quad \dot{v}_x = a_{px} - \frac{\mu}{r^3}x,$$

$$\dot{y} = v_y, \quad \dot{v}_y = a_{py} - \frac{\mu}{r^3}y, \tag{5.3}$$

$$\dot{z} = v_z, \quad \dot{v}_z = a_{pz} - \frac{\mu}{r^3}z,$$

$$r = \left(x^2 + y^2 + z^2\right)^{1/2}. \tag{5.4}$$

These equations can be solved by numerical methods, such as RK4 for example; see Section 5.4, which needs explicit representation of perturbations (a_{px}, a_{py}, a_{pz}). This kind of method is relatively slow, needs small step-size and then rounding off error accumulates.

5.1.3   Encke's Method

Cowell's Method integrates the sum of all accelerations; in Encke's Method only the difference in accelerations is integrated. First assume a reference (osculating) orbit whose position is ρ, and a true (perturbed) orbit is position $\mathbf{r}$. See Figure 5.1.

When $|\mathbf{r} - \rho| > Tolerance$, redefining the osculating orbit process is called rectification. Then we can have:

$$\ddot{\mathbf{r}} + \frac{\mu}{r^3}\mathbf{r} = \mathbf{a}_p \quad \text{and} \quad \ddot{\rho} + \frac{\mu}{\rho^3}\rho = 0. \tag{5.5}$$

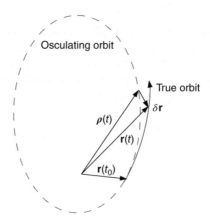

Figure 5.1   Osculating orbit and its true orbit

At time t_0, the osculating and true orbits agree with each other:

$$\mathbf{r}(t_0) = \boldsymbol{\rho}(t_0) \text{ and } \mathbf{v}(t_0) = \dot{\boldsymbol{\rho}}(t_0). \tag{5.6}$$

The deviation from the reference orbit at time t is:

$$\delta\mathbf{r} = \mathbf{r} - \boldsymbol{\rho} \Rightarrow \delta\ddot{\mathbf{r}} = \ddot{\mathbf{r}} - \ddot{\boldsymbol{\rho}} \tag{5.7}$$

$$\Rightarrow \delta\ddot{\mathbf{r}} = \mathbf{a}_p + \frac{\mu}{\rho^3}\left[\left(1 - \frac{\rho^3}{r^3}\right)\mathbf{r} - \delta\mathbf{r}\right]. \tag{5.8}$$

Small perturbations can cause numerical calculation difficulties. Let

$$2q = 1 - \frac{r^2}{\rho^2} \Rightarrow \frac{\rho^3}{r^3} = (1 - 2q)^{-3/2} \tag{5.9}$$

and

$$f = \frac{1}{q}\left[1 - (1 - 2q)^{-3/2}\right] \tag{5.10}$$

then

$$\delta\ddot{\mathbf{r}} = \mathbf{a}_p + \frac{\mu}{\rho^3}\left(fq\mathbf{r} - \delta\mathbf{r}\right). \tag{5.11}$$

For small perturbations:

$$q = -\frac{\rho_x\delta x + \rho_y\delta y + \rho_z\delta z}{\rho^2}. \tag{5.12}$$

f vs. q curves vs. binomial expansion for $1 - (1 - 2q)^{-3/2}$.

Encke Algorithm Summary

1. Given $\mathbf{r}(t_0) = \boldsymbol{\rho}(t_0), \mathbf{v}(t_0) = \dot{\boldsymbol{\rho}}(t_0), \delta\mathbf{r} = 0$.
2. Calculate $\delta\mathbf{r}(t_0 + \Delta t)$.
3. Calculate $\boldsymbol{\rho}(t_0 + \Delta t), q(t_0 + \Delta t), f(t_0 + \Delta t)$.
4. Calculate $\delta\mathbf{r}(t_0 + k\Delta t), k = step\#$.
5. If $\delta r/\rho > Tolerance$, Rectify $\Rightarrow \boldsymbol{\rho} = \mathbf{r}$.
6. $\dot{\boldsymbol{\rho}} = \mathbf{v}$, let $\mathbf{r} = \boldsymbol{\rho} + \delta\mathbf{r}$ and $\mathbf{v} = \dot{\boldsymbol{\rho}} + \delta\dot{\mathbf{r}}$.
7. Loop on time, back to (2).

5.1.4 Variation of Parameters (COEs)

For standard elements: $\mathbf{c} = [a, e, i, \Omega, \omega, M_0, T]$, we want to determine expressions for time derivatives of elements. As for Differential Corrections these are:

$$\mathbf{c} = \mathbf{c}_0 + \frac{\partial\mathbf{c}}{\partial t} \cdot \Delta t. \tag{5.13}$$

In the coordinate system $\hat{R}\hat{S}\hat{W}$ (perifocal rotated with velocity $\dot{\theta}$; See Figure 5.2) the perturbing force:

$$\mathbf{F} = m(F_r\hat{R} + F_s\hat{S} + F_w\hat{W}) \tag{5.14}$$

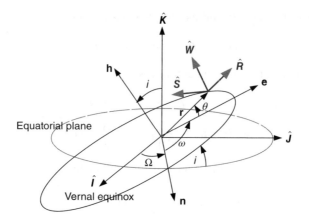

Figure 5.2   Perturbation in satellite coordinate system

then

$$\mathbf{r} = r\hat{R}, \text{ and } \mathbf{v} = \dot{\theta}\left(\frac{dr}{d\theta}\hat{R} + r\hat{S}\right). \tag{5.15}$$

da/dt can be derived by $h = r^2\dot{\theta} = \sqrt{\mu a(1 - e^2)}$:

$$\frac{da}{dt} = \frac{da}{d\mathcal{E}} \cdot \frac{d\mathcal{E}}{dt} = \left(\frac{\mu}{2\mathcal{E}^2}\right)\left[\dot{\theta}\left(\frac{dr}{d\theta}F_r + rF_s\right)\right] \tag{5.16}$$

$$\dot{\theta} = \sqrt{\frac{\mu}{a^3}\frac{a^2}{r^2}}\sqrt{1 - e^2}, \quad \frac{dr}{d\theta} = \frac{re\sin\theta}{1 + e\cos\theta}. \tag{5.17}$$

Then,

$$\frac{da}{dt} = \frac{2e\sin\theta}{n \cdot \sqrt{1 - e^2}}F_r + \frac{2a\sqrt{1 - e^2}}{nr}F_s. \tag{5.18}$$

By similar derivations, variations of other elements can be obtained:

$$\frac{de}{dt} = \frac{\sqrt{1 - e^2}\sin\theta}{na}F_r + \frac{\sqrt{1 - e^2}}{na^2e}\left[\frac{a^2(1 - e^2)}{r} - r\right]F_s \tag{5.19}$$

$$\frac{di}{dt} = \frac{r\cos u}{na^2\sqrt{1 - e^2}}F_w \tag{5.20}$$

$$\frac{d\Omega}{dt} = \frac{r\sin u}{na^2\sqrt{1 - e^2}\sin i}F_w \tag{5.21}$$

$$\frac{d\omega}{dt} = -\frac{\sqrt{1 - e^2}\cos\theta}{nae}F_r + \frac{p}{eh}\left[\sin\theta\left(1 + \frac{1}{1 + e\cos\theta}\right)\right]F_s$$
$$- \frac{r\cot i\sin u}{na^2\sqrt{1 - e^2}}F_w \tag{5.22}$$

$$\frac{dM_0}{dt} = -\frac{1}{na}\left(\frac{2r}{a} - \frac{1-e^2}{e}\cos\theta\right)F_r - \frac{1-e^2}{nae}\left(1 + \frac{r}{a(1-e^2)}\right)\sin\theta F_s$$

$$+ \frac{3\mu t}{2na^4}\cdot\frac{da}{dt}. \tag{5.23}$$

where u = angle from ascending node to satellite position $\mathbf{r}$, that is $u = \omega + \theta$.
These equations are the so-called Gauss form of variation of parameters.

COEs Algorithm:

1. At $t = t_0$ calculate orbital elements.
2. Calculate perturbation force and transform to $\hat{R}\hat{S}\hat{W}$ system.
3. Calculate 6 element derivatives.
4. Numerically integrate each equation.
5. Update orbital elements.
6. Calculate updated position and velocity vectors.
7. Repeat-Loop, back to (1).

5.2   Systematic Method to Derive VOP

Variation of parameter (VOP) equations of motion are a system of first-order differential equations describing the rates of change for the time-varying element [3].

$$\frac{d\mathbf{c}}{dt} = f(\mathbf{c}, t). \tag{5.24}$$

These elements are osculating elements. To derive the osculating element rates, we'll use the following notation:

$$\mathbf{r} = \mathbf{x}(a, e, i, \omega, \Omega, M, t) = \mathbf{x}(c_1, c_2, c_3, c_4, c_5, c_6, t) = \mathbf{x}(\mathbf{c}, t) \tag{5.25}$$

$$\mathbf{v} = \dot{\mathbf{x}}(a, e, i, \omega, \Omega, M, t) = \dot{\mathbf{x}}(c_1, c_2, c_3, c_4, c_5, c_6, t) = \dot{\mathbf{x}}(\mathbf{c}, t). \tag{5.26}$$

The six constant, $\mathbf{c}$, indicate that we can use any set of orbital elements (or more generally, any six independent constants of motion).

5.2.1   Variation of Orbital Elements

The Equation of Motion (EOM) Equation (2.15) for the unperturbed and perturbed systems (respectively) can be expressed as:

$$\ddot{\mathbf{x}}(\mathbf{c}, t) + \frac{\mu\mathbf{x}(\mathbf{c}, t)}{|\mathbf{x}(\mathbf{c}, t)|^3} = 0 \tag{5.27}$$

$$\ddot{\mathbf{x}}(\mathbf{c}, t) + \frac{\mu\mathbf{x}(\mathbf{c}, t)}{|\mathbf{x}(\mathbf{c}, t)|^3} = \mathbf{a}_{pert}, \tag{5.28}$$

where $\mathbf{a}_{pert}$ is the disturbing acceleration.

Since velocity is the position's derivative, we can also express the equations for velocity in these two systems:

$$\dot{\mathbf{x}}(\mathbf{c}, t) = \frac{d\mathbf{x}(\mathbf{c}, t)}{dt} = \frac{\partial \mathbf{x}(\mathbf{c}, t)}{\partial t} \tag{5.29}$$

$$\dot{\mathbf{x}}(\mathbf{c}, t) = \frac{d\mathbf{x}(\mathbf{c}, t)}{dt} = \frac{\partial \mathbf{x}(\mathbf{c}, t)}{\partial t} + \sum_{i=1}^{6} \frac{\partial \mathbf{x}(\mathbf{c}, t)}{\partial c_i} \frac{dc_i}{dt}. \tag{5.30}$$

Differentiate the position vector in the perturbed case Equation (5.28) twice to get the acceleration, using the following constraint for the perturbed system (condition of osculation):

$$\sum_{i=1}^{6} \frac{\partial \mathbf{x}(\mathbf{c}, t)}{\partial c_i} \frac{dc_i}{dt} \equiv 0. \tag{5.31}$$

The second derivative of the position vector gives us:

$$\frac{d^2 \mathbf{x}(\mathbf{c}, t)}{dt^2} = \frac{\partial^2 \mathbf{x}(\mathbf{c}, t)}{\partial t^2} + \sum_{i=1}^{6} \frac{\partial \dot{\mathbf{x}}(\mathbf{c}, t)}{\partial c_i} \frac{dc_i}{dt}. \tag{5.32}$$

We can now find the osculation element rates in terms of the original equations by substituting into Equation (5.27):

$$\frac{\partial^2 \mathbf{x}(\mathbf{c}, t)}{\partial t^2} + \sum_{i=1}^{6} \frac{\partial \dot{\mathbf{x}}(\mathbf{c}, t)}{\partial c_i} \frac{dc_i}{dt} + \frac{\mu \mathbf{x}(\mathbf{c}, t)}{|\mathbf{x}(\mathbf{c}, t)|^3} = \mathbf{a}_{pert}. \tag{5.33}$$

But the unperturbed EOM (Equation (5.27)) holds for any time and can be written:

$$\frac{\partial^2 \mathbf{x}(\mathbf{c}, t)}{\partial t^2} + \frac{\mu \mathbf{x}(\mathbf{c}, t)}{|\mathbf{x}(\mathbf{c}, t)|^3} = 0. \tag{5.34}$$

So, Equation (5.33) simplifies to:

$$\sum_{i=1}^{6} \frac{\partial \dot{\mathbf{x}}(\mathbf{c}, t)}{\partial c_i} \frac{dc_i}{dt} = \mathbf{a}_{pert}. \tag{5.35}$$

The above equation is incomplete because it gives three equation $\mathbf{a}_{pert}$ in the six elements rates dc_i/dt (i.e. 3 equations, 6 unknowns). The remaining equations come from condition of osculation in Equation (5.31).

In matrix notation, the complete system of equation gives us:

$$\begin{bmatrix} \sum_{i=1}^{6} \frac{\partial \mathbf{x}(\mathbf{c}, t)}{\partial c_i} \frac{dc_i}{dt} \\ \sum_{i=1}^{6} \frac{\partial \dot{\mathbf{x}}(\mathbf{c}, t)}{\partial c_i} \frac{dc_i}{dt} \end{bmatrix} = \begin{bmatrix} 0 \\ \mathbf{a}_{pert} \end{bmatrix}. \tag{5.36}$$

The solution to the Lagrange planetary equations of motion begins by taking the dot product of bottom vector (Equation (5.36)) with $\partial \mathbf{x}(\mathbf{c}, t)/\partial c_k$ and the top vector with $-\partial \dot{\mathbf{x}}(\mathbf{c}, t)/\partial c_k$, where $k = 1, 2, \cdots, 6$.

After replacing disturbing acceleration wrt gradient of disturbing potential function ($\mathbf{a}_{pert} \rightarrow \nabla\Phi$), the result of that dot product is:

$$\sum_{i=1}^{6} \frac{dc_i}{dt} \left(\frac{\partial \mathbf{x}(\mathbf{c}, t)}{\partial c_k} \cdot \frac{\partial \dot{\mathbf{x}}(\mathbf{c}, t)}{\partial c_i} - \frac{\partial \mathbf{x}(\mathbf{c}, t)}{\partial c_i} \cdot \frac{\partial \dot{\mathbf{x}}(\mathbf{c}, t)}{\partial c_k} \right) = \frac{\partial \Phi}{\partial c_k}. \tag{5.37}$$

Lagrange brackets are typically used for simplification:

$$[c_i, c_k] = \frac{\partial \mathbf{x}(\mathbf{c}, t)}{\partial c_k} \cdot \frac{\partial \dot{\mathbf{x}}(\mathbf{c}, t)}{\partial c_i} - \frac{\partial \mathbf{x}(\mathbf{c}, t)}{\partial c_i} \cdot \frac{\partial \dot{\mathbf{x}}(\mathbf{c}, t)}{\partial c_k}. \tag{5.38}$$

Using matrix operations and invoking Lagrange bracket notation, we can now rewrite the matrix equation for element rates:

$$\begin{pmatrix} [c_1, c_1] & \cdots & [c_1, c_6] \\ \vdots & & \vdots \\ [c_6, c_1] & \cdots & [c_6, c_6] \end{pmatrix} \left(\frac{d\mathbf{c}}{dt} \right) = \left(\frac{\partial \Phi}{\partial \mathbf{c}} \right). \tag{5.39}$$

Letting $\mathbf{L}$ represent the matrix of Lagrange brackets, the general form of the Lagrange Planetary equation becomes:

$$\dot{\mathbf{c}} = \mathbf{L}^{-1} \frac{\partial \Phi}{\partial \mathbf{c}}, \tag{5.40}$$

which reduces 15 equations using the identities $[c_i, c_i] = 0$, $[c_i, c_k] = -[c_k, c_i]$. Evaluation of the remaining brackets requires the partial derivatives of position and velocity, which need to be expressed as functions of the elements:

$$\frac{\partial \mathbf{x}(\mathbf{c}, t)}{\partial c_i} \quad \text{and} \quad \frac{\partial \dot{\mathbf{x}}(\mathbf{c}, t)}{\partial c_i}, \tag{5.41}$$

where $i = 1, 2, \cdots, 6$.

From Kaula (1966) [2], we obtain explicit dependence of position and velocity on the orbital elements using the transformation from the geocentric $(\hat{I}\hat{J}\hat{K})$ to apsidial $(\hat{P}\hat{Q}\hat{W})$ frame.

$$\mathbf{r}_{PQW} = \begin{bmatrix} r\cos\theta \\ r\sin\theta \\ 0 \end{bmatrix}, \qquad \dot{\mathbf{r}}_{PQW} = \begin{bmatrix} \dfrac{-na\sin\theta}{\sqrt{1-e^2}} \\ \dfrac{na(e+\cos\theta)}{\sqrt{1-e^2}} \\ 0 \end{bmatrix}.$$

Recall that $r = a(1 - e^2)/(1 + e\cos\theta)$ and Equation (2.100).

Letting $\mathbf{T}$ represent the transformation matrix between $\hat{P}\hat{Q}\hat{W}$ and $\hat{I}\hat{J}\hat{K}$, Equation (2.103):

$$\mathbf{x} = \mathbf{T}_{[IJK/PQW]} \mathbf{r}_{PQW}, \qquad \dot{\mathbf{x}} = \mathbf{T}_{[IJK/PQW]} \dot{\mathbf{r}}_{PQW}. \tag{5.42}$$

Previous relations show an important result that $\mathbf{T}$ is a function of only Ω, ω and i, whereas the position and velocity vectors are functions of a, e, θ. Consequently,

$$\frac{\partial(\mathbf{x}, \dot{\mathbf{x}})}{\partial(\Omega, \omega, i)} = \frac{\partial \mathbf{T}_{[IJK/PQW]}}{\partial(\Omega, \omega, i)} (\mathbf{r}_{PQW}, \dot{\mathbf{r}}_{PQW}) \tag{5.43}$$

$$\frac{\partial(\mathbf{x}, \dot{\mathbf{x}})}{\partial(a, e, M)} = \mathbf{T}_{[IJK/PQW]} \frac{\partial(\mathbf{r}_{PQW}, \dot{\mathbf{r}}_{PQW})}{\partial(a, e, M)}. \tag{5.44}$$

At periapsis, the $\hat{P}\hat{Q}\hat{W}$ position and velocity vectors become:

$$\mathbf{r}_{PQW} = \begin{bmatrix} a(1-e) \\ 0 \\ 0 \end{bmatrix}, \qquad \dot{\mathbf{r}}_{PQW} = \begin{bmatrix} 0 \\ na\sqrt{\dfrac{1+e}{1-e}} \\ 0 \end{bmatrix}.$$

Three cases arise in the solution:

Case 1:

$$[c_i, c_k] = na^2 \sqrt{1-e^2} \sum_{j=1}^{3} \left[\left(\frac{\partial T_{j1}}{\partial c_i} \frac{\partial T_{j2}}{\partial c_k} - \frac{\partial T_{j2}}{\partial c_k} \frac{\partial T_{j1}}{\partial c_i} \right) \right]. \tag{5.45}$$

Case 2:

$$[c_i, c_k] = \sum_{j=1}^{3} \left[a(1-e) \frac{\partial T_{j1}}{\partial c_i} \left(T_{j1} \frac{\partial v_P}{\partial c_k} + T_{j2} \frac{\partial v_Q}{\partial c_k} \right) \right.$$

$$\left. - \frac{na\sqrt{1-e^2}}{1-e} \frac{\partial T_{j2}}{\partial c_i} \left(T_{j1} \frac{\partial r_P}{\partial c_k} + T_{j2} \frac{\partial r_Q}{\partial c_k} \right) \right]. \tag{5.46}$$

Case 3:

$$[c_i, c_k] = \sum_{j=1}^{3} \left[T_{j1}^2 \left(\frac{\partial r_P}{\partial c_i} \frac{\partial v_P}{\partial c_k} - \frac{\partial r_P}{\partial c_k} \frac{\partial v_P}{\partial c_i} \right) + T_{j1}T_{j2} \left(\frac{\partial r_P}{\partial c_i} \frac{\partial v_Q}{\partial c_k} - \frac{\partial r_P}{\partial c_k} \frac{\partial v_Q}{\partial c_i} \right) \right.$$

$$\left. + T_{j1}T_{j2} \left(\frac{\partial r_Q}{\partial c_i} \frac{\partial v_P}{\partial c_k} - \frac{\partial r_Q}{\partial c_k} \frac{\partial v_P}{\partial c_i} \right) + T_{j2}^2 \left(\frac{\partial r_Q}{\partial c_i} \frac{\partial v_Q}{\partial c_k} - \frac{\partial r_Q}{\partial c_k} \frac{\partial v_Q}{\partial c_i} \right) \right]. \tag{5.47}$$

Kaula (1966) shows us an example of this. We start with the Lagrange bracket (using Case 1 in the above equation).

The partial derivatives are obtained by differentiating each component of the rotation matrix from $\hat{P}\hat{Q}\hat{W} \rightarrow \hat{I}\hat{J}\hat{K}$, Equation (2.103):

$$[\Omega, i] = \left[(-\sin\Omega\cos\omega - \cos\Omega\cos i\sin\omega)\sin\Omega\sin i\cos\omega \right.$$

$$- (\cos\Omega\cos\omega - \sin\Omega\cos i\sin\omega)\cos\Omega\sin i\cos\omega$$

$$- (\sin\Omega\sin\omega - \cos\Omega\cos i\cos\omega)\sin\Omega\sin i\sin\omega$$

$$\left. - (\cos\Omega\sin\omega + \sin\Omega\cos i\cos\omega)\cos\Omega\sin i\sin\omega \right] na^2\sqrt{1-e^2}$$

$$= \left[(-\sin^2\Omega\sin i\cos^2\omega) - (\cos^2\Omega\sin i\cos^2\omega) \right.$$

$$\left. - (\sin^2\Omega\sin i\sin^2\omega) - (\cos^2\Omega\sin i\sin^2\omega) \right] na^2\sqrt{1-e^2}$$

$$= -\left[\cos^2\omega(\sin^2\Omega + \cos^2\Omega) + \sin^2\omega(\sin^2\Omega + \cos^2\Omega) \right] na^2\sqrt{1-e^2}\sin i$$

$$= -[\cos^2\omega + \sin^2\omega] na^2\sqrt{1-e^2}\sin i$$

$$= -na^2\sqrt{1-e^2}\sin i. \tag{5.48}$$

The above simplification is straight forward.

There are only 12 nonzero results (Kaula, 1966), though 6 values differ only in sign from the relation $[\Omega, i] = -[\Omega, i]$ mentioned earlier. Similarly, we obtain the remaining unique Lagrange brackets in terms of classical elements. The following are the six brackets:

$$[\Omega, i] = -na^2 \sqrt{1 - e^2} \sin i \tag{5.49}$$

$$[\omega, a] = \frac{na\sqrt{1 - e^2}}{2} \tag{5.50}$$

$$[\Omega, a] = \frac{na\sqrt{1 - e^2}}{2} \cos i = [\omega, a] \cos i \tag{5.51}$$

$$[\omega, e] = \frac{-na^2}{1 - e^2} \tag{5.52}$$

$$[\Omega, e] = \frac{-na^2 e}{\sqrt{1 - e^2}} \cos i = [\omega, e] \cos i \tag{5.53}$$

$$[\omega, M_0] = -\frac{na}{2}. \tag{5.54}$$

From Equation (5.40), we find that $\mathbf{L} = [c_i, c_k]$ reduces to:

$$\mathbf{L} = \begin{bmatrix} 0 & 0 & 0 & [a, \Omega] & [a, \omega] & [a, M_0] \\ 0 & 0 & 0 & [e, \Omega] & [e, \omega] & 0 \\ 0 & 0 & 0 & [i, \Omega] & 0 & 0 \\ [\Omega, a] & [\Omega, e] & [\Omega, i] & 0 & 0 & 0 \\ [\omega, a] & [\omega, e] & 0 & 0 & 0 & 0 \\ [M_0, a] & 0 & 0 & 0 & 0 & 0 \end{bmatrix}, \tag{5.55}$$

which resembles the following when inverted:

$$\mathbf{L}^{-1} = \begin{bmatrix} 0 & 0 & 0 & 0 & 0 & \frac{1}{[a, M_0]} \\ 0 & 0 & 0 & 0 & \frac{1}{[\omega, e]} & \frac{[a, \omega]}{[M_0, a][\omega, e]} \\ 0 & 0 & 0 & \frac{1}{[\Omega, i]} & 0 & 0 \\ 0 & 0 & \frac{1}{[i, \Omega]} & 0 & 0 & 0 \\ 0 & \frac{1}{[e, \omega]} & 0 & 0 & 0 & 0 \\ \frac{1}{[M_0, a]} & \frac{[\omega, a]}{[a, M_0][e, \omega]} & \frac{[a, \Omega][e, \omega] + [e, \Omega][a, \omega]}{[i, \Omega][a, M_0][e, \omega]} & 0 & 0 & 0 \end{bmatrix}. \tag{5.56}$$

We can now solve Equation (5.40) to find the equations for the variation of parameters. For example:

$$\dot{\Omega} = \begin{bmatrix} 0 & 0 & \frac{1}{[i,\Omega]} & 0 & 0 & 0 \end{bmatrix} \begin{bmatrix} \partial\Phi/\partial a \\ \partial\Phi/\partial e \\ \partial\Phi/\partial i \\ \partial\Phi/\partial\Omega \\ \partial\Phi/\partial\omega \\ \partial\Phi/\partial M_0 \end{bmatrix}. \tag{5.57}$$

Then

$$\dot{\Omega} = \frac{1}{[i,\Omega]}\frac{\partial\Phi}{\partial i} = \frac{1}{-[\Omega,i]}\frac{\partial\Phi}{\partial i} \tag{5.58}$$

$$= \frac{1}{-na^2\sqrt{1-e^2}\sin i}\frac{\partial\Phi}{\partial i} \tag{5.59}$$

By Lagrangian VOP, Lagrange Planetary Equations can be obtained, which describe effects on orbital elements caused by perturbations, but only for conservative effects.

$$\frac{da}{dt} = \frac{2}{na}\frac{\partial\Phi}{\partial M_0} \tag{5.60}$$

$$\frac{de}{dt} = \frac{1-e^2}{na^2 e}\frac{\partial\Phi}{\partial M_0} - \frac{\sqrt{1-e^2}}{na^2 e}\frac{\partial\Phi}{\partial\omega} \tag{5.61}$$

$$\frac{di}{dt} = \frac{1}{na^2\sqrt{1-e^2}\sin i}\left(\cos i\frac{\partial\Phi}{\partial\omega} - \frac{\partial\Phi}{\partial\Omega}\right) \tag{5.62}$$

$$\frac{d\Omega}{dt} = \frac{1}{na^2\sqrt{1-e^2}\sin i}\frac{\partial\Phi}{\partial i} \tag{5.63}$$

$$\frac{d\omega}{dt} = \frac{\sqrt{1-e^2}}{na^2 e}\frac{\partial\Phi}{\partial e} - \frac{\cot i}{na^2\sqrt{1-e^2}}\frac{\partial\Phi}{\partial i} \tag{5.64}$$

$$\frac{dM_0}{dt} = -\frac{1-e^2}{na^2 e}\frac{\partial\Phi}{\partial e} - \frac{2}{na}\frac{\partial\Phi}{\partial a}. \tag{5.65}$$

Gauss also gives similar equations as a function of the perturbing forces F_R, F_S, F_W which can also handle nonconservative effects. See previous section.

5.2.2 Variation of r and v

In this section parameters to be varied are 6 components of $\mathbf{r}_0$ and $\mathbf{v}_0$. Now we want the derivatives of $\mathbf{r}_0$ and $\mathbf{v}_0$ wrt perturbation forces. From Kepler's Problem of f and g method in Section 2.5.4:

$$\mathbf{r} = f\mathbf{r}_0 + g\mathbf{v}_0 \quad \text{and} \quad \mathbf{v} = \dot{f}\mathbf{r}_0 + \dot{g}\mathbf{v}_0. \tag{5.66}$$

With VOP

$$\mathbf{r}_0 = F\mathbf{r} + G\mathbf{v} \quad \text{and} \quad \mathbf{v}_0 = \dot{F}\mathbf{r} + \dot{G}\mathbf{v}. \tag{5.67}$$

$F, G, \dot{F}, \dot{G}$ are the same as $f, g, \dot{f}, \dot{g}$ except t replaced by $-t$, X replaced by $-X$, and r and r_0 are interchanged. Then,

$$F = 1 - \frac{X^2}{r}C(z), \quad G = -\left[(t - t_0) - \frac{X^3}{\sqrt{\mu}}S(z)\right] \tag{5.68}$$

$$\dot{F} = \frac{\sqrt{\mu}X}{rr_0}[1 - zS(z)], \quad \dot{G} = 1 - \frac{X^2}{r_0}C(z) \tag{5.69}$$

$$z = \frac{X^2}{a} = \alpha X^2, \quad \alpha = \frac{1}{a} = \frac{2}{r} - \frac{v^2}{\mu}. \tag{5.70}$$

We will examine

$$\alpha \mathbf{r}_0 = \alpha F \mathbf{r} + \alpha G \mathbf{v} \tag{5.71}$$

$$\alpha \mathbf{v}_0 = \alpha \dot{F} \mathbf{r} + \alpha \dot{G} \mathbf{v}. \tag{5.72}$$

Assume $\mathbf{r}$ is constant and $\dot{\mathbf{v}}$ results from perturbing force, then

$$\frac{d\alpha}{dt} = -\frac{2}{\mu}\mathbf{v} \cdot \dot{\mathbf{v}} \tag{5.73}$$

$$\frac{d(\alpha \mathbf{r}_0)}{dt} = \frac{d(\alpha F)}{dt}\mathbf{r} + \frac{d(\alpha G)}{dt}\mathbf{v} + \alpha G\dot{\mathbf{v}} \tag{5.74}$$

$$\frac{d(\alpha \mathbf{v}_0)}{dt} = \frac{d(\alpha \dot{F})}{dt}\mathbf{r} + \frac{d(\alpha \dot{G})}{dt}\mathbf{v} + \alpha \dot{G}\dot{\mathbf{v}}. \tag{5.75}$$

then the derivatives in RHS can be obtained:

$$\frac{d(\alpha F)}{dt}, \frac{d(\alpha G)}{dt}, \frac{d(\alpha \dot{F})}{dt}, \frac{d(\alpha \dot{G})}{dt}. \tag{5.76}$$

Detailed discussion can be found in reference [1], Equations (9.4–45).

We must still determine dX/dt and dr_0/dt:

$$\alpha \frac{dr_0}{dt} - \text{Equation (9.4–48)}, \quad \alpha \frac{dX}{dt} - \text{Equation (9.4–49)}.$$

All these equations can be found in reference [1].

5.3   General Perturbations (GA)

Special perturbations: Give an answer to a specific problem with a given set of initial conditions which mainly uses numerical methods.
General perturbations (GP): Give an answer to a general problem, which mainly uses analytical methods.

5.3.1   An Example of GP

Let

$$\dot{x} - \alpha = \frac{v_s}{\left(1 + \mu_s x\right)^2}, \tag{5.77}$$

where α is a constant, μ_s, v_s small numbers. Consider right-hand side (RHS) as a perturbation. When there is no perturbation, $\dot{x} = \alpha \Rightarrow x = \alpha t + c$. Let $c = c(t)$ vary the parameter, then

$$\dot{x} = \alpha + \dot{c} \Rightarrow \dot{c} = v_s / \left[1 + \mu_s \left(\alpha t + c\right)\right]^2, \tag{5.78}$$

which is the perturbation. Expand $v_s / \left[1 + \mu_s \left(\alpha t + c\right)\right]^2$ in its power series. Get

$$\dot{c} = v_s \left[1 - 2\mu_s(\alpha t + c) + 3\mu_s^2(\alpha t + c)^2 + \ldots\right]. \tag{5.79}$$

Its first approximation, $\dot{c} = v_s$, is simple. Next $\dot{c} = v_s - 2\mu_s v_s(\alpha t + c)$, then $\dot{c} + 2\mu_s v_s c = v_s(1 - 2\mu_s \alpha t)$; this is the first order ordinary differential equation. Then we can get its integrating factor:

$$\left[e^{2\mu_s v_s t} c\right]' = v_s(1 - 2\mu \alpha t)e^{2\mu_s v_s t} = v_s e^{2\mu_s v_s t} - 2\mu_s v_s \alpha t e^{2\mu_s v_s t} \tag{5.80}$$

$$e^{2\mu_s v_s t} c = \frac{1}{2\mu_s} e^{2\mu_s v_s t} - \alpha t e^{2\mu_s v_s t} + \frac{\alpha}{2\mu_s v_s} e^{2\mu_s v_s t} \tag{5.81}$$

$$c = \frac{1}{2\mu_s} - \alpha t + \frac{\alpha}{2\mu_s v_s} \Rightarrow x = \frac{\alpha + v_s}{2\mu_s v_s}. \tag{5.82}$$

For orbital analysis, perturbation can be categorized as secular, short period, long period.

$$c = c_0 + \dot{c}_1(t - t_0) + K_1 \cos(2\omega) + K_2 \sin(2\theta + \omega) + K_3 \cos(2\theta). \tag{5.83}$$

In the RHS of the above equation the second term is called secular, the third term is long-periodic, the forth term is mixed-periodic, and the last term is short-periodic. The equation is known as "Possion series." K_i is the sum and product of polynomials in a, e, i.

5.3.2   General Perturbation Techniques

Osculating elements are orbital elements at an instant in time defined by their position and velocity. "Osculate" is Latin for "to kiss." The osculating orbit/ellipse "kisses" the trajectory at the prescribed instant. The osculating orbit is also the path the satellite would follow if the forces were instantaneously removed.

Mean elements are averaged over a period of time, which is useful for determining the satellite's long-term behavior.

Method of perturbations is a class of mathematical techniques for generating analytical solutions that describe the motion of a satellite subject to disturbing forces. It looks at small deviations to the solution for the unperturbed problem. One technique is the "Variation of Parameters," which looks at small variations of the normally constant orbital elements.

Mathematically, method of perturbations are

$$\frac{dc_i}{dt} = \varepsilon_1 f_i + \varepsilon_2 g_i, \quad 0 < \varepsilon_1 \ll 1, \quad 0 < \varepsilon_2 \ll 1 \tag{5.84}$$

$$c = c_0 + \varepsilon_1 \alpha_1(t) + \frac{\varepsilon_1^2}{2!}\alpha_2(t) + \varepsilon_2 \beta_1(t) + \frac{\varepsilon_2^2}{2!}\beta_2(t) + \frac{\varepsilon_1 \varepsilon_2}{2!}\gamma_2(t) + \cdots \tag{5.85}$$

Note that Lagrange planetary equations require partially disturbing potential wrt the orbital elements. It is easy to have $\Phi(L_{gc}, \lambda, r)$, where L_{gc} is geocentric latitude, λ is longitude, r is radius. But $\Phi(a, e, i, \Omega, \omega, M)$ are needed. Kaula (1966) developed an elegant solution to this problem (Equations (5.60)–(5.65)). We won't look at this, but let's look at the problem for just J_2 perturbation, as it is simple and useful.

$$R = -\frac{\mu J_2}{r}\left(\frac{R_\oplus}{r}\right)^2 \frac{3}{2}\left(\sin^2 L_{gc} - \frac{1}{3}\right). \tag{5.86}$$

Let's convert this to orbital elements $\sin L_{gc} = \sin i \sin(\omega + \theta)$, and $\sin^2(\alpha + \beta) = (1 - \cos 2(\alpha + \beta))/2$, thus

$$\Phi = -\frac{3\mu J_2}{2r}\left(\frac{R_\oplus}{r}\right)^2\left(\sin^2 i \sin^2(\omega + \theta) - \frac{1}{3}\right) \tag{5.87}$$

$$= -\frac{3\mu J_2}{2r}\left(\frac{R_\oplus}{r}\right)^2\left(\frac{\sin^2 i}{2} - \frac{\sin^2 i \cos 2(\omega + \theta)}{2} - \frac{1}{3}\right). \tag{5.88}$$

We want to isolate the secular variation, so we ignore long period ω and short period θ terms:

$$\tilde{\Phi} = -\frac{3}{2}n^2 R_\oplus^2 J_2 \left(\frac{a}{r}\right)^3\left(\frac{\sin^2 i}{2} - \frac{1}{3}\right). \tag{5.89}$$

r changes periodically over each revolution. Thus, we need to replace $(a/r)^3$ by its average over one period, since we want only the secular effect. As $dM = (r/a)dE$, $dE = r/(a\sqrt{1-e^2})d\theta$, $r = a(1-e^2)/(1 + e\cos\theta)$

$$\overline{\left(\frac{a}{r}\right)^3} = \frac{1}{2\pi}\int_0^{2\pi}\left(\frac{a}{r}\right)^3 dM = \frac{1}{2\pi}\int_0^{2\pi}\left(\frac{a}{r}\right)^3\frac{r}{a}dE \tag{5.90}$$

$$= \frac{1}{2\pi}\int_0^{2\pi}\left(\frac{a}{r}\right)^3\frac{r^2}{a^2\sqrt{1-e^2}}d\theta \tag{5.91}$$

$$= \frac{1}{2\pi(1-e^2)^{3/2}}\int_0^{2\pi}(1 + e\cos\theta)d\theta \tag{5.92}$$

$$= \frac{1}{(1-e^2)^{3/2}}. \tag{5.93}$$

Thus, replacing $(a/r)^3$ by its average gives:

$$\Phi_{AVG} = -\frac{3}{2}n^2 R_\oplus^2 J_2 \frac{1}{(1-e^2)^{3/2}} \left(\frac{\sin^2 i}{2} - \frac{1}{3} \right). \tag{5.94}$$

Thus

$$\frac{\partial \Phi_{AVG}}{\partial a} = \frac{\Phi_{AVG}}{n^2} \cdot 2n \cdot \frac{dn}{da} = \frac{-3\Phi_{AVG}}{a}. \tag{5.95}$$

Substituting into the Lagrange Planetary Equations gives the secular change of the orbital elements due to J_2, which is given from the Lagrange Planetary Equations due as $(J_2 = 0.00\,108\,262\,693)$

$$\dot{\Omega}_{SEC} = -\frac{3nR_\oplus^2 J_2}{2p^2} \cos i \tag{5.96}$$

$$\dot{\omega}_{SEC} = \frac{3nR_\oplus^2 J_2}{4p^2}(4 - 5\sin^2 i) \tag{5.97}$$

$$\dot{M}_{0SEC} = -\frac{3nR_\oplus^2 J_2 \sqrt{1-e^2}}{4p^2}(3\sin^2 i - 2). \tag{5.98}$$

The physical meaning of oblateness perturbations is

$$\ddot{\mathbf{r}} = -\frac{\mu \mathbf{r}}{r^3} + \Delta \mathbf{f} \tag{5.99}$$

$$\mathbf{r} \times \ddot{\mathbf{r}} = -\mathbf{r} \times \frac{\mu \mathbf{r}}{r^3} + \mathbf{r} \times \Delta \mathbf{f} \tag{5.100}$$

$$\frac{d}{dt}(\mathbf{r} \times \dot{\mathbf{r}}) = \dot{\mathbf{h}} = +\mathbf{r} \times \Delta \mathbf{f}, \tag{5.101}$$

which is a "secularly precessing ellipse." The equatorial bulge introduces a force component toward the equator, causing a regression of the node (for prograde orbits) and a rotation of periapse. Note: for $i < 90°$, $\dot{\Omega} < -0$; $i = 90°$, $\dot{\Omega} = 0$; $i > 90°$, $\dot{\Omega} > 0$. $\dot{\omega} = 0$ for $i = 63.4°$ or $116.6°$; this is called critical inclination for periapse precession. Gravity perturbation also affects geosynchronous orbits due to C_{22} perturbation. Large C_{22} perturbation is discussed in Chapter 6.

Sun synchronous orbits:

$$\dot{\Omega}_{SEC\,desired} = \frac{360°}{365.2\,421\,897} = 0.985\,647\,\text{deg/day}. \tag{5.102}$$

We can adjust a, e, i to accommodate this, for example $h = 800$ km, $e = 0.0$, $i = 98.6°$. Sun synchronous orbits's orbit plane remains at a constant angle $\Omega' s$ with respect to the Earth–Sun line. Its precession about the Earth is equal to the period of Earth's orbit about the Sun. See Figure 5.4.

Other zonal terms can be included also, for example [4]:

$$\dot{\omega}_{SEC} = \frac{3nJ_2R_\oplus^2}{4p^2}\left(4 - 5\sin^2 i\right)$$

$$+ \frac{3nJ_2^2R_\oplus^4}{128p^4}\left(56e^2 + (760 - 36e^2)\sin^2 i - (890 + 45e^2)\sin^4 i\right)$$

$$- \frac{15nJ_4R_\oplus^4}{128p^4}\left(64 - 72e^2 - (248 + 252e^2)\sin^2 i + (196 + 189e^2)\sin^4 i\right)$$

$$+ \frac{105nJ_6R_\oplus^6}{2048p^4}\begin{pmatrix} 256 + 960e^2 + 320e^4 \\ -(2048 + 6880e^2 + 2160e^4)\sin^2 i \\ +(4128 + 13080e^2 + 3960e^4)\sin^4 i \\ -(2376 + 14520e^2 + 2145e^4)\sin^6 i \end{pmatrix}. \tag{5.103}$$

We can also include periodic terms:

$$\Delta\omega_{SP} = \frac{3J_2R_\oplus^2}{2p^2}\begin{pmatrix} (2 - \frac{5}{2}\sin^2 i)(\theta - M + e\sin\theta) \\ +(1 - \frac{3}{2}\sin^2 i)(\frac{1}{e}(1 - \frac{1}{4}e^2)\sin\theta + \frac{1}{2}\sin(2\theta) + \frac{e}{12}\sin(3\theta)) \\ -\frac{1}{e}(\frac{1}{4}\sin^2 i + (\frac{1}{2} - \frac{15}{16}\sin^2 i)e^2)\sin(\theta + 2\omega) \\ +\frac{e}{16}\sin^2 \sin(\theta - 2\omega) - \frac{1}{2}(1 - \frac{5}{2}\sin^2 i)\sin(2\theta + 2\omega) \\ +\frac{1}{e}(\frac{7}{12}\sin^2 i - \frac{1}{6}(1 - \frac{19}{8}\sin^2 i)e^2\sin(3\theta + 3\omega) \\ +\frac{3}{8}\sin^2 \sin(4\theta + 2\omega) + \frac{e}{16}\sin^2 i\sin(5\theta + 2\omega) \end{pmatrix}. \tag{5.104}$$

In the equation elliptical expansion is used, see Appendix A. A similar equation for $\dot{\Omega}_{SEC}$ can also be found in reference [4].

5.3.3 Total Perturbation

The total perturbation value at $t + \Delta t$ includes initial value at t_0, secular perturbation, long period perturbation, short period perturbation (reference [4], Kozai and Brouwer method).
   The total perturbation

$$a = \bar{a} + \Delta a_{LP} + \Delta a_{SP} \tag{5.105}$$

$$e = e_0 + \Delta e_{LP} + \Delta e_{SP} \tag{5.106}$$

$$i = i_0 + \Delta i_{LP} + \Delta i_{SP} \tag{5.107}$$

$$\Omega = \Omega_0 + \dot{\Omega}\Delta t + \Delta\Omega_{LP} + \Delta\Omega_{SP} \tag{5.108}$$

$$\omega = \omega_0 + \dot{\omega}\Delta t + \Delta\omega_{LP} + \Delta\omega_{SP} \tag{5.109}$$

$$M = M_0 + n_0\Delta t + \Delta M_{LP} + \Delta M_{SP} \tag{5.110}$$

$$\bar{a} = a_0 \left(1 - \frac{3J_2R_\oplus^2}{4p^2}\left(2 - 3\sin^2 i\right)\right) \tag{5.111}$$

$$n_0^2 a_0^3 = \mu. \tag{5.112}$$

In Kaula's solutions, potential from a dependence on r, L, λ is first converted to a dependence on $a, e, i, \Omega, \omega, M$ and α_{gst}, then the perturbed orbit is propagated.

The semi-major axis, mean anomaly, and eccentricity may be propagated if $\dot{n}$ and $\ddot{n}$ are known. The mean anomaly is found from a Taylor Series expansion:

$$M = M_0 + n_0\Delta t + \frac{\dot{n}}{2}\Delta t^2 + \frac{\ddot{n}}{6}\Delta t^3. \tag{5.113}$$

$\dot{n}/2$ and $\ddot{n}/6$ are given in the TLEs and RINEX of GPS, see Chapter 7. $n_0 = \sqrt{\mu/a_0^3}$, $p_0 = a_0(1 - e^2)$, $M_0 = E_0 - e\sin E_0$; consider the J_2 perturbation:

$$a = a_0 - \frac{2a_0}{3n_0}\dot{n}\Delta t \tag{5.114}$$

$$e = e_0 - \frac{2(1 - e_0)}{3n_0}\dot{n}\Delta t \tag{5.115}$$

$$i = i_0 \tag{5.116}$$

$$\Omega = \Omega_0 - \frac{3n_0R_\oplus^2 J_2}{2p_0^2}\cos i\Delta t \tag{5.117}$$

$$\omega = \omega_0 + \frac{3n_0R_\oplus^2 J_2}{4p_0^2}(4 - 5\sin^2 i)\Delta t \tag{5.118}$$

$$M = M_0 + n_0\Delta t + \frac{\dot{n}}{2}\Delta t^2 + \frac{\ddot{n}}{6}\Delta t^3. \tag{5.119}$$

5.3.4 Analytical Perturbation Formulation

In general perturbation technique, first determine series expansion of perturbing acceleration, then integrate series expansion analytically term by term. Examine its major perturbations. For non spherical Earth, simplified gravitational potential is μ/r (sphere symmetric). Devise more accurate potential function ϕ. The most often used form is

$$\Phi = \frac{\mu}{r}\left[1 - \sum_{n=2}^{\infty} J_n\left(\frac{R_\oplus}{r}\right)^n P_n(\sin L)\right], \tag{5.120}$$

where J_n is coefficients to be decided (TBD); $R_\oplus$ is equatorial radius of the Earth; P_n is Legendre polynomials; L is geocentric latitude, $\sin L = z/r$.

5.3.4.1 Legendre Polynomials

These are a set of orthogonal polynomials on $[-1,1]$ with scalar product $(x = \sin L)$ given by:

$$\langle g, h \rangle = \int_{-1}^{1} g(x)h(x)dx \tag{5.121}$$

$$P_0(x) = 1, \quad P_1(x) = x, \tag{5.122}$$

$$\int_{-1}^{1} P_0(x)P_1(x)dx = \frac{1}{2}x^2 \Big|_{-1}^{1} = 0, \tag{5.123}$$

$$P_2(x) = x^2 - \frac{1}{3}, \quad P_3(x) = x^3 - \frac{3}{5}x. \tag{5.124}$$

$$\cdots \tag{5.125}$$

Generally normalize, so that $\langle g, g \rangle = 1$. Orthogonal set of the functions,

$$P_0(x) = 1, P_1(x) = x, \tag{5.126}$$

$$P_2(x) = 3/2(x^2 - 1/3), P_3(x) = 1/2(5x^3 - 3x), \tag{5.127}$$

$$P_4(x) = 1/8(35x^4 - 30x^2 + 3), \cdots \tag{5.128}$$

In general,

$$P_{k+1}(x) = \frac{(2k+1)xP_k(x) - kP_{k-1}(x)}{k+1}. \tag{5.129}$$

For **nonspherical perturbation**, normally text gives values $J_2 - J_7$. Note: J_2 dominates. Zonal harmonics are the variations which are dependent on mass distribution symmetrical about the north–south axis. Even numbered variations are symmetrical about the equatorial plane. Odd numbered variations are antisymmetrical. Tesseral depends on longitude and latitude. Sectorial depends on longitude only.

5.3.4.2 Atmospheric Drag

Perturbing acceleration:

$$\ddot{\mathbf{r}} = -\frac{1}{2}C_D \frac{A}{m}\rho v_a \dot{\mathbf{r}}_a, \tag{5.130}$$

where C_D is drag coefficient, A is cross-sectional area, m is vehicle mass, ρ = atmospheric density, v_a = vehicle speed, $\dot{\mathbf{r}}_a$ is inertial velocity $-\boldsymbol{\omega}_\oplus \times \mathbf{r}$

Two other perturbations are often considered. The acceleration due to solar radiation pressure is

$$\mathbf{a}_{SRP} = \frac{p_{SP}C_R A_*}{m_{SAT}} \frac{\mathbf{r}_{*SAT}}{|\mathbf{r}_{*SAT}|}, \tag{5.131}$$

where p_{SR} is solar pressure (mean solar flux)/(speed of light) $= 1353$ (W/m^2)/ c $\cong 4.51 \times 10^{-6}$ Ws/m^3 (N/m^2). C_R = satellite reflectivity (varies between 0 and 2.0), where 0 means body is translucent, no force; 1 means all radiation absorbed (black body), all force transmitted; 2 means all radiation is reflected, twice force is transmitted. A_* is satellite area exposed to the Sun. $\mathbf{r}_{*SAT}$ is a vector from satellite to Sun.

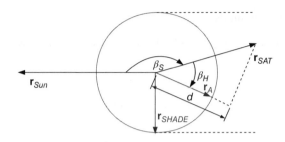

Figure 5.3   Shadow and visibility analysis of a satellite

In shadow, $\mathbf{a}_{SRP} = 0$. Normally, a cylindrical shadow model (Sun infinite distance from Earth) is used. The geometry of the entry and exit points for a satellite depends on the position of the Sun and the satellite. Here we assume that the shadow is a cylindrical cone beginning at a point on the Earth that's perpendicular to the Sun. Using the simple calculation:

$$\mathbf{r}_{SHADE} = 6378 \text{ km} \tag{5.132}$$

$$\cos \beta_S = \frac{\mathbf{r}_{SAT} \cdot \mathbf{r}_{Sun}}{r_{SAT} \, r_{Sun}}, \tag{5.133}$$

if $\beta_S > 90°$, and $r_{SAT} \sin \beta_S < r_{SHADE}$, then the satellite is in shadow. Similarly, if $\mathbf{r}_A$ is the position vector on the Earth and

$$\cos \beta_H = \frac{\mathbf{r}_A \cdot \mathbf{r}_{SAT}}{r_A \, r_{SAT}}, \tag{5.134}$$

if $d = r_{SAT} \cos \beta_H > r_A$, this means the satellite is above horizon. Also $\mathbf{r}_A$ is in darkness to view the satellite. See Figure 5.3. More complicated models take into account umbra/penumbra.

5.3.4.3   Third Body Perturbation

This section discusses orbit accelerations due to a third body.

In three-body geometry, the inertial coordinate system $\hat{I}\hat{J}\hat{Z}$ is used to measure all accelerations. The double subscripts are necessary to completely define the direction of each vector. The satellite is intended to be orbiting the Earth. More discussion is given in reference [4]. $\mathbf{r}_{\oplus SAT} = \mathbf{r}_{SAT} - \mathbf{r}_{\oplus}$, thus $\ddot{\mathbf{r}}_{\oplus SAT} = \ddot{\mathbf{r}}_{SAT} - \ddot{\mathbf{r}}_{\oplus}$ Based on Newton's Second Law and Law of Gravitation:

$$m_{\oplus} \ddot{\mathbf{r}}_{\oplus} = \sum \mathbf{F}_{g\oplus} = \frac{Gm_{\oplus} m_{SAT} \mathbf{r}_{\oplus SAT}}{r_{\oplus SAT}^3} + \frac{Gm_{\oplus} m_{\odot} \mathbf{r}_{\oplus \odot}}{r_{\oplus \odot}^3} \tag{5.135}$$

$$m_{SAT} \ddot{\mathbf{r}}_{SAT} = \sum \mathbf{F}_{gSAT} = -\frac{Gm_{\oplus} m_{SAT} \mathbf{r}_{\oplus SAT}}{r_{\oplus SAT}^3} - \frac{Gm_{\odot} m_{SAT} \mathbf{r}_{\odot SAT}}{r_{\odot SAT}^3}, \tag{5.136}$$

thus

$$\ddot{\mathbf{r}}_{\oplus SAT} = -\frac{Gm_{\oplus} \mathbf{r}_{\oplus SAT}}{r_{\oplus SAT}^3} - \frac{Gm_{\odot} \mathbf{r}_{\odot SAT}}{r_{\odot SAT}^3} - \frac{Gm_{SAT} \mathbf{r}_{\oplus SAT}}{r_{\oplus SAT}^3} - \frac{Gm_{\odot} \mathbf{r}_{\oplus \odot}}{r_{\oplus \odot}^3}. \tag{5.137}$$

And combining terms:

$$\ddot{\mathbf{r}}_{\oplus SAT} = -\frac{G(m_\oplus + m_{SAT})\mathbf{r}_{\oplus SAT}}{r_{\oplus SAT}^3} - Gm_\odot\left(\frac{\mathbf{r}_{\odot SAT}}{r_{\odot SAT}^3} - \frac{\mathbf{r}_{\odot\oplus}}{r_{\odot\oplus}^3}\right), \qquad (5.138)$$

in the right-hand side of the above equation, the first term is a two body term for Earth, the second term is the "direct" effect (acceleration of Sun on satellite), the third term is "indirect" effect (effect of Sun on Earth).

More generally

$$\ddot{\mathbf{r}}_{\odot SAT} = -\frac{G(m_\oplus + m_{SAT})\mathbf{r}_{\oplus SAT}}{r_{\oplus SAT}^3} \quad G\sum_{j=3}^{n} m_j\left(\frac{\mathbf{r}_{jSAT}}{r_{jSAT}^3} - \frac{\mathbf{r}_{j\oplus}}{r_{j\oplus}^3}\right). \qquad (5.139)$$

There are still other perturbations such as thrust etc. when $j = 3, ..., n$ they could be the Moon, the Sun, Jupiter, Saturn etc.

5.3.5 Gravity Potential

Total potential

$$\Phi = \frac{GM_\oplus}{r_2}\left[\sum_{l=0}^{\infty} -J_l\left(\frac{R_\oplus}{r_2}\right)^l P_l\left(\sin\phi_2\right)\right.$$

$$\left. + \sum_{l=2}^{\infty}\sum_{m=1}^{l}\left(\frac{R_\oplus}{r_2}\right)^l P_{lm}\left(\sin\phi_2\right)\left(C_{lm}\cos(m\lambda_2) + S_{lm}\sin(m\lambda_2)\right)\right], \qquad (5.140)$$

where

$$J_l \equiv \left(\frac{1}{M_\oplus R_\oplus^l}\right)\iiint \rho(r_1,\lambda,\phi_1)r_1^{l+2}\cos\phi_1 P_l(\sin\phi_1)dr_1 d\lambda_1 d\phi_1 \qquad (5.141)$$

$$C_{lm} \equiv \left(\frac{2}{M_\oplus R_\oplus^l}\right)\frac{(l-m)!}{(l+m)!}\iiint \rho(r_1,\lambda,\phi_1)r_1^{l+2}\cos(m\lambda_1)\cos\phi_1 P_{lm}(\sin\phi_1)dr_1 d\lambda_1 d\phi_1$$

$$S_{lm} \equiv \left(\frac{2}{M_\oplus R_\oplus^l}\right)\frac{(l-m)!}{(l+m)!}\iiint \rho(r_1,\lambda,\phi_1)r_1^{l+2}\sin(m\lambda_1)\cos\phi_1 P_{lm}(\sin\phi_1)dr_1 d\lambda_1 d\phi_1.$$

$R_\oplus$ is the Earth's radius; r_1 is the distance inside the Earth, where r_2 is the radius outside the Earth. Equation (5.120) is zonal perturbation potential. $P_{lm}(x)$ is the tesseral and sectorial function:

$$P_{lm} = (1-x^2)^{m/2}\frac{d^m}{dx^m}P_l(x),$$

when $x = \sin\phi_2$, P_0, P_1, P_2, P_3, P_4 can be found in Equations (5.126)–(5.128), and

$$P_{11} = \cos\phi_2, \quad P_{21} = 3\cos\phi_2\sin\phi_2, \quad P_{22} = 3\cos^2\phi_2 \qquad (5.142)$$

$$P_{31} = \frac{3}{2}\cos\phi_2(5\sin^2\phi_2 - 1), \quad P_{32} = 15\cos^2\phi_2\sin\phi_2, \qquad (5.143)$$

$$P_{33} = 15 \cos^3 \phi_2, \quad P_{41} = \frac{5}{2} \cos \phi_2 (7 \sin^3 -3 \sin \phi_2), \tag{5.144}$$

$$P_{42} = \frac{15}{2} \cos^2 \phi_2 (7 \sin^2 \phi_2 - 1), \quad P_{43} = 105 \cos^3 \phi_2 \sin \phi_2, \tag{5.145}$$

$$P_{44} = 105 \cos^4 \phi_2, \quad \cdots \tag{5.146}$$

The coefficients for Earth (WGS 84 model) are

$$J_0 = 1; \quad J_1 = 0, \quad J_2 = 1.0826300 \times 10^{-3}, \tag{5.147}$$

$$J_3 = -2.5321531 \times 10^{-6}, \quad J_4 = -1.6109876 \times 10^{-6} \tag{5.148}$$

$$C_{22} = 1.5747419 \times 10^{-6}, \quad S_{22} = -9.0237594 \times 10^{-7} \tag{5.149}$$

$$C_{31} = 2.1946736 \times 10^{-6}, \quad S_{31} = 2.7095717 \times 10^{-7} \tag{5.150}$$

$$C_{32} = 3.0968373 \times 10^{-7}, \quad S_{32} = -2.1212017 \times 10^{-7} \tag{5.151}$$

$$C_{33} = 1.0007897 \times 10^{-6}, \quad S_{33} = 1.9734562 \times 10^{-7} \tag{5.152}$$

$$C_{42} = 7.7809618 \times 10^{-8}, \quad S_{42} = 1.4663946 \times 10^{-7} \tag{5.153}$$

$$C_{44} = -3.9481643 \times 10^{-9}, \quad S_{44} = 6.540039 \times 10^{-9}. \tag{5.154}$$

We see that J_2 is much larger than other coefficients.

5.3.5.1 Real Application Models

For a **real-world application**, there are two steps:

Step 1: Orbital Period classification
If the orbit's period is less than 225 minutes, it is near Earth orbit;
   If period is not less than 225 minutes, it belongs to deep-space orbit. Note that for an orbit with 225 minute period circular orbit, its $r = 12\,250$ km.

Step 2: Choose one of the five math models
1. SGP (Simplified General Perturbations) is for Near-earth, in the gravitational model J_2 is included; drag effect's impact on mean motion is linear in time; long-period periodics with J_2, J_3 and osculating orbit are considered; short-period periodics just include J_2.
2. SGP4 is near-earth, enhanced gravitational model, power density atmospheric model included.
3. SDP4 is for deep-space; extension of SGP4; it models gravity effects of sun and moon and models sectoral and tesseral harmonics (emphasis on 12 and 24 hr period). Increases fidelity of short- and long-periodics.
4. SGP8 is also for near-earth; adds fidelity to gravitational and atmospheric models.
5. SDP8 is for deep-space; adds fidelity to deep-space effects; secular effects of drag and gravity.

5.3.5.2 Two Application Missions

Let's discuss low altitude-perturbative effects with Sun-synchronous orbit, and Moniya frozen orbit, which are designed by perturbation theory.

For the bulged effect of the Earth at the equator, C_{20} is of the order of 10^3 times as large as any of the other gravity coefficients $P_2(x) = 3/2(x^2 - 1/3), x = \sin \delta$ in Equation (5.129). The gravity potential becomes

$$\Phi_2 = \frac{\mu}{r^3} \left[C_{20} \left(1 - \frac{3}{2} \cos^2 \delta \right) \right]. \tag{5.155}$$

For convenience here, $r = r_2$, $C_{20} = -J_2 R^2$ in Equation (5.140).

To formulate an analytical approach to the problem the perturbing potential is averaged to find its long-term secular effects:

$$\bar{\Phi}_2 = \frac{\mu}{2\bar{a}^3(1 - \bar{e}^2)^{\frac{3}{2}}} \left[C_{20} \left(\frac{3}{2} \sin^2 \bar{i} - 1 \right) \right], \tag{5.156}$$

where the over-bars indicate that the orbital elements are now mean orbit elements. By submiting Equation (5.156) into Equations (5.60)–(5.65), the classical averaged equations for satellite motion about the Earth under the perturbation of C_{20} are:

$$\frac{di}{dt} = 0, \tag{5.157}$$

$$\frac{d\Omega}{dt} = \frac{3C_{20}n}{2a^2(1 - e^2)} \cos i, \tag{5.158}$$

$$\frac{d\omega}{dt} = -\frac{3C_{20}n}{2a^2(1 - e^2)^2} \left(2 - \frac{5}{2} \sin^2 i \right). \tag{5.159}$$

Equations (5.157)–(5.159) are the foundation to design a Sun-synchronous orbit and a frozen Molniya orbit, which are actually the same as Equations (5.96)–(5.97) for Ω and ω.

Since the Earth is not spherical, the center of gravity is not the same point of center of mass, $CG \neq CM$. Two principal effects exist. The first is the regression of line-of-nodes.

Still consider the Earth as spherically symmetrical, but now consider the oblateness of the Earth. The bulge at the equator is referred to as the J_2 effect. J_2 is a constant describing the size of the orbital bulge. The force caused by bulge is gravity which is conservative. Mechanical energy is constant, i.e. the orbit's semi-major axis can keep constant. And eccentricity and inclination do not change. The right ascension of the ascending node and argument of the perigee do change.

Nodal regression is the rotation of the orbital plane about the Earth's axis of rotation. The rate is a function of inclination, altitude which can be found by Equation (5.158). See Figure 3.1-5, in reference [1]. Nodes move westward for direct orbits, eastward for retrograde orbits. This is the basic idea to design a Sun-synchronous orbit using Equation (5.158).

Similarly the second effect is major axis rotation. For an orbit, its inclination is less then 63.4° or inclination is greater than 116.6°; its major axis rotates in the direction of satellite motion. For an orbit whose inclination is greater than 63.4° is less than 116.6°; its major axis rotates in the direction opposite that of the satellite. Its rate is a function of inclination, altitude which can be obtained by Equation (5.159) directly. See Figure 3.1-6, in reference [1] for

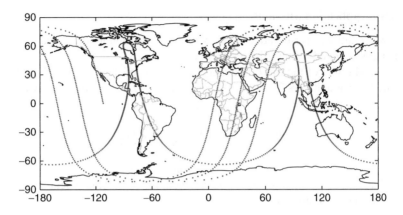

Figure 5.4   Examples of a Sun-synchronous orbit and a frozen Molniya orbit

more detail. And this is the reason that the periapsis is frozen in inertial space for a Molniya orbit. See Equation (5.159).

Molniya orbits have 12 hr periods; their inclination is 63.4°; Tundra orbits have the same inclination but their period is 24 hrs. Figure 5.4 shows the ground track of a Sun-synchronous orbit and a frozen Molniya orbit.

5.4   Numerical Methods

For differential equations of position and velocity, the Rungo-Kutta method can be used to solve them, such as the well-known RK4 expansion.

$$\dot{x} = f_1(\mathbf{r}, \mathbf{v}), \quad \dot{v}_x = g_1(\mathbf{r}, \mathbf{v}, \mathbf{a})$$

$$\dot{y} = f_2(\mathbf{r}, \mathbf{v}), \quad \dot{v}_y = g_2(\mathbf{r}, \mathbf{v}, \mathbf{a})$$

$$\dot{z} = f_3(\mathbf{r}, \mathbf{v}), \quad \dot{v}_z = g_3(\mathbf{r}, \mathbf{v}, \mathbf{a})$$

$$x(n+1) = x(n) + \frac{1}{6}\left(k_{11} + 2k_{12} + 2k_{13} + k_{14}\right)$$

$$v_x(n+1) = v_x(n) + \frac{1}{6}\left(l_{11} + 2l_{12} + 2l_{13} + l_{14}\right)$$

$$y(n+1) = y(n) + \frac{1}{6}\left(k_{21} + 2k_{22} + 2k_{23} + k_{24}\right)$$

$$v_y(n+1) = v_y(n) + \frac{1}{6}\left(l_{21} + 2l_{22} + 2l_{23} + l_{24}\right)$$

$$z(n+1) = z(n) + \frac{1}{6}\left(k_{31} + 2k_{32} + 2k_{33} + k_{34}\right)$$

$$v_z(n+1) = v_z(n) + \frac{1}{6}\left(l_{31} + 2l_{32} + 2l_{33} + l_{34}\right). \tag{5.160}$$

Start with known $\mathbf{r}_0$ and $\mathbf{v}_0$; let h be step-size,

$$k_{11} = h \cdot f_1(\mathbf{r}_0, \mathbf{v}_0),$$

$$l_{11} = h \cdot g_1(\mathbf{r}_0, \mathbf{v}_0, \mathbf{a})$$

$$k_{12} = h \cdot f_1(\mathbf{r}_0 + \frac{1}{2}k_{11}, \mathbf{v}_0 + \frac{1}{2}l_{11}),$$

$$l_{12} = h \cdot g_1(\mathbf{r}_0 + \frac{1}{2}k_{11}, \mathbf{v}_0 + \frac{1}{2}l_{11}, \mathbf{a})$$

$$k_{13} = \cdots$$

$$l_{13} = \cdots$$

$$\cdots \tag{5.161}$$

then $t = t_0 + h$, update x, y, z and $\dot{x}, \dot{y}, \dot{z}$; we have new $\mathbf{r}$ and $\mathbf{v}$.

There are some other numerical methods. **Adams-Moulton** is a multi-step method. It use both Predictor and Corrector.

1. Predict a solution value:

$$y(n+1)^{(0)} = y(n) + \frac{h}{24}\left[55f(n) - 59f(n-1) + 37f(n-2) - 9f(n-3)\right]. \tag{5.162}$$

2. Correct the solution value:

$$y(n+1) = y(n) + \frac{h}{24}\left[9f(n+1) + 19f(n) - 5f(n-1) + f(n-2)\right]. \tag{5.163}$$

5.5 Summary and Keywords

This chapter describes orbital perturbation by different methods, such as Cowell's method, Encke's method, variation of parameters, special and general perturbation techniques, gravity potential, including derivations and different application results.

Keywords: special perturbation, general perturbation, Cowell's method, Encle's method, variation of parameters, Lagrange brackets, secular perturbation, long period perturbation, short period perturbation, Legendre polynomials, gravity potential, atmospheric drag, third body perturbation, Sun synchronous orbit, frozen orbit, real application models.

Problems

5.1 An interplanetary spaceship consisting of science payload and a large solar sail has mass m. The sail is kept normal to the Sun and produces an acceleration from radiation pressure of:

$$\mathbf{a}_1 = \frac{2SA}{cm}\frac{\mathbf{r}}{r^3}.$$

where S is the solar constant, A is the sail area, c is the speed of light and r is the radius from the Sun. In this sketch, $\mathbf{a}_g$ is the gravitational acceleration between the Sun and the spacecraft.

Defining $\mu = GM_{\text{Sun}}$, what types of conic sections are possible when:

(a) $\mu > 2SA/c\,m$,
(b) $\mu = 2SA/c\,m$,
(c) $\mu < 2SA/c\,m$ (for 5 points extra credit),
(d) If in any of these cases the vehicle escapes, determine an analytic relation for escape velocity that is a function of the above parameters and the radius and speed at which the sail is unfurled.

5.2 Recall the equation for the acceleration of body 2 relative to body 1 for the N-body problem derived in class (Equation (5.139)). Here 1 means $\oplus$, 2 means SAT. Assuming constant mass and the absence of non gravitational forces, determine if the two-body problem is a good approximation when:

(a) m_j is very small, for all $j \geq 3$,
(b) r_{j2} and r_{j1} are very large, for all $j \geq 3$,
(c) $\mathbf{r}_{j2} \cong \mathbf{r}_{j1}$ and not large,
(d) $m_3 \cong m_1$.

5.3 The equation of motion for a satellite subjected to aerodynamic drag can be expressed as

$$\ddot{\mathbf{r}} = -\frac{\mu}{r^3}\mathbf{r} - qC_D A\frac{\dot{\mathbf{r}}}{\dot{r}},$$

where q is a function of $\mathbf{r}$ and $\dot{\mathbf{r}}$, and C_D and A are constants. Is specific angular momentum $(\mathbf{r} \times \mathbf{v})$ a constant? Justify your answer mathematically.

5.4 A spacecraft is initially in a circular orbit about the Earth. It experiences a constant, small drag deceleration along the orbital path, $f_s = -D$, while $f_r = f_w = 0$.

(a) Write the three equations which describe the perturbed motion of this spacecraft with respect to its Keplerian path (motion without drag).
(b) Solve these differential equations so that the x, y, and z components are expressed as a function of time in the ECI frame.
(c) Sketch or plot the in-plane portion of this motion.
(d) What about if the spacecraft is at the Earth–Moon L_4 point?

5.5 What is the inclination of a circular orbit with period of 100 minutes whose ground track moves eastward at 3 deg/day?

5.6 Consider three circular sun-synchronous orbits with radii $= 7000$ km, 8778 km and $10\,000$ km, respectively

(a) What are the inclinations of each of these three orbits?
(b) Assuming that each orbit has a 6 am ascending node, are these orbits ever eclipsed by the Earth's shadow?
(c) Determine the semi-major axis of a sun-synchronous orbit with $e = 0.3$ and an inertially fixed apsidal rotation.

5.7 One inclination for stationary perigee is 116.6°. If $r_p = 6700$ km, select values of a and e consistent with a sun synchronous orbit. Assume J_2 oblateness only.

5.8 Come up with a guidance law for a throttleable low thrust engine to change the eccentricity as quickly and efficiently as possible.

References

[1] Bate, R., Mueller, D., and White, J. (1971) *Fundamentals of Astrodynamics*. New York: Dover Publication, Inc.

[2] Kaula, W.M. (1966) *Theory of Satellite Geodesy*. Boston: Blaisdell.

[3] Battin, R.H. and Vaughan, R.J. (1987) *An Introduction to the Mathematics and Methods of Astrodynamics*. New York: AIAA.

[4] Vallado, D. (1997) *Fundamentals of Astrodynamics and Applications*. El Segundo CA: Microcosm Press.

6

Orbital Motion Around Asteroids

A relatively new problem in orbital mechanics is spacecraft (S/C) motion around asteroids. Asteroids are the minor planets that mainly circulate around the Sun between Mars and Jupiter. Their size varies from a few hundred kilometers to a few centimeters in diameter. In addition, comets travel more widely throughout the solar system, some of which pass near the Sun and the Earth. Space missions to send a spacecraft to orbit about an asteroid or comet have been carried out and are currently being prepared or proposed. Orbits around an asteroid or a comet are considerably different from the Keplerian orbits in the two-body problem. This is for two reasons: the irregular shape of an asteroid, and the rotation of the asteroid. These two factors are coupled and can cause the spacecraft orbit to experience large energy and angular momentum changes within short time periods [2], occurrences that are not seen for spacecraft motion around planetary bodies. Due to these effects the study of spacecraft motion about asteroidal bodies is new and challenging. Recent surveys of this problem can be found in references [25, 26].

Some data of a few well-known asteroids are show in Table 6.1. Note that Chinese CE-2 passed by Asteroid Toutatis at the end of 2012. See Figure 6.1.

6.1 Introduction

This chapter analyzes the characters of spacecraft orbital motion around asteroids. Three different methods such as averaging, elliptic expansion, and Poincaré map are used to study this problem for different aspects of motion including secular motion, resonance, and periodic orbits. The key to this problem is the stability of the orbit. Some of our conclusions obtained are summarized in the following sections.

Before we discuss the orbital motion around asteroids, let us first introduce a typical asteroid in the solar system. Its shape is simulated in Figure 6.2.

Castalia is a Near Earth Object (NEO), which has the potential to collide with Earth. Its nearest distance to Earth was 0.0251 AU once before. In the heliocentric frame, its $e_\odot = 0.483$, $a_\odot = 1.063$, $i_\odot = 8.9°$. Its rotation period is 4.07 hr. Its volume is 0.671 km^3, which corresponds to a mean radius 0.543 km. For describing positions about Castalia we use a body system whose origin is at the model's centroid and whose axes (x,y,z) correspond to the principal

Fundamental Spacecraft Dynamics and Control, First Edition. Weiduo Hu.
© 2015 John Wiley & Sons (Asia) Pte Ltd. Published 2015 by John Wiley & Sons (Asia) Pte Ltd.

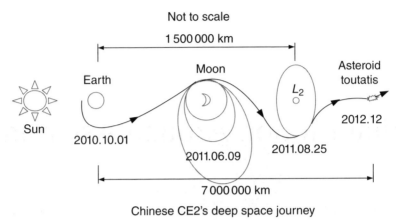

Figure 6.1   CE2 mission adapted from CAST

Table 6.1   Some data of a few well-known asteroids [3]

Asteroid Period name	Diameter (km)	Mass 10^{15}kg	Rotation T hrs	$a_\odot$ AU	$e_\odot$	$i_\odot$ deg	Period $T_\odot$ yrs
1 Ceres	960×932	870 000	9.075	2.767	0.0789	10.58	4.60
2 Pallas	$570 \times 525 \times 482$	318 000	7.811	2.774	0.2299	34.84	4.61
3 Juno	240	20 000	7.210	2.669	0.2579	12.97	4.36
4 Vesta	530	300 000	5.342	2.362	0.0895	7.14	3.63
243 Ida	58×23	100	4.633	2.861	0.0451	1.14	4.84
Mathilde	$66 \times 48 \times 46$	103.3	417.7	2.646	0.2660	6.71	4.31
433 Eros	$33 \times 13 \times 13$	6.69	5.270	1.458	0.2229	10.83	1.76
Gaspra	$19 \times 12 \times 11$	10	7.042	2.209	0.1738	4.10	3.29
Icarus	1.4	0.001	2.273	1.078	0.8269	22.86	1.12
Castalia	1.8×0.8	0.0005	4.07	1.063	0.4831	8.89	1.10
Toutatis	$4.6 \times 2.4 \times 1.9$	0.05	130.	2.512	0.6339	0.47	3.98

axes of smallest, intermediate, and largest moment of inertia, respectively. The model's rotation pole lies along the z axis and its shape fits into the bounding box centered at its center of mass. $-0.762 \leq x \leq 0.851$, $-0.462 \leq y \leq 0.519$, $-0.444 < z < 0.382$, its density is approximate $1.7 \leq \rho \leq 2.5$ g/cm^3. The total mass $M = 1.4091 \times 10^{12}$ kg. Its gravity coefficients of a harmonic expansion are shown in Table 6.2. These data of the asteroid Castalia are from reference [9]. We can see that C_{20} and C_{22} are much larger than others coefficients.

6.2   Problem Formulation

For the general two-body problem, if we assume the spacecraft is a point mass, its position vector in inertial coordinates satisfies the equation of motion,

$$\ddot{\mathbf{r}}_i = \mathbf{F}, \qquad (6.1)$$

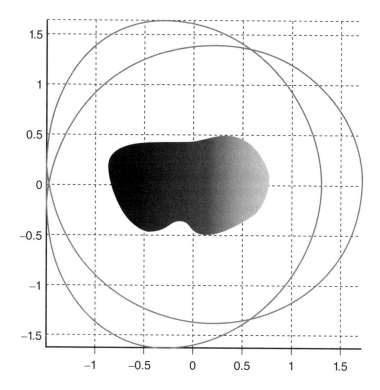

Figure 6.2 The intricate scale simulated shape of asteroid Castalia with a $\omega_T:n = 5:3$ periodic orbit [29]

Table 6.2 The gravity coefficients for Castalia [9]

Order	Degree	C coefficient	S coefficient
0	0	1.0	–
1	0	0.0	–
1	1	0.0	0.0
2	0	−0.110 298	–
2	1	0.0	0.0
2	2	0.156 733	0.0
3	0	−0.015 112	–
3	1	−0.037 935	0.001 211
3	2	0.006 325	0.000 616
3	3	0.020 568	−0.013 715
4	0	0.036 630	–
4	1	0.002 706	0.000 407
4	2	−0.051 363	0.003 949
4	3	0.006 140	−0.001 747
4	4	0.050 334	−0.006 839

where $\mathbf{r}_i$, $\mathbf{F}$ or denoted as $\vec{r}_i$, $\vec{F}$ are the position and acceleration vectors. The ideal case occurs when there is no perturbation, $\mathbf{F} = -\mu\mathbf{r}_i/r_i^3$, meaning that the attracting body is a point mass (or a sphere with uniform mass distribution), with its principle moments of inertia equal, $I_{xx} = I_{yy} = I_{zz}$. Then the spacecraft orbit is a classical Keplerian orbit and its semi-major axis a, eccentricity e, inclination i, longitude of ascending node Ω, argument of periapsis ω are all constant. These elements are well-defined in many text books.

When the attracting body is an oblate body rotating about its axis of symmetry (such as a planet), the spacecraft orbit is perturbed from its ideal Keplerian orbits, but its acceleration $\mathbf{F}$ has no relation to the main body's rotation. Since the shape of an oblate body is symmetric about its rotation axis, the relation between their moments of inertia is $I_{xx} = I_{yy} < I_{zz}$. Spacecraft motion around an oblate body has been widely studied since the 1960s [4, 5]. This is the well-known J_2 (i.e. C_{20} only) perturbation problem. Normally, for spacecraft motion around an oblate body, the orbit's semi-major axis a, eccentricity e, and inclination i are constant on average, but its longitude of ascending node Ω and argument of periapsis ω have secular variations. Some of these results are used to design satellite orbits for specific applications, such as Sun-synchronous orbits.

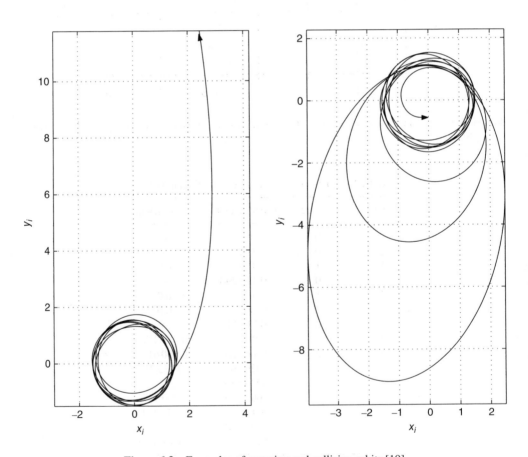

Figure 6.3 Examples of escaping and collision orbits [19]

Table 6.3 The summary of application and background

Central body-shape	Sphere	Oblate	Irregular
Moments of inertia	$I_{xx} = I_{yy} = I_{zz}$	$I_{xx} = I_{yy} < I_{zz}$	$I_{xx} < I_{yy} < I_{zz}$
Gravity coefficients	$C_{20} = C_{22} = 0$	$C_{20} < 0, C_{22} = 0$	$C_{20} < 0, C_{22} > 0$
Relation to ω_T	No	No	Yes
Motion constants	a, e, i, Ω, ω	$\bar{a}, \bar{e}, \bar{i}$	No

For spacecraft motion around an asteroid whose moments of inertia are, in general, different from each other, $I_{xx} < I_{yy} < I_{zz}$, the acceleration **F** has a close relationship with the asteroid's rotation, making the study of motion around an asteroid completely different from the case of motion around an oblate body. Not only will Ω and ω change, but i, and even a and e, may also change drastically over time. For some cases these changes are rapid, making the spacecraft motion unpredictable to some degree; this unpredictability is an instance of chaotic motion, which exists in many dynamic systems [6, 7]. Here we define "chaotic" motion as sensitive dependence on initial conditions. Thus, the problem of spacecraft motion around an asteroid is not only a practical engineering problem, but also a theoretical problem in dynamical systems as well.

Table 6.3 gives a summary of the discussion in this section.

6.3 Motion Equations

There are some approaches that can be used to model the acceleration **F**. One is to model the asteroid as an ellipsoid [8], and calculate its gravity potential and derivatives whose first order partial derivatives are the acceleration **F**. The second method represents the gravity force potential using a polyhedron model [10]. The third method is to use the spherical harmonics expansion of the gravity potential. Spacecraft motions for a number of specific asteroids are calculated and analyzed. In these studies, the polyhedron and spherical harmonics expansion gravity models are used.

In this chapter we focus on the general problem of spacecraft motion around asteroids by modeling the asteroid perturbation potential with the two most significant gravity coefficients, C_{20} and C_{22}. The model is simple, but can describe the main character of spacecraft motion about an asteroid. One of these characteristics is that the spacecraft motion is also affected by the central body's rotation. Since our gravity field is rotating with the asteroid, some aspects of our problem have similarities to the three-body problem, which is discussed comprehensively in book [11] and has been studied for the last one hundred years.

For definiteness we will specify the frame so that $C_{20} \leq 0$ and $C_{22} \geq 0$, implying that the principal moments of inertia are ordered as $I_{xx} \leq I_{yy} \leq I_{zz}$ with axes x, y and z. The relations between the gravity coefficients and the principal moments of inertia of the body (normalized by the body mass) are [12, 13]:

$$C_{20} = -\frac{1}{2} \left(2I_{zz} - I_{xx} - I_{yy} \right), \tag{6.2}$$

$$C_{22} = \frac{1}{4} \left(I_{yy} - I_{xx} \right). \tag{6.3}$$

Let $\mathbf{r} = [x, y, z]^T$, $\boldsymbol{\omega} = [0, 0, \omega_T]^T$ both expressed in body-fixed coordinate. The scalar form of the above equation is:

$$\ddot{x} - 2\omega_T \dot{y} = \omega_T^2 x - \frac{\mu x}{r^3} + \frac{\partial U}{\partial x}, \tag{6.4}$$

$$\ddot{y} + 2\omega_T \dot{x} = \omega_T^2 y - \frac{\mu y}{r^3} + \frac{\partial U}{\partial y}, \tag{6.5}$$

$$\ddot{z} = -\frac{\mu z}{r^3} + \frac{\partial U}{\partial z}, \tag{6.6}$$

where U is perturbation potential. These equations are defined in the body-fixed coordinate frame whose rotation rate is a constant ω_T, and thus includes the effect of coriolis and centripetal accelerations. Here we assume that the asteroid is in uniform rotation about its maximum moment of inertia (the z-axis) with a rotation rate ω_T and a corresponding rotation period $T = 2\pi/\omega_T$. More complete discussions of these equations as applied to motions about an asteroid have been reported in reference [8].

6.4 Progress and Some Conclusions

We use three different methods such as averaging, elliptic expansion, Poincaré map to study this problem for different aspects of motion including secular motion, resonance and periodic orbits. The key to this problem is the stability of the orbit. Some of our conclusions obtained are summarized here. We see that by using these three different methods, averaging, resonance analysis and periodic orbit computation, we can reveal different aspects of the nature of spacecraft motion in a uniformly rotating, second degree and order gravity field. We also find that they are closely related, and that some results in our secular and resonant analyzes can be used to discuss the existence and stability of periodic orbits.

6.4.1 Secular Motion

We use averaging methods to remove the rapidly changing variables and keep the slowly changing variables that describe an orbit's secular motion. We study the averaged motion for three cases: when the central body does not rotate, rotates slowly, and rotates rapidly. For the slow rotation cases, we derive the averaged Lagrange planetary equations for this system and use them, in conjunction with the Jacobi integral, to give a complete description of orbital motion. We show that, under the averaging assumptions, that is, with the size-shape stability condition, the problem is completely integrable and can be reduced to quadratures. For the case of no rotation these quadratures can be expressed in terms of elliptic functions and integrals. For the rapid rotation case, motion can be reduced to the well-known oblate body perturbation problem. One application of the secular motion analysis is to discuss the condition of periodic orbits, for example the 1:2 periodic family discussed in reference [15]. Two papers [17, 16] have been published from this part.

One interesting result is the similarity of angular momentum vector to free rigid body motion.

The averaged dynamics of the angular momentum are:

$$\frac{dh_x}{dt} = [B(1 - \sigma)h_z + \omega_T]h_y, \tag{6.7}$$

$$\frac{dh_y}{dt} = -(Bh_z + \omega_T)h_x, \tag{6.8}$$

$$\frac{dh_z}{dt} = B\sigma h_x h_y, \tag{6.9}$$

where B is value and σ are value determined by the orbit and the mass properties of the central body. When $\omega_T = 0$, Equations (6.7)–(6.9) are completely the same with the free rigid body motion Equations (9.181)–(9.183) in structure.

There are six equilibria which correspond to the two unstable frozen orbits, the two stable frozen orbits, the direct orbit and the retrograde orbit. See Figure 6.4.

It can be seen that there are some similarities between the stability conditions of torque free rigid body rotational motion and our problem of a particle moving around a rigid body, especially when $\omega_T = 0$. For rigid body motion, rotation is stable when the rigid body rotates about its largest or smallest moment of inertia. For our problem, the particle's motion is stable when it circles the axis of the largest or smallest moment of inertia of the central body. There are fundamental reasons for this similarity, as the two kinds of motions are both determined by the mass distribution of the central body which is described by the moments of inertia. This

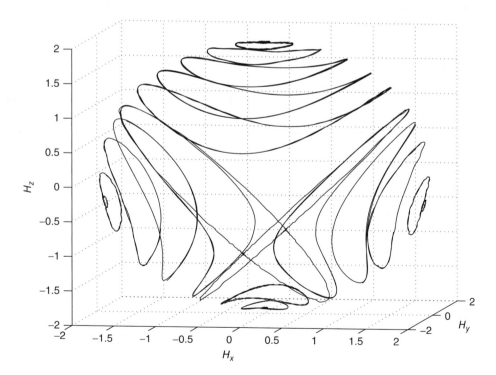

Figure 6.4 Numerical verification of secular motion by angular momentum calculations in 3-D space for a range of different initial conditions [16]

similarity can provide insight into stable and unstable orbits in the second degree and order gravity field.

We see the averaged Equations (6.7–6.9) completely describe the orbital angular momentum motion in Figure 6.4. Particularly interesting are the "separatrices" between the libration and circulation cases – heteroclinic connections between the unstable frozen orbits for the mean elements in our averaged motion analysis. The orbit normal can cross these lines without violating any fundamental uniqueness properties of the full solution. Thus, it is possible for orbital motion in the real system to jump between circulation and libration, potentially defining a region of chaotic motion in terms of the orientation of the orbital plane.

6.4.2   Resonance

We analyze the short-period motion which is closely related to resonance. The study of resonance can answer some of the size–shape stability problems which are proposed in references [15, 19]. Resonance in our system exists because of the longitudinal dependent tesseral C_{22}, and due to commensurability between orbital mean motion n and the asteroid rotation ω_T. Specifically, we use elliptic expansions to derive the Fourier expansions of semi-major axis $a(t)$ and eccentricity $e(t)$ for nominal orbits in our system, that is the frequencies, amplitudes and phase angles of the Fourier expansion of $a(t), e(t)$ are estimated. With the Fourier expansions, we can explain the dynamics of the semi-major axis for near circular orbits. Additionally, by averaging along near resonant orbits, we can explain why $a(t)$ and $e(t)$ sometimes experience large changes over short time spans. Together with the resonant integral analysis [2], we can estimate a stability index for our size-shape stability analysis, proposed in reference [15]. Additionally, the Fourier expansion of $a(t)$ and $e(t)$ can be used to analyze the periodic orbits in reference [15], for example the 1:2 periodic family. Some numerical analyses and verifications are also given. In Figure 6.5 the orbits are usually size–shape stable in the lower-right region of the plot, corresponding to small gravity coefficients, large orbit radii, and slow rotation; in the upper-left region the orbits are usually size–shape unstable, corresponding to large gravity coefficients, small orbit radii, and fast rotation. We note that there are some unstable orbits in the stable orbit region and that there are also some isolated regions of stable orbits in the unstable region.

We note that at resonances between the mean motion and the asteroid rotation, which occur at $(a_0/r_s)^{3/2} = 1:2; 1:1; 3:2; 2:1; 5:2$ etc., the stability boundary is near a local minimum. This implies that at these resonances stable orbits only exist when the parameter $(I_{zz} - I_{xx})/r_s^2$ is small and the perturbation is weak.

6.4.3   Periodic Orbits

Periodic orbits around asteroids are also discussed. First, we analyze the four equilibrium points in the rotating 2nd degree and order gravity field. Two of them are always unstable; the other two can be stable or unstable, their stability determined by the central body rotation rate ω_T, the two gravity coefficients, C_{20} and C_{22}, and the asteroid's mass.

If we define the $r_s = (\mu/\omega_T^2)^{1/3}$ as synchronous orbit, then the stability condition [19] is

$$r_s^2 + C_{20} - 162C_{22} > 0. \tag{6.10}$$

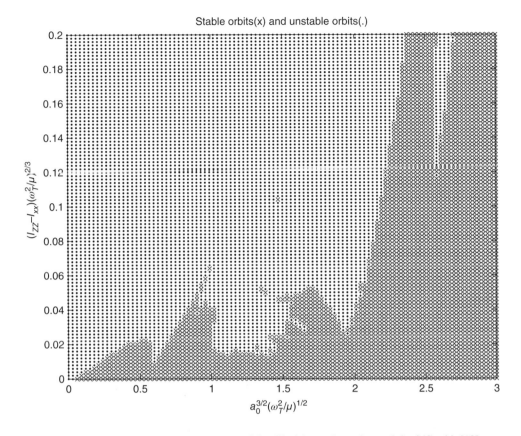

Figure 6.5   The stability regions vs. asteroid's ellipticity and rotation and the S/C orbit [19]

So Equation 6.10 can be used as a criterion to determine the stability of the synchronous orbit. We see that when the central body rotates fast enough, or C_{22} is large enough, the above equation is not satisfied. For the Earth, with C_{22} small and r_s large, the equilibrium $(0, y_{eq}, 0)$ is stable, as is well known [14]. And also we can find y_{eq} [19]

$$|y_{eq}| \approx r_s + \frac{-C_{20}/2 - 3C_{22}}{r_s} < r_s. \tag{6.11}$$

Next we compute families of periodic orbits. In our searching procedure, we remove the two unity eigenvalues from the state transition matrix (STM) to find a robust, nonsingular Poincaré map to solve for the periodic orbits. The algorithm converges well, especially for stable periodic orbits. As we give the initial variations a Jacobi constraint, the procedure automatically searches for a periodic orbit at a fixed energy level. By calculating the STM for the periodic orbit, we find the characteristic multipliers which are the indices of stability of the periodic orbit. These are used to test the stability of the periodic orbits. Then, by using a relatively automatic searching procedure for the periodic orbits, we find five basic families of periodic orbits and discuss their existence and stability at different central body rotation rates. We call these five basic families the near direct, far direct, 1:1 resonant, 1:2 resonant and retrograde periodic

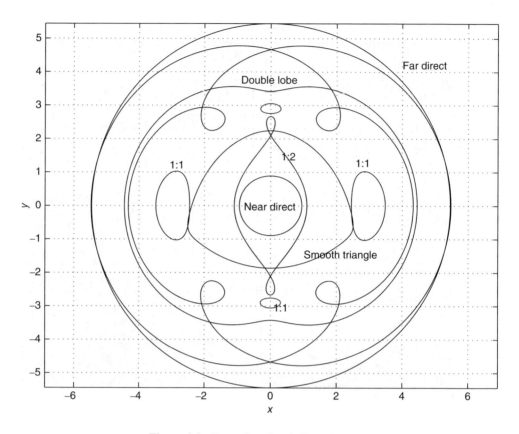

Figure 6.6   Examples of periodic orbits [18]

orbits. There are also other "special" periodic orbits that lie between these basic periodic orbit families. See Figure 6.6. A paper [18] has been published from this discussion.

6.5 Conclusions

This chapter studies spacecraft orbital motion about a rotating 2nd degree and order gravity field, with its main application to orbital motion about an asteroid. These dynamical systems are nonintegrable in general. Our goal is to understand how the dynamics of this system change as a function of the central body's rotation rate and mass distribution. To carry out this analysis we use three different approaches: averaging, resonance analysis, and periodic orbit computation. By using these analyses, we can better understand the whole character of a spacecraft's motion around an asteroid, which is much different from the character of spacecraft motion around the Earth, or other planets in the solar system.

6.6 Summary and Keywords

This chapter introduces the relatively new problem, i.e. spacecraft orbital motion around an asteroid, including its characters, methods, and conclusions obtained mainly by the author.

Keywords: asteroid, orbital motion, secular motion, resonant orbit, periodic orbit, chaotic motion, Poincaré map, equilibrium points, orbital stability, averaging method.

Acknowledgments

The work described here was partially supported by the TMOD Technology Program with a grant from the Jet Propulsion Laboratory, California Institute of Technology which is under contract with the NASA.

Problems

6.1 Why is spacecraft orbital motion around an asteroid different from its motion around the Earth?

6.2 How many equilibrium points in the rotating asteroid frame about spacecraft orbital motion in it? Are they stable or unstable?

References

[1] Hu, W., Scheeres, D J., *et al.* (2008) The characteristics of near asteroid orbital dynamics and its implication to mission analysis (in Chinese), *Progress in Astronomy*, **28**(3): 193–208.

[2] Scheeres, D.J. (1999) The effect of C_{22} on orbit energy and angular momentum, *Celestial Mechanics and Dynamical Astronomy*, 1999, **73**: 339–348.

[3] http://sbn.psi.edu/pds

[4] Garfinkel, B. (1958) On the motion of a satellite of an oblate planet, *The Astronomical Journal*, **63**: 88–96.

[5] Kozai Y. (1959) The motion of a close Earth satellite, *The Astronomical Journal*, **64**: 367–377.

[6] Roy, A.E. Steves, B.A. (1995) From Newton to Chaos, *NATO ASI Series B: Physics*, 336.

[7] Ferraz-Mello, S. (1999) Slow and fast diffusion in asteroid-belt resonances, *Celestial Mechanics and Dynamical Astronomy*, **73**: 25–37.

[8] Scheeres, D.J. (1994) Dynamics about uniformly rotating triaxial ellipsoids: applications to asteroids, *Icarus*, **110**: 225–238.

[9] Scheeres, D.J., Ostro, S.J., Hudson, R.S., and Werner, R.A. (1996) Orbits close to asteroid 4769 Castalia, *Icarus*, **121**: 67–87.

[10] Werner, R.A. and Scheeres, D.J. (1997) Exterior gravitation of a polyhedron derived and compared with harmonic and mascon gravitation representation of asteroid 4769 Castalia, *Celestial Mechanics and Dynamical Astronomy*, **65**: 313–44.

[11] Szebehely, V. (1967) *Theory of Orbits*. New York: Academic Press Inc.

[12] Danby, J.M.A. (1992) *Fundamentals of Celestial Mechanics*, 2nd edn. Richmond, VA: Willmann-Bell.

[13] Taff, L.G. (1985) *Celestial Mechanics: A Computational Guide for the Practitioner*, New York: John Wiley & Sons, Inc.

[14] Kaula, W.M. (1966) *Theory of Satellite Geodesy*. Boston: Blaisdell.

[15] Hu, W. (2002) *Orbital Motion in Uniformly Rotating Second Degree and Order Gravity Fields*, Ph.D. Dissertation of The University of Michigan.

[16] Hu, W. and Scheeres, D.J. (2000) Spacecraft motion about slowly rotating asteroids, *AIAA Journal of Guidance, Control and Dynamics*, 2002, **25**(4): 765–75.
Previous version appeared at *AAS/AIAA Space Flight Mechanics Meeting[C]*, Clearwater, Florida.

[17] Scheeres, D.J. and Hu, W. (2001) Secular motion in a 2nd degree and order gravity field with no rotation, *Celestial Mechanics and Dynamical Astronomy*, **79**: 183–200.

[18] Hu, W. and Scheeres, D.J. (2001) Periodic orbits in rotating 2nd degree and order gravity fields, *Chinese Journal of Astronomy and Astrophysics*, 2008, **1**: 08–18.
Previous version appared at *9th International Space Conference of Pacific-basin Societies (ISCOPS)[C]*, Pasadena, California.

[19] Hu, W. and Scheeres, D.J. (2004) Numerical determination of stability regions for orbital motion in uniformly rotating second degree and order gravity fields, *Planetary and Space Science*, **52**: 685–92.

[20] Magri, C., Ostro, S.J., Scheeres, D.J., *et al.* (2007) Radar observations and a physical model of asteroid 1580 Betulia, *Icarus*, **186**(1): 152–77.

[21] Mysen, E, *et al.* Olsen, O., and Aksnes, K. (2006) Chaotic gravitational zones around a regularly shaped complex rotating body, *Planetary and Space Science*, **54**(8): 750–60.

[22] Olsen, O. (2006) Orbital resonance widths in a uniformly rotating second degree and order gravity field, *Astronomy & Astrophysics*, **449**(2): 821–6.

[23] Lara, M., Palacian, J.F., and Russell, R.P. (2010) Mission design through averaging of perturbed Keplerian system: the paradigm of an Enceladus orbit, *Celestial Mechanics and Dynamical Astronomy*, **108**: 1–12.

[24] Delsate, N. (2011) Analytical and numerical study of the ground-track resonances of Dawn orbiting Vesta, *Planetary and Space Science*, **59**(13): 1372–83.

[25] Scheeres, D.J. (2012) Orbital mechanics about asteroids and comets, *AIAA Journal of Guidance Control Dynamics*, **35**(3): 987–97.

[26] Scheeres, D.J. (2012) *Orbital Motion in Strongly Perturbed Environments: Applications to Asteroid, Comet and Planetary Satellite Orbiter* London: Springer-Praxis.

[27] Wang, Y. and Xu, S. (2013) Attitude stability of a spacecraft on a stationary orbit around an asteroid subject to gravity gradient torque, *Celestial Mecnanics and Dynamical Astronomy*, **115**(4): 333–52.

[28] Herrera, E., Palmer, P.L., and Roberts, R.M. (2012) Modeling the fravitational potential of a non-spherical asteroid, *AIAA Journal of GCD*, **36**(3): 790–8.

[29] Hu, W. and Scheeres, D. (2014), Averaging analyses for spacecraft orbital motion around asteroids, *Acta Mechanica Sinica*, **30**: 294–300.

7

Application

This chapter introduce some examples of practical applications of orbital mechanics in other fields.

7.1 Two Line Elements (TLE)

Once an orbit has been determined for a particular space object, we might want to communicate that information to another person who is interested in tracking the object. A standard format for communicating the orbit of an Earth-orbiting object is the two-line element set, or TLE. An example TLE for a Russian spy satellite, Molniya (See Figure 7.6) is:

```
MOLNIYA 1-93
1 28163U 04005A   05111.15254065  .00000265  00000-0  10000-3 0  4278
2 28163  62.9152 143.9979 7233471 287.8575  24.1954  2.00601438  8577
```

The orbital elements of this satellite are: Period: $P = 24/2.00\,601\,438 = 11.964$ hr $= 43\,070$ sec ; semi-major axis: $a = (\mu P^2/(2\pi)^2)^{1/3} = 26\,557$ km; eccentricity: $e = 0.7\,233\,471$; inclination: $i = 62.9152° = 1.0981$; longitude of ascending node: $\Omega = 143.9979° = 2.5132$; argument of periapsis: $\omega = 287.8575° = 5.0241$; mean anomaly: $M = 24.1954° = 0.42\,229$; eccentric anomaly: $E = 60.1375° = 1.0496$, true anomaly: $\theta = 110.6264° = 1.9308$. We can also know that the satellite was launched in 2004, the epoch time is April 21, 3 hr 39 min 39.5 sec in 2005.

Note that the TLE actually has three lines.

Documentation for the TLE format is available at the website http://celestrak.com and is reproduced here for your convenience. This website also includes current TLEs for many satellites.

```
Data for each satellite consists of three lines in the following format:
AAAAAAAAAAAAAAAAAAAAAAAAA
1 NNNNNU NNNNNAAA NNNNN.NNNNNNNN +.NNNNNNNN +NNNNN-N +NNNNN-N N NNNNN
2 NNNNN NNN.NNNN NNN.NNNN NNNNNNN NNN.NNNN NNN.NNNN NN.NNNNNNNNNNNNNN
```

Fundamental Spacecraft Dynamics and Control, First Edition. Weiduo Hu.

Line 0 is a twenty-four character name (to be consistent with the name length in the NORAD SATCAT).

Lines 1 and 2 are the standard Two-Line Orbital Element Set Format identical to that used by NORAD and NASA. The format description is:

Line 1
Column   Description
01   Line Number of Element Data
03-07   Satellite Number
08   Classification (U=Unclassified)
10-11   International Designator (Last two digits of launch year)
12-14   International Designator (Launch number of the year)
15-17   International Designator (Piece of the launch)
19-20   Epoch Year (Last two digits of year)
21-32   Epoch (Day of the year and fractional portion of the day(UT))
34-43   First Time Derivative of the Mean Motion
45-52   Second Time Derivative of Mean Motion (decimal point assumed)
54-61   BSTAR drag term (decimal point assumed)
63   Ephemeris type
65-68   Element number
69   Checksum (Modulo 10)
         (Letters, blanks, periods, plus signs = 0; minus signs = 1)

Line 2
Column   Description
01   Line Number of Element Data
03-07   Satellite Number
09-16   Inclination [Degrees]
18-25   Right Ascension of the Ascending Node [Degrees]
27-33   Eccentricity (decimal point assumed)
35-42   Argument of Perigee [Degrees]
44-51   Mean Anomaly [Degrees]
53-63   Mean Motion [Revs per day(solar day)]
64-68   Revolution number at epoch [Revs]
69   Checksum (Modulo 10)

All other columns are blank or fixed.

Some examples of TLEs are listed here, which include many Chinese satellites:
```
HST
1 20580U 90037B   13215.73048880  .00001283  00000-0  78873-4 0  2954
2 20580  28.4704 321.1166 0003165 132.2805  22.7802 15.03811562 76868
TIANGONG 1
1 37820U 11053A   13217.07032766  .00050899  00000-0  38205-3 0  1696
2 37820  42.7772  63.0939 0007334 222.8519 275.3274 15.71755641106343
JB-3 2 (ZY 2B)
1 27550U 02049A   13217.18334491  .00008672  00000-0  18623-3 0  2013
2 27550  97.1009 203.8213 0006285 152.1080  38.3708 15.44905968600471
ISS (ZARYA)
1 25544U 98067A   13217.18208943  .00003855  00000-0  75048-4 0  3307
2 25544  51.6490 225.5716 0003644 271.0398 177.9490 15.50171497842306
```

```
CZ-4C DEB
1 39214U 13037G   13216.99308881  .00000306  00000-0  63518-4 0   494
2 39214  98.0302 234.3020 0012043 249.5218 110.4695 14.66742893  2028
FENGYUN 1D
1 27431U 02024B   13217.00239089  .00000142  00000-0  10595-3 0  4614
2 27431  98.8205 180.0705 0016218 101.0204 259.2791 14.09405435577478
FENGYUN 2E
1 33463U 08066A   13216.57431583 -.00000342  00000-0  10000-3 0  2579
2 33463   1.3513  50.4207 0002501 188.2151  25.7309  1.00278693 16971
ZHONGXING-6B
1 31800U 07031A   13216.85476446 -.00000373  00000-0  10000-3 0  7266
2 31800   0.0125  50.9522 0002064  78.3520 247.5273  1.00270891 22352
BEIDOU G3
1 36590U 10024A   13216.88972757 -.00000366  00000-0  10000-3 0  8625
2 36590   1.5102  27.9434 0002144 311.5060  44.9878  1.00270244 11653
BEIDOU IGSO 2
1 37256U 10068A   13216.46236078 -.00000251  00000-0  10000-3 0  6919
2 37256  54.6249 327.5580 0021243 186.8086  85.6373  1.00284607  9676
CHINASAT 10 (ZX 10)
1 37677U 11026A   13216.58836015 -.00000363  00000-0  10000-3 0  6049
2 37677   0.0436 258.5084 0004346 284.2759  92.8372  1.00270597  7864
GPS BIIF-4  (PRN 27)
1 39166U 13023A   13215.57467135 -.00000037  00000-0  10000-3 0   552
2 39166  55.0347  96.8018 0001768 358.3936   1.6698  2.00564904  1597
COSMOS 2478 (746)
1 37938U 11071A   13216.43334432 -.00000016  00000-0  00000+0 0  5004
2 37938  64.8941 112.6675 0012738 241.8671 313.7450  2.13102482 13237
GALILEO-FM4 (GSAT0104)
1 38858U 12055B   13214.93651530  .00000059  00000-0  00000+0 0  1151
2 38858  55.2062 231.3463 0003697 246.3410 113.6744  1.70473562  5018
MOLNIYA 1-93
1 28163U 04005A   13215.76876521 -.00000137  00000-0  10000-3 0  2015
2 28163  64.6412 118.1934 7147207 250.3246  24.6898  2.00599880 69308
IRIDIUM 98 [S]
1 27451U 02031B   13216.21859954  .00000106  00000-0  25647-4 0  9020
2 27451  86.4411 252.9183 0004669 131.4465 271.2314 14.42946045588453
```

7.2   GPS RINEX

GPS Nominal Constellation consists of 24 operational satellites in six orbits planes, four satellites in each plane. The near circular orbits are 20 200 km above the Earth at an inclination of 55 degrees with a near 12 hr period.

7.2.1   Distribution of Ephemerides

The broadcast ephemeris is decoded by all GPS receivers and for geodetic receivers the software that converts the receiver binary to an exchange format outputs an ASCII version.
The exchange format: Receiver Independent Exchange format (RINEX) has a standard for the broadcast ephemeris [2].

7.2.2   A GPS RINEX Navigation File

```
      2.10              N:  GPS NAV DATA                          RINEX VERSION / TYPE
JPS2RIN 1.07            RUN BY               04-SEP-01 13:20      PGM / RUN BY / DATE
build October 30, 2000 (c) Topcon Positioning Systems           COMMENT
     .2235D-07    .2235D-07   -.1192D-06   -.1192D-06            ION ALPHA
     .1290D+06    .4915D+05   -.1966D+06    .3277D+06            ION BETA
     .465661287308D-08 .168753899743D-13    319488       1130   DElTA-UTC: A0,A1,T,W
     13                                                          LEAP SECONDS
                                                                 END OF HEADER
 1 01  9  4 10   0  0.0  .193310435861D-03  .147792889038D-11  .000000000000D+00
     .166000000000D+03 -.248750000000D+02  .474519765653D-08 -.311447558422D+01
    -.136531889439D-05  .504714518320D-02  .268220901489D-05  .515374749756D+04
     .208800000000D+06  .968575477600D-07  .195770043888D+01 -.260770320892D-07
     .965370522140D+00  .329406250000D+03 -.172314458485D+01 -.830927468607D-08
    -.173221501085D-09  .000000000000D+00  .113000000000D+04  .000000000000D+00
     .400000000000D+01  .000000000000D+00 -.325962901115D-08  .422000000000D+03
     .208799000000D+06
 4 01  9  4 10   0  0.0  .576881226152D-03 -.187583282241D-10  .000000000000D+00
     .195000000000D+03 -.250312500000D+02  .413231498465D-08  .238621932869D+01
    -.114925205708D-05  .545852934010D-02  .113416463137D-04  .515378430557D+04
     .208800000000D+06  .540167093277D-07 -.132884154953D+00  .147148966789D-06
     .973307813857D+00  .167062500000D+03 -.414280964991D+00 -.772817905253D-08
    -.375015620906D-09  .000000000000D+00  .113000000000D+04  .000000000000D+00
     .200000000000D+01  .000000000000D+00 -.605359673500D-08  .451000000000D+03
     .208799000000D+06
```

We can obtain much information from the Navigation Date File. For example, GPS satellite 1, at year: 2001, month: 9, date: 4, hour: 10, minute: 0, second: 0.0, its orbital elements are $a = 2.6561e + 007$m, $e = 5.0471e - 003$, $i_0 = 9.6537e - 001$, $\Omega_0 = 1.9577e + 000$, $\omega = -1.7231e + 000$, $M_0 = -3.1144e + 000$. Then its position in ECI system is $[1.4789e + 007, 7.3347e + 006, 2.0977e + 007]$.

7.3   Remote Observation

7.3.1   Orbital Coverage

See Figure 7.1. The satellite coverage area can be calculated:

$$A = \int_0^{\lambda_{max}} 2\pi(R_E \sin \lambda)(R_E d\lambda) = 2\pi R_E^2(1 - \cos \lambda_{max}), \tag{7.1}$$

where R_E is the Earth mean radius, $\lambda_{max} = \arccos(R_E/(R_E + H))$, H is the height of the satellite, $D_{max}^2 = (R_E + H)^2 - R_E^2$, $\tan \rho = R_E/D_{max}$.

Interested readers can find the coverage correspondence and differences between the 2D figure in Figure 7.2 and the 3D figure in Figure 7.3 centered at point C, where the λ_{max} is $30°, 60°, 81.3°$, and near $90°$, respectively. Note that $\lambda_{max} = 81.3°$ is the geosynchronous S/C's coverage angle. Here the the coverage angle centered at Beijing (Point A) and Sydney (Point B) is $\lambda_{max} = 20°$.

Table 7.1   Navigation message file: header section description [2]

HEADER LABEl (Columns 61–80)	DESCRIPTION	FORMAT
RINEX VERSION/TYPE	- Format version (2) - File type ('N' for Navigation data)	I6,14X, A1,19X
PGM/RUN BY/DATE	- Name of program creating current file - Name of agency creating current file - Date of file creation	A20, A20, A20
COMMENT*	Comment line(s)	A60
ION ALPHA*	Ionosphere parameters A0-A3 of almanac	2X,4D12.4
ION BETA*	Ionosphere parameters B0-B3 of almanac	2X,4D12.4
DElTA-UTC*: A0,A1, T,W	Almanac parameters to compute time in UTC A0,A1: terms of polynomial T : reference time for UTC data W : UTC reference week number	3X,2D19.12, 2I9
LEAP SECONDS*	Delta time due to leap seconds	I6
END OF HEADER*	Last record in the header section. Records marked with * are optional	60X

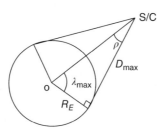

Figure 7.1   The satellite coverage geometry

7.3.2   *Ground Track*

Sub-satellite Point (SSP) on Earth's surface defined by an imaginary line connecting the satellite and the Earth's center.

Ground track is the trace of a S/C's path over the surface of the earth, i.e. trace of sub-points over time.

The ground track is important for a S/C looking down on the Earth, which helps to determine which orbit will best maximize mission effectiveness of satellite. It also helps to determine the over flight/proximity of hostile satellites with respect to key areas.

A great circle is a circle that cuts through the center of the Earth (spherical trig). Orbits trace out a great circle. See great circle example in Figure 7.3 $a = \widehat{CB}$, $b = \widehat{AC}$, $c = \widehat{AB}$. Figure 7.2

Table 7.2 Ephemeris data definitions [2]

	Ephemeris date definitions
M_0	Mean anomaly at reference time
Δn	Mean motion difference from computed value
e	Eccentricity
$\sqrt{a}$	Square root of the semi-major axis
Ω_0	Longitude of ascending node of orbit plane at weekly epoch
i_0	Inclination angle at reference time
ω	Argument of perigee
$\dot{\Omega}$	Rate of right ascension
IDOT	Rate of inclination angle
C_{uc}	Amplitude of the cosine harmonic correction term to the argument of latitude
C_{us}	Amplitude of the sine harmonic correction term to the argument of latitude
C_{rc}	Amplitude of the cosine harmonic correction term to the orbit radius
C_{rs}	Amplitude of the sine harmonic correction term to the orbit radius
C_{ic}	Amplitude of the cosine harmonic correction term to the angle of inclinations
C_{is}	Amplitude of the sine harmonic correction term to the angle of inclination
t_{oe}	Reference time ephemeris
IODE	Issue of date (ephemeris)

is Mercator Projection of Figure 7.3, as Mercator Projection is a flat map projection of the spherical Earth.

The tracing of orbits appears as a sine wave on Mercator Projection. The Earth rotates $360/24 = 15°$/hour. Orbit traces shift to the west.

To calculate SSP, compute S/C position vector $r = [r_1, r_2, r_3]^T$ in ECI first. Secondly determine Greenwich Sidereal Time (GST) α_g at epoch with α_{g0}. See Figure 2.17 and Figure 7.4. Then the latitude is

$$\delta_s = \sin^{-1} \frac{r_3}{r} = \arcsin(\sin i \sin(\omega + \theta)). \tag{7.2}$$

And longitude is (when $r_1 > 0$)

$$L_s = \arctan\left(\frac{r_2}{r_1}\right) - \alpha_g \tag{7.3}$$

$$= \Omega + \arctan(\tan(\omega + \theta) \cos i). \tag{7.4}$$

When $r_1 < 0$, $\pi + \arctan\left(r_2/r_1\right) - \alpha_g$. Note that propagate position vector in the "the usual way," propagate GST using $\alpha_g = \alpha_{g0} + \omega_E(t - t_0)$, where ω_E is the angular velocity of the Earth.

Let ΔN (node displacement) be the angular difference between the ascending node for two successive orbits. For direct orbits with period less than 24 hours, nodal decrease (westward) is $360-\Delta N$. If S/C period is $P = (360° - \Delta N)/(15°/\text{hr})$, calculate semi-major axis a from relationship $P = 2\pi\sqrt{a^3/\mu}$. Orbital inclination can also be determined from ground track. For a direct orbit, it is the highest latitude; for a retrograde orbit it is $180°$ minus highest latitude.

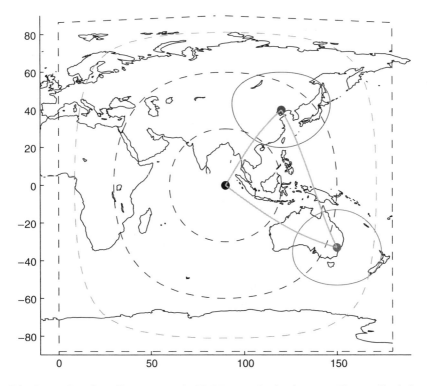

Figure 7.2 Examples of satellite coverage in 2D Mercator Projection map. The satellite is just above the equator with different heights

A circular orbit has symmetrical ground track. An elliptical orbit has non-symmetrical ground track. A S/C moves fastest at perigee, so spreads out ground track, whereas it moves slowest at apogee, so compresses ground track. See Figure 7.6. Molniya orbits are highly elliptical; position of apogee is over a desired location. It seems that Molniya "hang" over the Earth at apogee and move faster than the earth at perigee.

If i is inclination, β is launch azimuth, ϕ_0 is latitude of the launch site, then based on the spherical triangle relation (see Equation (A.35) Appendix A; see also Figure 7.5).

$$\cos i = -\cos 90° \cos \beta + \sin 90° \sin \beta \cos \phi_0$$

$$\Rightarrow \cos i = \sin \beta \cos \phi_0. \tag{7.5}$$

Hence inclination of the orbit determines the maximum latitude the ground track will reach. Assuming no maneuvers after launch, launch sites will determine inclination – more on this in launch considerations. Injection point will determine where the ground track will start.

In Figure 7.6 shows the orbits of a Molniya, an ISS, a near GEO orbit near the longitude 100° and inclination of near 14°. A GPS orbit is also plotted, where the Molniya and GPS ground tracks are closed repeated curves as their period is 12 hrs which is half a day on the Earth. Interested reader can find the correspondence between these orbits and their ground tracks in the figure.

Table 7.3 Navigation message file: data record description [2]

OBS. RECORD	DESCRIPTION	FORMAT
PRN / EPOCH / SV CLK	– Satellite PRN number	I2,
	– Epoch: Toc - Time of Clock	
	year (2 digits)	5I3,
	month	
	day	
	hour	
	minute	
	second	F5.1,
	– SV clock bias (seconds)	3D19.12
	– SV clock drift (sec/sec)	
	– SV clock drift rate (sec/sec2)	
BROADCAST ORBIT – 1	– IODE Issue of Data, Ephemeris	3X,4D19.12
	– Crs (meters)	
	– Delta n (radians/sec)	
	– M0 (radians)	
BROADCAST ORBIT – 2	– Cuc (radians)	3X,4D19.12
	– e Eccentricity	
	– Cus (radians)	
	– sqrt(A) (sqrt(m))	
BROADCAST ORBIT – 3	– Toe Time of Ephemeris	3X,4D19.12
	(sec of GPS week)	
	– Cic (radians)	
	– OMEGA (radians)	
	– CIS (radians)	
BROADCAST ORBIT – 4	– i0 (radians)	3X,4D19.12
	– Crc (meters)	
	– omega (radians)	
	– OMEGA DOT (radians/sec)	
BROADCAST ORBIT – 5	– IDOT (radians/sec)	3X,4D19.12
	– Codes on L2 channel	
	– GPS Week # (to go with TOE)	
	– L2 P data flag	
BROADCAST ORBIT – 6	– SV accuracy	3X,4D19.12
	– SV health (MSB only)	
	– TGD (seconds)	
	– IODC Issue of Data, Clock	
BROADCAST ORBIT – 7	– Transmission time of message	3X,4D19.12
	(sec of GPS week, derived e.g.	
	from Z – count in Hand Over Word (HOW)	
	– spare	
	– spare	
	– spare	

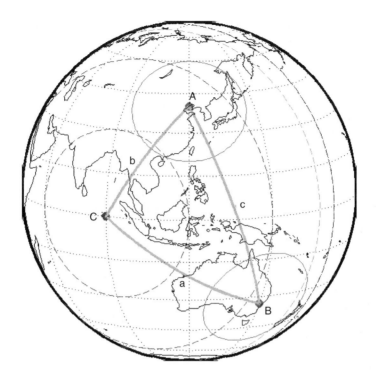

Figure 7.3   Examples of satellite coverage in three dimensions, the spherical triangles are on great arcs

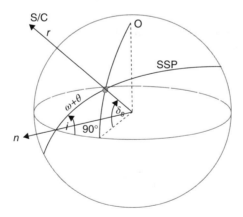

Figure 7.4   The sub-satellite point (SSP) geometry

7.3.3   Ground Station

Orbital determination, elevation angle, and visibility are closely related problems.

Elevation angle, El, is measured up from horizon to target as discussed in Section 2.4.4.

Minimum elevation angle is typically based on the performance of an antenna or sensor.

Table 7.4 The computation method. Many variables and equations here are introduced in Chapter 5, such as Equtions (5.115)–(5.120) [2]

Equations	
Algorithm	**Description**
$\mu = 3.986\,004\,418 \times 10^{14}$ meters^3/sec^2	WGS84(GPS) gravitational parameter
$\dot{\Omega}_e = 7.2\,921\,151\,467 \times 10^{-5}$ rad/sec	WGS84(GPS) Earth rotation rate
$n_o = \sqrt{\mu/a^3}$	Computed mean motion
$t_j = t - t_{oe}$	Time from reference epoch
$n = n_o + \Delta n$	Corrected mean motion
$M_j = M_o + nt_j$	Mean anomaly
$M_j = E_j - e\sin E_j$	Kepler's equation
$\tan v_j = \sqrt{1-e^2}\sin E_j/(\cos E_j - e)$	True anomaly
$\phi_j = v_j = \omega$	Argument of latitude
$\delta u_j = C_{us}\sin 2\phi_j + C_{uc}\cos 2\phi_j$	Argument of latitude correction
$\delta r_j = C_{rs}\sin 2\phi_j + C_{rc}\cos 2\phi_j$	Radius correction
$\delta i_j = C_{is}\sin 2\phi_j + C_{ic}\cos 2\phi_j$	Inclination correction
$u_j = \phi_j + \delta u_j$	Corrected argument of latitude
$r_j = a(1 - e\cos E_j) + \delta r_j$	Corrected radius
$i_j = i_o + \delta i_j + \mathrm{IDOT}t_j$	Corrected inclination
$x'_j = r_j\cos u_j$	Position in orbital plane
$y'_j = r_j\sin u_j$	
$\Omega = \Omega_o + (\dot{\Omega} - \dot{\Omega}_e)t_j - \dot{\Omega}_e t_{oe}$	Corrected longitude of ascending node
$x_j = x'_j\cos\Omega_j - y'_j\sin i_j\sin\Omega_j$	Earth-fixed
$y_j = x'_j\sin\Omega_j + y'_j\cos i_j\sin\Omega_j$	Coordinates
$z_j = y'_j\sin i_j$	

If $t_j > 302,400$ s, subtract $604\,800$ s from t_j. If $t_j < -302\,400$ s, add $604\,800$ s to t_j.

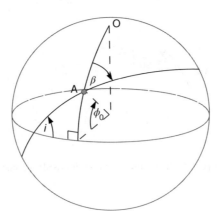

Figure 7.5 The launch site geometry

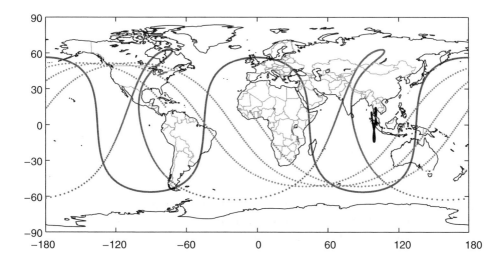

Figure 7.6   Examples of ground track

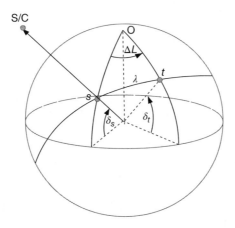

Figure 7.7   The SSP and target geometry

Coverage area A is determined by Equation (7.1). Angle η is called the nadir angle. The angle $\rho = \eta_{\max}$ is called the apparent Earth radius. When $El = 0$, $\rho = \eta_{\max} = \pi/2 - \lambda_{\max}$, that is, the line from the satellite to the target is at a tangent to the Earth.

For a target with known position vector, λ is easily computed by the side law of cosine in Appendix A.

$$\cos \lambda = \cos \delta_s \cos \delta_t \cos \Delta L + \sin \delta_s \sin \delta_t, \tag{7.6}$$

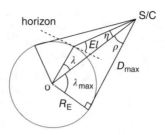

Figure 7.8   The satellite and ground target geometry

where δ_t is the latitude of ground station or target, and δ_s is latitude of the sub-satellite point. ΔL is the longitude difference between SSP and ground target. See Figure 7.7. And $\eta + \lambda + El = \pi/2$, the distance D is from target to satellite, $r_{sat} = R_E + H$ is the satellite position. See Figure 7.8.

$$D = R_E \frac{\sin \lambda}{\sin \eta}, \tag{7.7}$$

$$\tan \rho = \frac{R_E}{D_{\max}}, \tag{7.8}$$

$$\sin \rho = \frac{R_E}{r_{sat}} = \frac{R_E}{R_E + H} = \frac{\sin \eta}{\sin(90° + El)}. \tag{7.9}$$

Then

$$\tan \eta = \frac{R_E \cdot \sin \lambda}{r_{SAT} - R_E \cos \lambda} = \frac{\sin \rho \sin \lambda}{1 - \sin \rho \cos \lambda}. \tag{7.10}$$

The condition for visibility is $El > 0$. This is also the problem of predicting satellite rise and set times. A simple analysis of visibility is given in Figure 5.3.

In Figure 7.9, the ground station is near Beijing. The satellite orbit is a circular orbit. Its altitude $H = 360$ km, inclination $i = 45°$.

Figure 7.9 and Figure 7.10 are the same orbit from the view of different frames. An example of coordinate transformation between ECEF and THS may be found in Section 2.4.4.

7.4   Summary and Keywords

This chapter introduces some practical application examples which are directly related to the knowledge or method of orbital mechanics, such as two line element (TLE), GPS RINEX, orbital coverage, ground track, with many actual examples.

Keywords: two line element (TLE), GPS navigation message, coverage, sub-satellite point, ground track, satellite visibility.

Figure 7.9   Satellite visibility at a ground station near Beijing in geographic frame (ECEF)

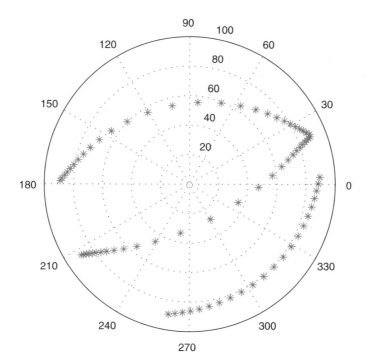

Figure 7.10   Satellite's visibility in Figure 7.9 in the topocentric frame. In the figure the radius axis is $90°-El$, the direction angle equals $-Az + 90°$, El, Az are defined in Section 2.4.4

Problems

7.1   What is satellite coverage area when the satellite's height is zero and infinity?

7.2   The following two-line element set (TLE) is for a recently launched space station
called TIANGONG 1

```
TIANGONG 1
1 37820U 11053A   13217.07032766  .00050899  00000-0  38205-3 0 1696
2 37820  42.7772  63.0939 0007334 222.8519 275.3274 15.71755641106343
```

For this problem, you should ignore Earth oblateness effects, as well as the "perturbation"
terms in the TLE ($\dot{n}, \ddot{n}$, and B^*).

(a) What are the date and the Eastern Standard Time of epoch?
(b) What are the orbital elements of the satellite? (a, e, i, Ω, ω, and v_0, with a in km, and
angles in degrees).
(c) What are the latitude and longitude of the sub-satellite point (SSP) at epoch? Where
is this point?
(d) What is the area, in km^2, of the instantaneous access area (IAA) at epoch? What
fraction of the theoretical maximum IAA (half the Earth) is this?
(e) If the minimum elevation angle is $El_{min} = 10°$, then what is the "reduced" IAA at
epoch?
(f) What are the latitude and longitude of the SSP the next time (after epoch) the satellite
passes through apoapsis?

7.3   What would be the effect on the ground track of a geosynchronous satellite if:

(a) $e > 0$ but the orbital period did not change?
(b) i changed but all other parameters remained the same?
(c) the orbital period was reduced?
(d) $e > 0, i > 0$, other elements remained the same?
(e) $e > 0, i > 0, \omega > 0$ other elements remained the same?

Sketch the resulting ground tracks to illustrate your answer.

References

[1] Vallado, D. (1997) *Fundamentals of Astrodynamics and Applications*. El Segundo, CA: Microcosm
Press.
[2] GPS Joint program office, Navstar GPS Space Segment/Navigation User Interfaces, IS-GPS-200,
Revision D, Dec. 2004. CA, USA.

Part Two

Attitude Dynamics

8

Rigid Body Kinematics

From this chapter, we begin to introduce attitude kinematics and dynamics, attitude determination and control.

In this chapter we discuss the description of attitude kinematics using reference frames, rotation matrices, Euler parameters, Euler angles, and quaternion.

Recall the fundamental dynamics equations law of linear momentum and angular momentum

$$\mathbf{f} = \frac{d\mathbf{p}}{dt}, \quad \mathbf{g} = \frac{d\mathbf{h}}{dt}, \tag{8.1}$$

where $\mathbf{f}$ is force, $\mathbf{g}$ is torque, $\mathbf{p}$ is linear momentum and $\mathbf{h}$ is angular momentum. For both equations, we must relate momentum to kinematics.

Before we introduce attitude kinematics, we recall Translational Kinematics. Newton's Second Law can be written in first-order state-vector form as

$$\dot{\mathbf{r}} = \mathbf{p}/m, \quad \dot{\mathbf{p}} = \mathbf{f}. \tag{8.2}$$

Here, $\mathbf{p}$ is the linear momentum, defined by the kinematics equation; that is, $\mathbf{p} = m\mathbf{v} = m\dot{\mathbf{r}}$.

Thus, the kinematics differential equation allows us to integrate the velocity to compute the position. The kinetics differential equation allows us to integrate the applied force to compute the linear momentum.

In general, $\mathbf{f} = \mathbf{f}(\mathbf{r}, \mathbf{p})$. Consider $\mathbf{f} = m\mathbf{a}$ expressed in an inertial frame: $m\ddot{\mathbf{r}} = \mathbf{f} \Leftrightarrow [m\ddot{r}_1, m\ddot{r}_2, m\ddot{r}_3] = [f_1, f_2, f_3]$. Equivalently, $\dot{\mathbf{p}} = \mathbf{f} \Leftrightarrow [\dot{p}_1, \dot{p}_2, \dot{p}_3] = [f_1, f_2, f_2]$, and $\dot{\mathbf{r}} = \mathbf{p}/m \Leftrightarrow [\dot{r}_1, \dot{r}_2, \dot{r}_3] = [p_1/m, p_2/m, p_3/m]$. We need to develop rotational equations equivalent to the translational kinematics equations.

8.1 Attitude Notions

8.1.1 Attitude

Spacecraft Attitude is the orientation of the spacecraft with respect to other objects. Spacecraft Attitude Determination is the process of measuring and calculating the spacecraft attitude, i.e,

Fundamental Spacecraft Dynamics and Control, First Edition. Weiduo Hu.
© 2015 John Wiley & Sons (Asia) Pte Ltd. Published 2015 by John Wiley & Sons (Asia) Pte Ltd.

Table 8.1   The summary of linear and angular motion

$\mathbf{f} = \mathrm{d}\mathbf{p}/\mathrm{d}t$	$\mathbf{g} = \mathrm{d}\mathbf{h}/\mathrm{d}t$
Linear momentum	Angular momentum
= mass × velocity	= inertia × angular velocity
$\mathrm{d}/\mathrm{d}t$ (linear momentum)	$\mathrm{d}/\mathrm{d}t$ (angular momentum)
= applied forces	= applied torques
$\mathrm{d}/\mathrm{d}t$(position)	$\mathrm{d}/\mathrm{d}t$ (attitude)
= linear momentum/mass	= "angular momentum/inertia"

"attitude knowledge." Spacecraft Attitude Control is the generation of moments and torques required to place the spacecraft in the appropriate attitude.

S/C pointing is a mission's requirements for a variety of purposes. In scientific sensors, it is used for deep space observation of considerable time (Hubble), comet identification, scan Earth (ground imaging: Nadir swath Mapping, True scan), scan Earth limb (Aurora Borealis), space physics (Electron Density), Sun physics, and launch imaging.

It is used in communications such as geosynchronous communications satellites which must point antennae for communications gain, ground station communications for Uplink/Downlink, and inter-satellite communications.

It is also use for power, i.e, solar arrays must point towards the Sun.

And in a thermal system those thermally sensitive components must not be in direct sunlight.

Sensitive instruments which light sensitive components must not point at the Sun (star cameras, IR sensors, etc.).

It is also used for orbit transfer, in which orbit transfer burns must be made in the correct direction, spacecraft Rendezvous Docking port and collar alignment, end of life spacecraft disposal, final spacecraft disposal (de-orbit for LEO, higher orbit for GEO).

Attitude is defined by three parameters. Typically, the relative orientation of two orthogonal, co-origination vector triads. It requires two (single axis) or three (three axis) parameters (degrees-of-freedom) to define.

Single Axis Attitude means that one axis is pointing along a single vector; the motion around that axis is undefined. The examples are: Spin-stabilized spacecraft and Single axis sensors (Ladar pencil beam).

Three Axis Attitude means that one axis may be pointing along a single vector, but the other two are also fixed. A complete definition of this kind of orientation requires three parameters. This approach is the most common and will be typically assumed for this book. Typically, we define the attitude as the relationship between a frame rigidly attached to the spacecraft's body (Body Frame) and either the Earth-Centered Inertial (ECI) frame or the Orbital frame, occasionally the Earth-Centered Earth-Fixed (ECEF) frame.

8.1.2   Frames

Here we introduce some coordinate, some of which may be listed already in Chapter 1, but they are defined in the field of attitude, maybe in the same name or by a different name.

For ECI (Earth-Centered Inertial) also called Geocentric Reference Frame (GCRF), its $\hat{X}$ axis points towards a fixed point in space, First Point of Aries, that is, a vector through the Sun from Earth center on March 21. Its $\hat{Z}$ axis is aligned along Earth's spin axis (North Pole), wobbles slightly, must define epoch. Then its $\hat{Y}$ axis makes a right-handed orthogonal frame. It is inertially fixed. It is the $\hat{I}\hat{J}\hat{K}$ which was defined in Chapter 1.

For ECEF (Earth-Centered Earth-Fixed) also known as Internally Terrestrial Reference (ITRF), its $\hat{X}$ axis is through Greenwich meridian at the equator. Its $\hat{Z}$ axis is aligned along the Earth's spin axis (North Pole), and its $\hat{Y}$ axis makes a right-handed orthogonal frame. It is noninertial frame which rotates with the Earth.

In an Orbital (roll, pitch, yaw), noninertial coordinate system, its $\hat{Z}$ axis points inward to the Earth's center (focus), $\hat{Y}$ axis is perpendicular to orbit plane, opposite direction of angular momentum, and its $\hat{X}$ axis makes a right-handed orthogonal frame which points roughly in the direction of motion for circular orbits. The orbital frame is similar to the frame $\hat{R}\hat{S}\hat{W}$ which was introduced in Chapter 1 with a transformation matrix (see Equation (8.116)).

Body frame (rigidly attached to main S/C body) is often aligned along the principal axes of inertia.

It is normally denoted that reference frames are triads of mutually orthogonal unit vectors.

The definition of the frames are: the Inertial Frame, $\mathcal{F}_i : \hat{i}_1, \hat{i}_2, \hat{i}_3$; the Orbital Frame, $\mathcal{F}_o : \hat{o}_1, \hat{o}_2, \hat{o}_3$; the Body Frame: $\mathcal{F}_b : \hat{b}_1, \hat{b}_2, \hat{b}_3$. See Figure 8.1. These axes are mutually perpendicular in each set, which satisfies the following relation as $\hat{i}_i$ for example.

$$\hat{i}_i \cdot \hat{i}_j = \begin{cases} 1 & \text{if } i = j \\ 0 & \text{if } i \neq j, \end{cases} \tag{8.3}$$

which means $\hat{i}_i \cdot \hat{i}_j = \delta_{ij}$. For this right-handedness of reference frames:

$$\begin{array}{lll} \hat{i}_1 \times \hat{i}_1 = 0 & \hat{i}_1 \times \hat{i}_2 = \hat{i}_3 & \hat{i}_1 \times \hat{i}_3 = -\hat{i}_2 \\ \hat{i}_2 \times \hat{i}_1 = -\hat{i}_3 & \hat{i}_2 \times \hat{i}_2 = 0 & \hat{i}_2 \times \hat{i}_3 = \hat{i}_1 \\ \hat{i}_3 \times \hat{i}_1 = \hat{i}_2 & \hat{i}_3 \times \hat{i}_2 = -\hat{i}_1 & \hat{i}_1 \times \hat{i}_3 = 0. \end{array} \tag{8.4}$$

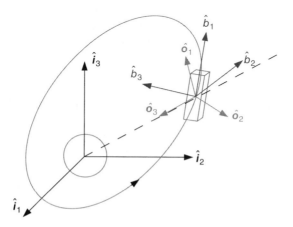

Figure 8.1 The inertial $\mathcal{F}_i$, orbital $\mathcal{F}_o$ and body $\mathcal{F}_b$ frames

This means

$$\hat{i}_i \times \hat{i}_j = \varepsilon_{ijk}\hat{i}_k, \tag{8.5}$$

where

$$\hat{i}_i \cdot \hat{i}_j = \begin{cases} 1 & \text{for } i,j,k \text{ an even permutation of } 1,2,3 \\ -1 & \text{for } i,j,k \text{ an odd permutation of } 1,2,3 \\ 0 & \text{otherwise (i.e.,if any repetitions occur).} \end{cases} \tag{8.6}$$

The same thing happens in orbital and body frames.

Or even more succinctly as

$$\{\hat{i}\} \times \{\hat{i}\}^T = \begin{Bmatrix} 0 & i_3 & -i_2 \\ -i_3 & 0 & i_1 \\ i_2 & -i_1 & 0 \end{Bmatrix} = -\{\hat{i}\}^\times. \tag{8.7}$$

The $[]^\times$ notation defines a skew-symmetric 3×3 matrix whose 3 unique elements are the components of the 3×1 matrix $[]$

The same notation applies if the components of the 3×1 matrix $[]$ are scalars instead of vectors:

$$\mathbf{a} = \begin{bmatrix} a_1 \\ a_2 \\ a_3 \end{bmatrix} \Rightarrow \mathbf{a}^\times = \begin{bmatrix} 0 & -a_3 & a_2 \\ a_3 & 0 & -a_1 \\ -a_2 & a_1 & 0 \end{bmatrix}. \tag{8.8}$$

The "skew-symmetry" property is satisfied since

$$(\mathbf{a}^\times)^T = -\mathbf{a}^\times. \tag{8.9}$$

8.1.3   Vector

Reference frame is dextral triad of orthonormal unit vectors. A vector is an abstract mathematical object with two properties: direction and length.

Vectors used in this book include, for example, position, velocity, acceleration, force, momentum, torque, angular velocity. Vectors can be expressed in any reference frame.

Keep in mind that the term "state vector" refers to a different type of object – specifically, a state vector is generally a column matrix collecting all the system states.

Vector can be expressed as a linear combination of the unit vectors, but it must be clear about which reference frame is used. Orthogonal means the base vectors are perpendicular (orthogonal) to each other, and have unit length (normalized).

A vector can be uniquely represented as having a magnitude and direction. A vector is independent of any frame.

Frequently we collect the components of the vector into a matrix. $\mathbf{v} = [v_1, v_2, v_3]^T$. Note the absence of an overarrow here by using a bold character.

When we can easily identify the associated reference frame, we use the simple notation above; however, when multiple reference frames are involved, we use a subscript to make the connection clear.

Examples: $\mathbf{v}_i$ denotes a vector in $\mathcal{F}_i$, $\mathbf{v}_o$ denotes a vector in $\mathcal{F}_o$, $\mathbf{v}_b$ denotes a vector in $\mathcal{F}_b$.

See Figure 8.2. A vector can be represented in any frame as:

$$\mathbf{v} = v_1^i \hat{i}_1 + v_2^i \hat{i}_2 + v_3^i \hat{i}_3. \tag{8.10}$$

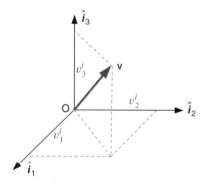

Figure 8.2   A vector in inertial frame $\mathcal{F}_i$

In this frame, the vector can be equivalently represented as a 3×1 matrix of scalars.

$$\mathbf{v}^i \equiv \begin{bmatrix} v_1^i & v_2^i & v_3^i \end{bmatrix}^T . \tag{8.11}$$

The scalars in the matrix are found as

$$v_1^i = \mathbf{v} \cdot \hat{i}_1, \quad v_2^i = \mathbf{v} \cdot \hat{i}_2, \quad v_3^i = \mathbf{v} \cdot \hat{i}_3.$$

Matrix multiplication arises frequently in dynamics and control, and an interesting application involves the 3×1 matrix of a vector's components and the 3×1 "matrix" of a frame's base vector.

The same vector can be represented in multiple frames as:

$$\mathbf{v} = v_1^i \hat{i}_1 + v_2^i \hat{i}_2 + v_3^i \hat{i}_3 \Rightarrow \begin{bmatrix} v_1^i & v_2^i & v_3^i \end{bmatrix}^T \equiv \mathbf{v}^i \tag{8.12}$$

$$\mathbf{v} = v_1^o \hat{o}_1 + v_2^o \hat{o}_2 + v_3^o \hat{o}_3 \Rightarrow \begin{bmatrix} v_1^o & v_2^o & v_3^o \end{bmatrix}^T \equiv \mathbf{v}^o \tag{8.13}$$

$$\mathbf{v} = v_1^b \hat{b}_1 + v_2^b \hat{b}_2 + v_3^b \hat{b}_3 \Rightarrow \begin{bmatrix} v_1^b & v_2^b & v_3^b \end{bmatrix}^T \equiv \mathbf{v}^b . \tag{8.14}$$

For the same vector, the 3×1 matrixes differ between frames, i.e. normally

$$\begin{bmatrix} v_1^i & v_2^i & v_3^i \end{bmatrix}^T \neq \begin{bmatrix} v_1^o & v_2^o & v_3^o \end{bmatrix}^T \neq \begin{bmatrix} v_1^b & v_2^b & v_3^b \end{bmatrix}^T . \tag{8.15}$$

The 2-norm (length) of each matrix is the same

$$\| \mathbf{v} \| = \| \mathbf{v}^i \| = \| \mathbf{v}^o \| = \| \mathbf{v}^b \| . \tag{8.16}$$

8.2   Attitude Parameters

8.2.1   *Direct Cosine Matrix*

Suppose we know components in body frame and we want to know components in inertial frame:

$$\mathbf{v} = v_1^i \hat{i}_1 + v_2^i \hat{i}_2 + v_3^i \hat{i}_3 = v_1^b \hat{b}_1 + v_2^b \hat{b}_2 + v_3^b \hat{b}_3. \tag{8.17}$$

Now we can find v_1^i by taking the dot product between $\mathbf{v}$ and $\hat{i}_1$, or:

$$\mathbf{v}_1^i = \hat{i}_1 \cdot \mathbf{v} = v_1^b(\hat{i}_1 \cdot \hat{b}_1) + v_2^b(\hat{i}_1 \cdot \hat{b}_2) + v_3^b(\hat{i}_1 \cdot \hat{b}_3). \tag{8.18}$$

Similarly, v_2^i and v_3^i can be found:

$$\mathbf{v}_2^i = \hat{i}_2 \cdot \mathbf{v} = v_1^b(\hat{i}_2 \cdot \hat{b}_1) + v_2^b(\hat{i}_2 \cdot \hat{b}_2) + v_3^b(\hat{i}_2 \cdot \hat{b}_3) \tag{8.19}$$

$$\mathbf{v}_3^i = \hat{i}_3 \cdot \mathbf{v} = v_1^b(\hat{i}_3 \cdot \hat{b}_1) + v_2^b(\hat{i}_3 \cdot \hat{b}_2) + v_3^b(\hat{i}_3 \cdot \hat{b}_3). \tag{8.20}$$

Notice that the terms in the parentheses are not dependent on the vector components of $\mathbf{v}$, just on the orientation between frames $\mathcal{F}_i$ and $\mathcal{F}_b$.

This means that the matrix representation for $\mathbf{v}$ in $\mathcal{F}_i$ is linearly related to the matrix representation for $\mathbf{v}$ in $\mathcal{F}_b$:

$$\begin{bmatrix} v_1^i \\ v_2^i \\ v_3^i \end{bmatrix} = \begin{bmatrix} \hat{i}_1 \cdot \hat{b}_1 & \hat{i}_1 \cdot \hat{b}_2 & \hat{i}_1 \cdot \hat{b}_3 \\ \hat{i}_2 \cdot \hat{b}_1 & \hat{i}_2 \cdot \hat{b}_2 & \hat{i}_2 \cdot \hat{b}_3 \\ \hat{i}_3 \cdot \hat{b}_1 & \hat{i}_3 \cdot \hat{b}_2 & \hat{i}_3 \cdot \hat{b}_3 \end{bmatrix} \begin{bmatrix} v_1^b \\ v_2^b \\ v_3^b \end{bmatrix} = \begin{bmatrix} r_{11} & r_{12} & r_{13} \\ r_{21} & r_{22} & r_{23} \\ r_{31} & r_{32} & r_{33} \end{bmatrix} \begin{bmatrix} v_1^b \\ v_2^b \\ v_3^b \end{bmatrix} \tag{8.21}$$

$$\Rightarrow [v]^i = \mathbf{R}^{ib} [v]^b . \tag{8.22}$$

See Figure 8.3. The equation $[v]^i = \mathbf{R}^{ib} [v]^b$ is a linear system of the form $\mathbf{Ax} = \mathbf{b}$. Thus, to determine the components in the inertial frame, we need to determine $\mathbf{R}^{ib}$ and solve the linear system. The components of $\mathbf{R}^{ib}$ are:

$$r_{jk}^{ib} = \hat{i}_j \cdot \hat{b}_k = \cos(\hat{i}_k, \hat{b}_j) \tag{8.23}$$

Thus the components of $\mathbf{R}$ are the direction cosines and the rotation matrix is also known as the Direction Cosine Matrix (DCM)

Note also that

$$r_{jk}^{ib} = \hat{i}_j \cdot \hat{b}_k = r_{kj}^{bi} \quad \text{or} \quad \mathbf{R}^{ib} = (\mathbf{R}^{bi})^T. \tag{8.24}$$

Also

$$[v]^i = \mathbf{R}^{ib} [v]^b , \quad [v]^b = \mathbf{R}^{bi} [v]^i , \quad [v]^b = \mathbf{R}^{bi}\mathbf{R}^{ib} [v]^b , \tag{8.25}$$

$$\mathbf{R}^{bi}\mathbf{R}^{ib} = 1, \quad \mathbf{R}^{bi} = \mathbf{R}^{ib-1} . \tag{8.26}$$

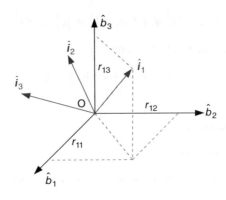

Figure 8.3   Direct cosine matrix between inertial $\mathcal{F}_i$ and body frame $\mathcal{F}_b$

The inverse of a rotation matrix is simply its transpose:

$$\mathbf{R}^{ib} = \mathbf{R}^{biT} = \mathbf{R}^{bi-1}. \tag{8.27}$$

A rotation matrix is an "orthonormal" matrix: its rows and columns are components of mutually orthogonal unit vectors.

Rotation about the $\hat{i}_3$ axis:

$$r_{jk}^{bi} = \hat{b}_j \cdot \hat{i}_k = \cos(\hat{b}_j, \hat{i}_k) \tag{8.28}$$

$$r_{11}^{bi} = \hat{b}_1 \cdot \hat{i}_1 = \cos(\hat{b}_1, \hat{i}_1) = \cos\theta \tag{8.29}$$

$$r_{12}^{bi} = \hat{b}_1 \cdot \hat{i}_2 = \cos(\hat{b}_1, \hat{i}_2) = \sin\theta \tag{8.30}$$

$$r_{21}^{bi} = \hat{b}_2 \cdot \hat{i}_1 = \cos(\hat{b}_2, \hat{i}_1) = -\sin\theta \tag{8.31}$$

$$r_{22}^{bi} = \hat{b}_2 \cdot \hat{i}_2 = \cos(\hat{b}_2, \hat{i}_2) = \cos\theta \tag{8.32}$$

$$r_{33}^{bi} = \hat{b}_3 \cdot \hat{i}_3 = \cos(\hat{b}_3, \hat{i}_3) = \cos(0) = 1 \tag{8.33}$$

$$r_{13}^{bi} = \hat{b}_1 \cdot \hat{i}_3 = \cos(\hat{b}_1, \hat{i}_3) = \cos(90°) = 0. \tag{8.34}$$

The rotation matrix $\mathbf{R}$ represents the attitude. It has 9 numbers, but they are not independent. There are 6 constraints on the 9 elements of a rotation matrix.

$$\mathbf{R}_i \cdot \mathbf{R}_j = \begin{cases} 1 & \text{if } i=j \\ 0 & \text{if } i \neq j \end{cases}. \tag{8.35}$$

$\mathbf{R}_i, \mathbf{R}_j$ is a row or column of the rotation matrix. This rotation has 3 degrees of freedom. There are many different sets of parameters that can be used to represent or parameterize rotations, including Euler angles, Euler parameters (aka quaternions), Rodrigues parameters (aka Gibbs vectors), modified Rodrigues parameters.

8.2.2   Euler Angles

Leonhard Euler(1707–1783) reasoned that the rotation from one frame to another can be visualized as a sequence of three simple rotations about base vectors.

Each rotation is through an angle (Euler angle) about a specified axis. Let's consider the rotation from $\mathcal{F}_i$ to $\mathcal{F}_b$ using three Euler angles, θ_1, θ_2, and θ_3.

The first rotation is about the axis $\hat{i}_3$, through angle θ_1. The resulting frame is denoted $\mathcal{F}_{i'}$ or $\{i'\}$. See Figure 8.4.

$$\mathbf{R}^{i'i} = \mathbf{R}_3(\theta_1) = \begin{bmatrix} \cos\theta_1 & \sin\theta_1 & 0 \\ -\sin\theta_1 & \cos\theta_1 & 0 \\ 0 & 0 & 1 \end{bmatrix} \tag{8.36}$$

$$\Rightarrow \mathbf{v}_{i'} = \mathbf{R}_3(\theta_1)\mathbf{v}_i. \tag{8.37}$$

The second rotation is about the axis $\hat{i}'_2$, through angle θ_2. The resulting frame is denoted $\mathcal{F}_{i''}$ or $\{i''\}$.

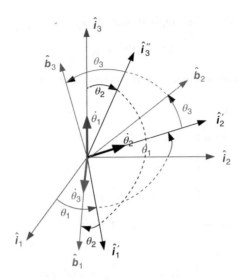

Figure 8.4   Euler angles and angle velocity for 3-2-1 sequence

The rotation matrix notation for the "simple" rotations is $\mathbf{R}_i(\theta_j)$ denotes a rotation about the i-axis. The subscript on $\mathbf{R}$ defines which simple rotation axis is used, and the subscript j on θ defines which of the three angles in the Euler sequence it is:

$$\mathbf{R}^{i''i'} = \mathbf{R}_2(\theta_2) = \begin{bmatrix} \cos\theta_2 & 0 & -\sin\theta_2 \\ 0 & 1 & 0 \\ \sin\theta_2 & 0 & \cos\theta_2 \end{bmatrix} \tag{8.38}$$

$$\Rightarrow \mathbf{v}_{i''} = \mathbf{R}_2(\theta_2)\mathbf{v}_{i'} = \mathbf{R}_2(\theta_2)\mathbf{R}_3(\theta_1)\mathbf{v}_i. \tag{8.39}$$

The third rotation is a "1" rotation, through angle θ_3. The resulting frame is the desired body frame, denoted $\mathcal{F}_b$ or $\{\hat{b}\}$.

$$\mathbf{R}^{bi''} = \mathbf{R}_1(\theta_3) = \begin{bmatrix} 1 & 0 & 0 \\ 0 & \cos\theta_3 & \sin\theta_3 \\ 0 & -\sin\theta_3 & \cos\theta_3 \end{bmatrix} \tag{8.40}$$

$$\Rightarrow \mathbf{v}_b = \mathbf{R}_1(\theta_3)\mathbf{v}_{i''} = \mathbf{R}_1(\theta_3)\mathbf{R}_2(\theta_2)\mathbf{R}_3(\theta_1)\mathbf{v}_i. \tag{8.41}$$

We have performed a 3-2-1 rotation from $\mathcal{F}_i$ to $\mathcal{F}_{i''}$

$$\mathbf{R}^{bi} = \mathbf{R}_1(\theta_3)\mathbf{R}_2(\theta_2)\mathbf{R}_3(\theta_1) \tag{8.42}$$

$$= \begin{bmatrix} 1 & 0 & 0 \\ 0 & c\theta_3 & s\theta_3 \\ 0 & -s\theta_3 & c\theta_3 \end{bmatrix} \begin{bmatrix} c\theta_2 & 0 & -s\theta_2 \\ 0 & 1 & 0 \\ s\theta_2 & 0 & c\theta_2 \end{bmatrix} \begin{bmatrix} c\theta_1 & s\theta_1 & 0 \\ -s\theta_1 & c\theta_1 & 0 \\ 0 & 0 & 1 \end{bmatrix} \tag{8.43}$$

$$= \begin{bmatrix} c\theta_1 c\theta_2 & s\theta_1 c\theta_2 & -s\theta_2 \\ -s\theta_1 c\theta_3 + c\theta_1 s\theta_2 s\theta_3 & c\theta_1 c\theta_3 + s\theta_1 s\theta_2 s\theta_3 & c\theta_2 s\theta_3 \\ s\theta_1 s\theta_3 + c\theta_1 s\theta_2 c\theta_3 & s\theta_1 s\theta_2 c\theta_3 - c\theta_1 s\theta_3 & c\theta_2 c\theta_3 \end{bmatrix}. \tag{8.44}$$

We can select arbitrary values of the three angles and compute a rotation matrix.
For example, let $\theta_1 = \pi/6, \theta_2 = \pi/4$, and $\theta_3 = \pi/3$, then

$$\mathbf{R} = \begin{bmatrix} 0.6124 & 0.3536 & -0.7071 \\ 0.2803 & 0.7392 & 0.6124 \\ 0.7392 & -0.5732 & 0.3536 \end{bmatrix}. \tag{8.45}$$

Conversely, given a rotation matrix, we can extract the Euler angles. Choose the easy one first:

$$R_{13} : -\sin\theta_2 = -0.7071 \Rightarrow \theta_2 = \sin^{-1} 0.7071 = 0.7854 \approx \pi/4 \tag{8.46}$$

$$R_{11} : \cos\theta_1 \cos\theta_2 = 0.6124 \rightarrow \theta_1 - 0.5236 \approx \pi/6 \tag{8.47}$$

$$R_{23} : \cos\theta_2 \sin\theta_3 = 0.6124 \Rightarrow \theta_3 = 1.0472 \approx \pi/3. \tag{8.48}$$

Quadrant checks are generally necessary for the second and third calculations. For example, as $R_{12} > 0, \cos\theta_2 > 0$, then $\sin\theta_1 > 0$

We have just developed a (3-2-1) rotation, but there are other possibilities.

There are 3 choices for the first rotation, 2 choices for the second rotation, and 2 choices for the third rotation, so there are $3 \times 2 \times 2 = 12$ possible Euler angle sequences.

The Euler angle rotation sequences are

$$(1\text{-}2\text{-}3)(1\text{-}3\text{-}2)(2\text{-}3\text{-}1)(2\text{-}1\text{-}3)(3\text{-}1\text{-}2)(3\text{-}2\text{-}1)$$
$$(1\text{-}2\text{-}1)(1\text{-}3\text{-}1)(2\text{-}3\text{-}2)(2\text{-}1\text{-}2)(3\text{-}1\text{-}3)(3\text{-}2\text{-}3).$$

The sequence in the first row is sometimes called the asymmetric rotation sequence, and the sequence in the second row is called the symmetric sequence.

The roll-pitch-yaw sequence is an asymmetric sequence (3-2-1), whereas the $\Omega - i - \omega$ sequence (3-1-3) is a symmetric sequence.

In the 3-2-1 sequence (which we used earlier), θ_1 is the yaw angle ψ, θ_2 is the pitch angle θ, and θ_3 is the roll angle ϕ.

The 1-2-3 sequence leads to

$$\mathbf{R}^{bi} = \mathbf{R}_3(\theta_3)\mathbf{R}_2(\theta_2)\mathbf{R}_1(\theta_1) \tag{8.49}$$

$$= \begin{bmatrix} c\theta_3 & s\theta_3 & 0 \\ -s\theta_3 & c\theta_3 & 0 \\ 0 & 0 & 1 \end{bmatrix} \begin{bmatrix} c\theta_2 & 0 & -s\theta_2 \\ 0 & 1 & 0 \\ s\theta_2 & 0 & c\theta_2 \end{bmatrix} \begin{bmatrix} 1 & 0 & 0 \\ 0 & c\theta_1 & s\theta_1 \\ 0 & -s\theta_1 & c\theta_1 \end{bmatrix} \tag{8.50}$$

$$= \begin{bmatrix} c\theta_2 c\theta_3 & c\theta_1 s\theta_3 + s\theta_1 s\theta_2 c\theta_3 & s\theta_1 s\theta_3 - c\theta_1 s\theta_2 c\theta_3 \\ -c\theta_2 s\theta_3 & c\theta_1 c\theta_3 - s\theta_1 s\theta_2 s\theta_3 & c\theta_1 s\theta_2 s\theta_3 + s\theta_1 c\theta_3 \\ s\theta_2 & -s\theta_1 c\theta_2 & c\theta_1 c\theta_2 \end{bmatrix}, \tag{8.51}$$

where θ_1 is the roll angle ϕ, θ_2 is the pitch angle θ, and θ_3 is the yaw angle ψ.

Note that the two matrices are not the same, and the rotations do not commute. However, if we assume that the angles are small (appropriate for many vehicle dynamics problems), then the approximations of the two matrices are equal. For small angles, $\cos\theta \approx 1$ and $\sin\theta = \theta$.

For 3-2-1 sequence,

$$\mathbf{R}^{bi} \approx \begin{bmatrix} 1 & \theta_1 & -\theta_2 \\ -\theta_1 & 1 & \theta_3 \\ \theta_2 & -\theta_3 & 1 \end{bmatrix}. \tag{8.52}$$

In the 3-2-1 sequence (which we used earlier), θ_1 is the yaw angle, θ_2 is the pitch angle, and θ_3 is the roll angle.

For 1-2-3 sequence,

$$\mathbf{R}^{bi} \approx \begin{bmatrix} 1 & \theta_3 & -\theta_2 \\ -\theta_3 & 1 & \theta_1 \\ \theta_2 & -\theta_1 & 1 \end{bmatrix}. \tag{8.53}$$

In the 1-2-3 sequence (which we used earlier), θ_1 is the roll angle, θ_2 is the pitch angle, and θ_3 is the yaw angle. For all sequence,

$$\mathbf{R}^{bi} \approx 1 - \boldsymbol{\theta}^{\times} \Rightarrow \mathbf{R}^{bi} = 1 - \begin{bmatrix} \text{roll} \\ \text{pitch} \\ \text{yaw} \end{bmatrix}^{\times}. \tag{8.54}$$

In summary, the subsequent transformations are important:

1. Given a sequence, say (3-1-2) for example, derive the Euler angle representation of $\mathbf{R}$. Be sure to get the order correct.
2. Given a sequence and some values for the angles, compute the numerical values of $\mathbf{R}$. Be sure to know the difference between degrees and radians.
3. Given the numerical values of R, extract numerical values of the Euler angles associated with a specified sequence. Be sure to make appropriate quadrant checks, and to check your answer.

8.3   Differential Equations of Kinematics

Given the velocity of a point and initial conditions for its position, we can compute its position as a function of time by integrating the differential equation $\dot{\mathbf{r}} = \mathbf{v}^b \equiv \mathbf{v}_b$.

We now need to develop the equivalent differential equations for the attitude when the angular velocity is known.

We use one frame-to-frame at a time, just as we did for developing rotation matrices. See Figure 8.4. (3-2-1) rotation from $\mathcal{F}_i$ to $\mathcal{F}_{i'}$ to $\mathcal{F}_{i''}$ to $\mathcal{F}_b$. 3-rotation from $\mathcal{F}_i$ to $\mathcal{F}_{i'}$ about $\hat{i}_3 \equiv \hat{i}'_3$ through θ_1. The angular velocity of $\mathcal{F}_{i'}$ with respect to $\mathcal{F}_i$ is

$$\boldsymbol{\omega}^{i'i} = \dot{\theta}_1 \hat{i}_3 = \dot{\theta}_1 \hat{i}'_3. \tag{8.55}$$

We can express $\boldsymbol{\omega}^{i'i}$ in any frame, but $\mathcal{F}_i$ and $\mathcal{F}_{i'}$ are especially simple:

$$\boldsymbol{\omega}^{i'i}_i = [0 \ 0 \ \dot{\theta}_1]^T \tag{8.56}$$

$$\boldsymbol{\omega}^{i'i}_{i'} = [0 \ 0 \ \dot{\theta}_1]^T. \tag{8.57}$$

Keep the notation in mind: $\boldsymbol{\omega}^{i'i}_i$ is the angular velocity of $\mathcal{F}_{i'}$ with respect to $\mathcal{F}_i$, expressed in $\mathcal{F}_i$.

2-rotation from $\mathcal{F}_{i'}$ to $\mathcal{F}_{i''}$ about $\hat{i}'_2 \equiv \hat{i}''_2$ through θ_2

The angular velocity of $\mathcal{F}_{i''}$ with respect to $\mathcal{F}_{i'}$ is

$$\boldsymbol{\omega}^{i''i'} = \dot{\theta}_2 \hat{i}'_2 = \dot{\theta}_2 \hat{i}''_2. \tag{8.58}$$

We can express $\boldsymbol{\omega}^{i''i'}$ in any frame, but $\mathcal{F}_{i'}$ and $\mathcal{F}_{i''}$ are especially simple:

$$\boldsymbol{\omega}^{i''i'}_{i'} = [0 \ \dot{\theta}_2 \ 0]^T \tag{8.59}$$

$$\boldsymbol{\omega}^{i''i'}_{i''} = [0 \ \dot{\theta}_2 \ 0]^T. \tag{8.60}$$

Keep the notation in mind: $\boldsymbol{\omega}^{i''i'}_{i'}$ is the angular velocity of $\mathcal{F}_{i''}$ with respect to $\mathcal{F}_{i'}$, expressed in $\mathcal{F}_{i'}$.

1-rotation from $\mathcal{F}_{i''}$ to $\mathcal{F}_b$ about $\hat{i}''_1 \equiv \hat{b}_1$ through θ_3

The angular velocity of $\mathcal{F}_{ib}$ with respect to $\mathcal{F}_{i''}$ is

$$\boldsymbol{\omega}^{bi''} = \dot{\theta}_3 \hat{i}''_1 = \dot{\theta}_3 \hat{\mathbf{b}}_1. \tag{8.61}$$

We can express $\boldsymbol{\omega}^{bi''}$ in any frame, but $\mathcal{F}_{i''}$ and $\mathcal{F}_b$ are especially simple:

$$\boldsymbol{\omega}^{bi''}_{i''} = [\dot{\theta}_3 \ 0 \ 0]^T \tag{8.62}$$

$$\boldsymbol{\omega}^{bi''}_{b} = [\dot{\theta}_3 \ 0 \ 0]^T. \tag{8.63}$$

Keep the notation in mind: $\boldsymbol{\omega}^{bi''}_{b}$ is the angular velocity of $\mathcal{F}_b$ with respect to $\mathcal{F}_{i''}$, expressed in $\mathcal{F}_b$.

Angular velocities are vectors and add like vectors:

$$\boldsymbol{\omega}^{bi} = \boldsymbol{\omega}^{bi''} + \boldsymbol{\omega}^{i''i'} + \boldsymbol{\omega}^{i'i}. \tag{8.64}$$

We have expressed these three vectors in different frames; to add them together, we need to express all of them in the same frame.

Typically, we want $\boldsymbol{\omega}^{bi}$, so we need to rotate the 3×1 matrices we have just developed into $\mathcal{F}_b$. We have:

$$\boldsymbol{\omega}^{i'i}_{i} = \boldsymbol{\omega}^{i'i}_{i'} = [0 \ 0 \ \dot{\theta}_1]^T \Rightarrow \text{need } \mathbf{R}^{bi} = \mathbf{R}_1(\theta_3)\mathbf{R}_2(\theta_2)\mathbf{R}_3(\theta_1) \tag{8.65}$$

$$\boldsymbol{\omega}^{i''i'}_{i'} = \boldsymbol{\omega}^{i''i'}_{i''} = [0 \ \dot{\theta}_2 \ 0]^T \Rightarrow \text{need } \mathbf{R}^{bi'} = \mathbf{R}_1(\theta_3)\mathbf{R}_2(\theta_2) \tag{8.66}$$

$$\boldsymbol{\omega}^{bi''}_{i''} = \boldsymbol{\omega}^{bi''}_{b} = [\dot{\theta}_3 \ 0 \ 0]^T \Rightarrow \text{need } \mathbf{R}^{bi''} = \mathbf{R}_1(\theta_3). \tag{8.67}$$

We previously developed all these rotation matrices, so we just need to apply them and add the results.

Carry out the matrix multiplications and additions. Then obtain

$$\boldsymbol{\omega}^{bi} = \mathbf{R}^{bi''} \boldsymbol{\omega}^{bi''}_b + \mathbf{R}^{bi'} \boldsymbol{\omega}^{i''i'}_{i''} + \mathbf{R}^{bi} \boldsymbol{\omega}^{i'i}_{i} \tag{8.68}$$

$$= \begin{bmatrix} \dot{\theta}_3 - \sin\theta_2\dot{\theta}_1 \\ \cos\theta_3\dot{\theta}_2 + \cos\theta_2\sin\theta_3\dot{\theta}_1 \\ -\sin\theta_3\dot{\theta}_2 + \cos\theta_2\cos\theta_3\dot{\theta}_1 \end{bmatrix} \tag{8.69}$$

$$= \begin{bmatrix} -\sin\theta_2 & 0 & 1 \\ \cos\theta_2\sin\theta_3 & \cos\theta_3 & 0 \\ \cos\theta_2\cos\theta_3 & -\sin\theta_3 & 0 \end{bmatrix} \begin{bmatrix} \dot{\theta}_1 \\ \dot{\theta}_2 \\ \dot{\theta}_3 \end{bmatrix} \tag{8.70}$$

$$= \mathbf{S}(\boldsymbol{\theta})\dot{\boldsymbol{\theta}} \tag{8.71}$$

or

$$\dot{\boldsymbol{\theta}} = \mathbf{S}^{-1}(\boldsymbol{\theta})\boldsymbol{\omega} = \begin{bmatrix} 0 & \sin\theta_3/\cos\theta_2 & \cos\theta_3/\cos\theta_2 \\ 0 & \cos\theta_3 & -\sin\theta_3 \\ 1 & \sin\theta_3\sin\theta_2/\cos\theta_2 & \cos\theta_3\sin\theta_2/\cos\theta_2 \end{bmatrix}\begin{bmatrix} \dot{\omega}_1 \\ \dot{\omega}_2 \\ \dot{\omega}_3 \end{bmatrix}. \tag{8.72}$$

For this Euler angle set (3-2-1), the Euler rates go to infinity when $\cos\theta_2 \to 0$. The reason is that near $\theta_2 = \pi/2$ the first and third rotations are indistinguishable.

For the "symmetric" Euler angle sequences (3-1-3, 2-1-2, 1-3-1, etc.) the singularity occurs when $\theta_2 = 0$ or π.

For the "asymmetric" Euler angle sequences (3-2-1, 2-3-1, 1-3-2, etc.) the singularity occurs when $\theta_2 = \pi/2$ or $3\pi/2$.

This kinematic singularity is a major disadvantage of using Euler angles for large-angle motion.

There are attitude representations that do not have a kinematic singularity, but 4 or more scalars are required.

If the Euler angles and rates are small, then $\sin\theta_i \approx \theta_i$ and $\cos\theta_i \approx 1$:

$$\boldsymbol{\omega}_b^{bi} \approx \begin{bmatrix} -\theta_2 & 0 & 1 \\ \theta_3 & 1 & 0 \\ 1 & -\theta_3 & 0 \end{bmatrix}\begin{bmatrix} \dot{\theta}_1 \\ \dot{\theta}_2 \\ \dot{\theta}_3 \end{bmatrix} \approx \begin{bmatrix} \dot{\theta}_3 \\ \dot{\theta}_2 \\ \dot{\theta}_1 \end{bmatrix}. \tag{8.73}$$

Suppose you want to invert the $n \times n$ matrix $\mathbf{A}$, with elements A_{ij} The elements of the inverse are

$$A_{ij}^{-1} = \frac{C_{ij}}{|\mathbf{A}|}, \tag{8.74}$$

where C_{ji} is the cofactor, computed by multiplying the determinant of the $(n-1) \times (n-1)$ minor matrix obtained by deleting the j_{th} row and i_{th} column from $\mathbf{A}$, by $(-1)^{i+j}$.

This formulation is absolutely unsuitable for calculating matrix inverses in numerical work, especially with larger matrices, since it is computationally expensive.

One normally uses LU decomposition instead. Elementary row reduction is essentially LU decomposition.

Note: In most cases, we do not need the inverse anyway; we need the solution to a linear system.

Let us invert the matrix $\mathbf{S}(\boldsymbol{\theta})$ in Equation (8.71):

$$\mathbf{S}(\boldsymbol{\theta}) = \begin{bmatrix} -\sin\theta_2 & 0 & 1 \\ \cos\theta_2\sin\theta_3 & \cos\theta_3 & 0 \\ \cos\theta_2\cos\theta_3 & -\sin\theta_3 & 0 \end{bmatrix}. \tag{8.75}$$

In row reduction, we augment the matrix with the identity matrix:

$$\begin{bmatrix} -s\theta_2 & 0 & 1 & 1 & 0 & 0 \\ c\theta_2 s\theta_3 & c\theta_3 & 0 & 0 & 1 & 0 \\ c\theta_2 c\theta_3 & -s\theta_3 & 0 & 0 & 0 & 1 \end{bmatrix} \tag{8.76}$$

and apply simply row-reduction operations to transform the left 3×3 block to the identity, leaving the inverse in the right 3×3 block.

In this case, we must swap the first row with one of the other two rows, say the 3rd row, which amounts to a permutation by:

$$\mathbf{P} = \begin{bmatrix} 0 & 0 & 1 \\ 0 & 1 & 0 \\ 1 & 0 & 0 \end{bmatrix}. \tag{8.77}$$

Note that $\mathbf{P}^{-1} = \mathbf{P}$.

$$\begin{bmatrix} c\theta_2 c\theta_3 & -s\theta_3 & 0 & 0 & 0 & 1 \\ c\theta_2 s\theta_3 & c\theta_3 & 0 & 0 & 1 & 0 \\ -s\theta_2 & 0 & 1 & 1 & 0 & 0 \end{bmatrix}. \tag{8.78}$$

Multiply row 2 by $\sin\theta_3 / \cos\theta_3$ and add to row 1.
Multiply row 1 by $-\sin\theta_3 \cos\theta_3$ and add to row 2.

$$\begin{bmatrix} c\theta_2 c\theta_3 + c\theta_2 s\theta_3 \tan\theta_3 & 0 & 0 & 0 & \tan\theta_3 & 1 \\ 0 & c\theta_3 & 0 & 0 & 1 - \sin^2\theta_3 & -\sin\theta_3 \cos\theta_3 \\ -s\theta_2 & 0 & 1 & 1 & 0 & 0 \end{bmatrix}. \tag{8.79}$$

Divide row 1 by $c\theta_2 c\theta_3 + c\theta_2 s\theta_3 \tan\theta_3 = \cos\theta_2 / \cos\theta_3$ and simplify.
Multiply resulting row 1 by $s\theta_2$, add to row 3, and simplify.
Divide row 2 by $c\theta_3$ and simplify

$$\begin{bmatrix} 1 & 0 & 0 & 0 & s\theta_3/c\theta_2 & c\theta_3/c\theta_2 \\ 0 & 1 & 0 & 0 & c\theta_3 & -s\theta_3 \\ 0 & 0 & 1 & 1 & s\theta_3 \tan\theta_2 & c\theta_3 \tan\theta_2 \end{bmatrix}. \tag{8.80}$$

8.3.1 Typical Problem Involving Angular Velocity and Attitude

Given initial conditions for the attitude (in any form), and a time history of angular velocity, compute $\mathbf{R}$ or any other attitude representation as a function of time.

This requires integration of one of the sets of differential equations involving angular velocity.

Use a rotation from $\mathcal{F}_i$ to $\mathcal{F}_b$ using two Euler angles, α, and $\pi/2 - \delta$. See Figure 8.5. The first (3) rotation is about the $\hat{i}_3$ axis, through angle α.

The second (2) rotation is about the $\hat{E}$ axis through $\pi/2 - \delta$. The three unit vectors $\hat{S}\hat{E}\hat{Z}$ (see Chapter 1) have derivatives:

$$\dot{\hat{S}} = \dot{\delta}\,\hat{Z} + \sin\delta\,\dot{\alpha}\,\hat{E} \tag{8.81}$$

$$\dot{\hat{E}} = -\cos\delta\,\dot{\alpha}\,\hat{Z} - \sin\delta\,\dot{\alpha}\,\hat{S} \tag{8.82}$$

$$\dot{\hat{Z}} = \cos\delta\,\dot{\alpha}\,\hat{E} - \dot{\delta}\,\hat{S}. \tag{8.83}$$

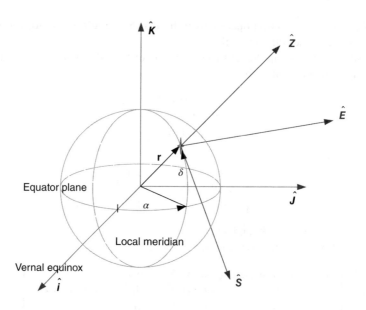

Figure 8.5   Rotation from $\hat{I}\hat{J}\hat{K} \rightarrow \hat{S}\hat{E}\hat{Z}$

8.4   Euler's Theorem

The most general motion of a rigid body with a fixed point is a rotation about a fixed axis.

The axis, denoted $\mathbf{a}$ or $\hat{\mathbf{a}}$ is called the eigenaxis or Euler axis. The angle of rotation Φ is called the Euler angle or the principal Euler angle. See Figure 8.6.

$$\mathbf{a}^b = \mathbf{R}^{bi}\mathbf{a}^i = \mathbf{a}^i = [a_1, a_2, a_3]^T \tag{8.84}$$

$$\mathbf{R}^{bi} = \cos\Phi\mathbf{1} + (1 - \cos\Phi)\mathbf{a}\mathbf{a}^T - \sin\Phi\mathbf{a}^\times \tag{8.85}$$

$$\mathbf{a}^\times = \begin{bmatrix} 0 & -a_3 & a_2 \\ a_3 & 0 & -a_1 \\ -a_2 & a_1 & 0 \end{bmatrix}. \tag{8.86}$$

Let us see some observations regarding $\mathbf{a}$ and Φ.

Since $\mathbf{R}^{bi}\mathbf{a} = \mathbf{a}$, the Euler axis is the eigenvector of $\mathbf{R}$ associated with the eigenvalue 1. We can check this result:

$$\mathbf{R}\mathbf{a} = \left[\cos\Phi\mathbf{1} + (1 - \cos\Phi)\mathbf{a}\mathbf{a}^T - \sin\Phi\mathbf{a}^\times\right]\mathbf{a} \tag{8.87}$$

$$= \cos\Phi\mathbf{1}\mathbf{a} + (1 - \cos\Phi)\mathbf{a}\mathbf{a}^T\mathbf{a} - \sin\Phi\mathbf{a}^\times\mathbf{a} \tag{8.88}$$

$$= \cos\Phi\mathbf{a} + \mathbf{a}\mathbf{a}^T\mathbf{a} - \cos\Phi\mathbf{a}\mathbf{a}^T\mathbf{a} - \sin\Phi\mathbf{a}^\times\mathbf{a} \tag{8.89}$$

$$= \cos\Phi\mathbf{a} + \mathbf{a} - \cos\Phi\mathbf{a} - \sin\Phi\mathbf{a}^\times\mathbf{a}(\mathbf{a}^T\mathbf{a} = 1) \tag{8.90}$$

$$= \cos\Phi\mathbf{a} + \mathbf{a} - \cos\Phi\mathbf{a}(\mathbf{a}^\times\mathbf{a} = 0) \tag{8.91}$$

$$= \mathbf{a}. \tag{8.92}$$

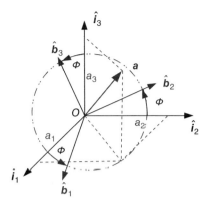

Figure 8.6   The geometry of Euler theorem

Thus every rotation matrix has an eigenvalue that is equal to $+1$. This fact justifies the term eigenaxis for the Euler axis. This parameterization requires four parameters.

We can extract $\mathbf{a}$ and Φ from $\mathbf{R}$. Just as we need to be able to compute Euler angles from a given rotation matrix, we need to be able to compute the Euler axis and Euler angle:

$$\Phi = \cos^{-1}\left[\frac{1}{2}(\text{trace }\mathbf{R} - 1)\right] \tag{8.93}$$

$$\mathbf{a}^\times = \frac{1}{2\sin\Phi}\left(\mathbf{R}^T - \mathbf{R}\right). \tag{8.94}$$

Again for the example in Section 8.2.2, its $\Phi = 1.2105$, $\mathbf{a} = [0.6335, 0.7728, 0.0391]^T$.

One can show that the kinematics differential equations for $\mathbf{a}$ and Φ are:

$$\dot{\Phi} = \mathbf{a}^T\boldsymbol{\omega} \tag{8.95}$$

$$\dot{\mathbf{a}} = \frac{1}{2}\left[\mathbf{a}^\times - \coth(\Phi/2)\mathbf{a}^\times\mathbf{a}^\times\right]. \tag{8.96}$$

So this system of equations also has kinematic singularities, at $\Phi = 0$ and $\Phi = 2\pi$.

8.4.1   Another Four-Parameter Set

The Euler parameter set, also known as a quaternion, is a four-parameter set with some advantages over the Euler axis/angle set:

$$\mathbf{q} = \mathbf{a}\sin\frac{\Phi}{2} \tag{8.97}$$

$$q_4 = \cos\frac{\Phi}{2}. \tag{8.98}$$

The vector component, $\mathbf{q}$, is a 3×1, whereas the scalar component, q_4, is, well, a scalar. The quaternion is denoted by $\bar{\mathbf{q}} = [\mathbf{q}^T, q_4]^T$, a 4×1 matrix.

To compute the rotation matrix using the quaternion:

$$\mathbf{R} = (q_4^2 - \mathbf{q}^T\mathbf{q})\,\mathbf{1} + 2\mathbf{q}\mathbf{q}^T - 2q_4\mathbf{q}^\times \tag{8.99}$$

$$= \begin{bmatrix} q_1^2 - q_2^2 - q_3^2 + q_4^2 & 2(q_1q_2 + q_3q_4) & 2(q_1q_3 - q_2q_4) \\ 2(q_1q_2 - q_3q_4) & -q_1^2 + q_2^2 - q_3^2 + q_4^2 & 2(q_2q_3 + q_1q_4) \\ 2(q_1q_3 + q_2q_4) & 2(q_2q_3 - q_1q_4) & -q_1^2 - q_2^2 + q_3^2 + q_4^2 \end{bmatrix}. \tag{8.100}$$

To compute the quaternion using the rotation:

$$q_4 = \pm\frac{1}{2}\sqrt{1 + \text{trace}\,\mathbf{R}} \tag{8.101}$$

$$\mathbf{q} = \frac{1}{4q_4}\begin{bmatrix} r_{23} - r_{32} \\ r_{31} - r_{13} \\ r_{12} - r_{21} \end{bmatrix}. \tag{8.102}$$

Again for the example in Section 8.2.2, $\bar{\mathbf{q}} = [0.3604, 0.4397, 0.0223, 0.8224]^T$.
As with Euler angles, we are frequently interested in small attitude motions.
If Φ is small, then $\mathbf{q} = \mathbf{a}\sin(\Phi/2) \approx \mathbf{a}\Phi/2$, and $q_4 \approx 1$.
Hence for small Φ:

$$\mathbf{R} = (q_4^2 - \mathbf{q}^T\mathbf{q})\mathbf{1} + 2\mathbf{q}\mathbf{q}^T - 2q_4\mathbf{q}^\times \tag{8.103}$$

$$\approx (1 - 0)\mathbf{1} + 2(0) - 2\mathbf{q}^\times \tag{8.104}$$

$$\approx 1 - 2\mathbf{q}^\times. \tag{8.105}$$

Compare this expression with the previously developed $\mathbf{R} \approx \mathbf{1} - \boldsymbol{\theta}^\times$ for Euler angles. We see for 3-2-1 rotation, $q_1 = \theta_3/2$, $q_2 = \theta_2/2$, $q_3 = \theta_1/2$,
Small rotations are commutative:

$$\mathbf{R}^{cb}\mathbf{R}^{ba} \approx [\mathbf{1} - 2\mathbf{q}_2^\times][\mathbf{1} - 2\mathbf{q}_1^\times] \approx \mathbf{1} - 2\mathbf{q}_2^\times - 2\mathbf{q}_1^\times. \tag{8.106}$$

Differential equations $\dot{\bar{\mathbf{q}}}$:

$$\dot{\bar{\mathbf{q}}} = \frac{1}{2}\begin{bmatrix} \mathbf{q}^\times + q_4\mathbf{1} \\ -\mathbf{q}^T \end{bmatrix}\boldsymbol{\omega} = \mathbf{Q}(\bar{\mathbf{q}})\boldsymbol{\omega}. \tag{8.107}$$

Note that there is no kinematic singularity with these differential equations.

8.4.2 Summary and Extension of Kinematics Notation

Several equivalent methods of describing attitude or orientation are: rotation matrix, denoted as DCM, which is the vectors of one frame expressed in the other; it is also the dot products of vectors of one frame with those of the other.

Euler angles have 12 different sets, so getting the order right is required! Euler axis/angle are unit vector and angle. Euler parameters = quaternions: unit 4 ×1.

It is important to be able to compute one from the other for any given representation. There are other attitude representations.

We have seen direction cosines, Euler angles, Euler angle/axis, and quaternions. Two other common representations are: Euler-Rodriguez parameters, denoted as $\mathbf{p}$:

$$\mathbf{p} = \mathbf{a} \tan \frac{\Phi}{2} \tag{8.108}$$

$$\mathbf{R} = \mathbf{1} + \frac{2}{1 + \mathbf{p}^T \mathbf{p}} (\mathbf{p}^\times \mathbf{p}^\times - \mathbf{p}^\times) \tag{8.109}$$

$$\bar{\mathbf{q}} = \frac{1}{\sqrt{1 + \mathbf{p}^T \mathbf{p}}} \begin{bmatrix} \mathbf{p} \\ 1 \end{bmatrix} \tag{8.110}$$

$$\dot{\mathbf{p}} = \frac{1}{2} (\mathbf{p} \mathbf{p}^T + \mathbf{1} + \mathbf{p}^\times) \boldsymbol{\omega}. \tag{8.111}$$

And Modified Rodriguez parameters, denoted as $\boldsymbol{\sigma}$:

$$\boldsymbol{\sigma} = \mathbf{a} \tan \frac{\Phi}{4} \tag{8.112}$$

$$\mathbf{R} = \frac{1}{1 + \boldsymbol{\sigma}^T \boldsymbol{\sigma}} \left[\left(1 - (\boldsymbol{\sigma}^T \boldsymbol{\sigma})^2\right) \mathbf{1} + 2 \boldsymbol{\sigma} \boldsymbol{\sigma}^T - 2 \left(1 - (\boldsymbol{\sigma}^T \boldsymbol{\sigma})^2\right) \boldsymbol{\sigma}^\times \right] \tag{8.113}$$

$$\dot{\boldsymbol{\sigma}} = \frac{1}{2} \left[\mathbf{1} - \boldsymbol{\sigma}^\times + \boldsymbol{\sigma} \boldsymbol{\sigma}^T - \frac{1 + \boldsymbol{\sigma}^T \boldsymbol{\sigma}}{2} \mathbf{1} \right]. \tag{8.114}$$

Again for the example in Section 8.2.2, $\sigma = [0.1978, 0.2413, 0.0122]^T$.

An Illustrative Example

Let us develop the rotation matrix relating the Earth-centered inertial (ECI), that is, $\hat{I}\hat{J}\hat{K}$ frame $\mathcal{F}_i$ and the orbital frame $\mathcal{F}_o$. We consider the case of an elliptical orbit, with right ascension of the ascending node (or RAAN), Ω, inclination, i, and argument of latitude, $u = \omega + \theta$. Recall that argument of latitude is the angle frame the ascending node to the position of the satellite, and is especially useful for circular orbits, since argument of periapsis, ω, is not defined for circular orbits. See Figure 8.7.

We denote the ECI frame $(\mathcal{F}_i)$ by $\{\hat{i}\}$, and the orbital frame $(\mathcal{F}_o)$ by $\{\hat{o}\}$. Intermediate frames are designated using primes, as the Euler angle development above. We use a "3-1-3" sequence as follows: Begin with a "3" rotation about the inertial $\hat{i}_3$ axis through the RAAN, Ω. This rotation is followed by a "1" rotation about the $\hat{i}'_1$ axis through the inclination, i. The last rotation is another "3" rotation about the $\hat{i}''_3$ axis through the argument of latitude, u.

We denote the resulting reference frame by $\{\hat{o}'\}$, since it is not quite the desired orbital reference frame. Recall that the orbital reference frame for a circular orbit has its three vectors aligned as follows: $\{\hat{o}_1\}$ is in the direction of the orbital velocity vector (the $\mathbf{v}$ direction), $\{\hat{o}_2\}$ is in the direction opposite to the orbit normal (the $-\hat{h}(\hat{W})$ direction), and $\{\hat{o}_3\}$ is the nadir direction (or the $-\hat{r}(-\hat{R})$ direction). However, the frame resulting frame the (3-1-3) rotation developed above has its unit vectors aligned in the $\mathbf{r}, \mathbf{v}$ and $\mathbf{h}$ directions, respectively.

Now it is possible to go back and choose angles so that the (3-1-3) rotation gives the desired orbital frame; however, it is instructive to see how to use two more rotations to get frame the $\{\hat{o}'\}$ frame to the $\{\hat{o}\}$. Specifically, if we perform another "3" rotation about $\{\hat{o}'_3\}$ through 90° and a "1" rotation about $\{\hat{o}''_1\}$ through 270°, we arrive at the desired orbital reference

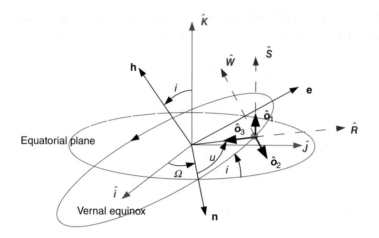

Figure 8.7   From inertial $\mathcal{F}_i$ to orbital frame $\mathcal{F}_o$ and $\hat{R}\hat{S}\hat{W}$ frame

frame. These final two rotations lead to an interesting rotation matrix, that is, from $\hat{R}\hat{S}\hat{W}$ to $\hat{O}_1\hat{O}_2\hat{O}_3(\mathcal{F}_o)$, see Figure 8.7:

$$\mathbf{R}^{oo'} = \mathbf{R}_1(270°)\mathbf{R}_3(90°) \tag{8.115}$$

$$= \begin{bmatrix} 1 & 0 & 0 \\ 0 & 0 & -1 \\ 0 & 1 & 0 \end{bmatrix} \begin{bmatrix} 0 & 1 & 0 \\ -1 & 0 & 0 \\ 0 & 0 & 1 \end{bmatrix} = \begin{bmatrix} 0 & 1 & 0 \\ 0 & 0 & -1 \\ -1 & 0 & 0 \end{bmatrix}. \tag{8.116}$$

Careful study of this rotation matrix reveals that its effect is to move the second row to the first row. Negate the third row and move it to the second row, and negate the first row and move to the third row. So, the rotation matrix that takes vectors from the inertial frame to the orbital frame is

$$\mathbf{R}^{oi} = \mathbf{R}^{oo'}\mathbf{R}_3(u)\mathbf{R}_1(i)\mathbf{R}_3(\Omega), \tag{8.117}$$

which, when expanded, gives

$$\mathbf{R}^{oi} = \begin{bmatrix} -suc\Omega - cucis\Omega & -sus\Omega + cucic\Omega & cusi \\ -sis\Omega & sic\Omega & -ci \\ cuc\Omega + sucis\Omega & -cus\Omega - sucic\Omega & -susi \end{bmatrix}. \tag{8.118}$$

where $u = \omega + \theta$. This rotation matrix is not resulted from attitude motion; it is the rotation to Euler angles, not of the Euler angles. If the transformation is from Inertial Frame to Perifocal Frame, then the true anomaly $\theta = 0$.

Knowing the 9 numbers in the rotation matrix, we can compute the Euler angles.

$$i = \cos^{-1}(-R_{23}), \quad u = \tan^{-1}(-R_{33}/R_{13}), \quad \Omega = \tan^{-1}(-R_{21}/R_{22}) \tag{8.119}$$

Quadrant checks are imperative. It's better to use $\texttt{atan2(y,x)}$ in MATLAB®.

8.5 Attitude Determination

8.5.1 Sensors

Attitude is measured by measuring the attitude of a celestial object that has a known orientation in the inertial coordinate system. One measured vector has only two degrees of freedom. Roll along the axis is undefined. Two measured vectors have four pieces of information, and only three are required for a unique solution. Hence just two independent measurements are needed.

Some items can be measured. The first is the direction from S/C to the sun, which must be out of umbra and the Sun must be in the S/C sun sensor field of view (FOV) also. The second direction that can be measured is the direction to the center of the Earth from S/C. From HEO, such as GEO, the Earth can be assumed a centroid, but for LEO, two points should be considered, i.e. Earth limb. Lighting conditions make the limb difficult to see in visual bands. Earth limb (CO_2) is very visible in infrared (IR).

Another item that can be measured is the direction to a star or set of stars. A single star (star tracker) limits attitude motion as its FOV is small. Multiple stars (star field imager) need a star map and image processing.

The Earth's magnetic field can also be measured. But it varies significantly in orbit and rotates with the Earth. Models of field accuracy vary.

The Moon can also be measured too; it has wide variation in attitude and requires multiple sensors, and/or wide FOV.

Items you can measure can be the angular velocity vector, which does not tell actual attitude, but variations from an attitude can be found by integration methods when it is used in conjunction with other sensors, such as radio beacons.

Narrow band and intensity maps allow ground measurement of spacecraft attitude.

When using differential GPS signals, their accuracy is dependent on relative separation of receiving antennas. And when velocity of ~8 km/s, doppler effects appear, and also satellites in view change rapidly. There are also some legal issues.

Table 8.2 The comparison of sensors adapted from reference [1]

Sensor	Accuracy	Characteristics and applicability
Magneto-meters	1.0° (5000 km alt) 5.0° (200 km alt)	Attitude measured relative to Earth's local magnetic field. Magnetic field uncertainties and variability dominate accuracy. Usable below ~ 6000 km, economical, orbital dependent
Earth-sensor	0.05° (GEO) 0.1° (LEO)	Horizon uncertainties dominate accuracy. Highly accurate units use scanning. orbit dependent, poor in yaw, relatively expensive
Sun sensor	0.01°	FOV: ±30°, simple, reliable, intermittent use, low cost
Star sensor	2″ (arc-sec)	FOV: ±6°, heavy, complex, expensive, very accurate
Gyroscopes	0.001 deg/hr	Normal use involves periodically resetting reference. Rate only, power cost, can be heavy and expensive
Directional antennas	0.01° to 0.5°	Typically 1% of the antenna beamwidth

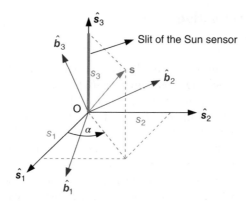

Figure 8.8   The frame of Sun sensor $\mathcal{F}_s$

Let us introduce more about the two most frequently used sensors for S/C.

8.5.1.1 Sun Sensor

See Figure 8.8. It is the frame of Sun sensor $\mathcal{F}_s$. What the Sun sensor measures is the angle α,

$$\alpha = \arctan\left(\frac{s_2}{s_1}\right), \tag{8.120}$$

where $\mathbf{s}_s = [s_1, s_2, s_3]^T$ are the components of the Sun direction in the $\mathcal{F}_s$. If $\mathbf{s}_i = [\cos \delta_s \cos \alpha_s, \cos \delta_s \sin \alpha_s, \sin \delta_s]^T$ is the Sun direction in $\mathcal{F}_i$, δ_s, α_s are the declination and right ascension of the Sun in celestial sphere, and $\mathbf{s}_s$ in $\mathcal{F}_s$, then

$$\mathbf{s}_s = \mathbf{R}^{sb} \mathbf{R}^{bo} \mathbf{R}^{oi} \mathbf{s}_i, \tag{8.121}$$

where $\mathbf{R}^{sb}$ is the fixed rotation matrix from body from $\mathcal{F}_b$ to the Sun sensor frame $\mathcal{F}_s$.

$$\mathbf{R}^{sb} = \begin{bmatrix} r_{ax} & r_{ay} & r_{az} \\ r_{bx} & r_{by} & r_{bz} \\ r_{cx} & r_{cy} & r_{cz} \end{bmatrix}. \tag{8.122}$$

Then component s_1, s_2 in Figure 8.8 can be obtained

$$s_1 = [r_{ax} r_{ay} r_{az}] \begin{bmatrix} 1 & 0 & 0 \\ 0 & c\phi & s\phi \\ 0 & -s\phi & c\phi \end{bmatrix} \begin{bmatrix} c\theta & 0 & -s\theta \\ 0 & 1 & 0 \\ s\theta & 0 & c\theta \end{bmatrix} \begin{bmatrix} c\psi & s\psi & 0 \\ -s\psi & c\psi & 0 \\ 0 & 0 & 1 \end{bmatrix} \begin{bmatrix} s_{ox} \\ s_{oy} \\ s_{oz} \end{bmatrix} \tag{8.123}$$

$$= [r_{ax} r_{ay} r_{az}] \begin{bmatrix} 1 & 0 & 0 \\ 0 & c\phi & s\phi \\ 0 & -s\phi & c\phi \end{bmatrix} \begin{bmatrix} c\theta & 0 & -s\theta \\ 0 & 1 & 0 \\ s\theta & 0 & c\theta \end{bmatrix} \begin{bmatrix} s_{ox} & s_{oy} & 0 \\ s_{oy} & -s_{ox} & 0 \\ 0 & 0 & s_{oz} \end{bmatrix} \begin{bmatrix} c\psi \\ s\psi \\ 1 \end{bmatrix} \tag{8.124}$$

$$s_2 = [r_{bx} r_{by} r_{bz}] \begin{bmatrix} 1 & 0 & 0 \\ 0 & c\phi & s\phi \\ 0 & -s\phi & c\phi \end{bmatrix} \begin{bmatrix} c\theta & 0 & -s\theta \\ 0 & 1 & 0 \\ s\theta & 0 & c\theta \end{bmatrix} \begin{bmatrix} c\psi & s\psi & 0 \\ -s\psi & c\psi & 0 \\ 0 & 0 & 1 \end{bmatrix} \begin{bmatrix} s_{ox} \\ s_{oy} \\ s_{oz} \end{bmatrix} \tag{8.125}$$

$$= [r_{bx} r_{by} r_{bz}] \begin{bmatrix} 1 & 0 & 0 \\ 0 & c\phi & s\phi \\ 0 & -s\phi & c\phi \end{bmatrix} \begin{bmatrix} c\theta & 0 & -s\theta \\ 0 & 1 & 0 \\ s\theta & 0 & c\theta \end{bmatrix} \begin{bmatrix} s_{ox} & s_{oy} & 0 \\ s_{oy} & -s_{ox} & 0 \\ 0 & 0 & s_{oz} \end{bmatrix} \begin{bmatrix} c\psi \\ s\psi \\ 1 \end{bmatrix}, \tag{8.126}$$

where $s_o = [s_{ox}, s_{oy}, s_{oz}]^T = \mathbf{R}^{oi} s_i$, by Equation (8.120),

$$k_1 \cos \psi + k_2 \sin \psi + k_3 = 0, \tag{8.127}$$

where

$$\begin{bmatrix} k_1 \\ k_2 \\ k_3 \end{bmatrix} = \begin{bmatrix} s_{ox} & s_{oy} & 0 \\ s_{oy} & -s_{ox} & 0 \\ 0 & 0 & s_{oz} \end{bmatrix} \begin{bmatrix} c\theta & 0 & s\theta \\ 0 & 1 & 0 \\ -s\theta & 0 & c\theta \end{bmatrix} \begin{bmatrix} 1 & 0 & 0 \\ 0 & c\phi & -s\phi \\ 0 & s\phi & c\phi \end{bmatrix} \tag{8.128}$$

$$\cdot \left\{ \begin{bmatrix} s_{bx} \\ s_{by} \\ s_{bz} \end{bmatrix} - \tan\alpha \begin{bmatrix} s_{ax} \\ s_{ay} \\ s_{az} \end{bmatrix} \right\}, \tag{8.129}$$

then the yaw angle can be solved:

$$\psi = -\arctan \frac{k_3}{\sqrt{k_1^2 + k_2^2}} - \arctan \frac{k_1}{k_2} \tag{8.130}$$

Normally Equation (8.161) is a nonlinear equation. It is difficult to use directly. Its linearized equation of Equation (8.120) is

$$\alpha = \alpha_0 + \frac{\partial \alpha}{\partial \phi}\phi + \frac{\partial \alpha}{\partial \theta}\theta + \frac{\partial \alpha}{\partial \psi}(\psi - \psi_0). \tag{8.131}$$

By proper choice the $\mathbf{R}^{sb}$, such as unit matrix, as calculated by the angle between the Sun's direction and the Earth nadir direction, i.e. $\cos \theta_{se} = r_{33}$ in $\mathbf{R}^{oi}$, λ_s is the angle from x_o to y_o in $\mathcal{F}_o$, the Sun's direction in $\mathcal{F}_o$ is

$$s_o = \begin{bmatrix} \sin \theta_{se} \cos \lambda_s \\ \sin \theta_{se} \sin \lambda_s \\ \cos \theta_{se} \end{bmatrix}. \tag{8.132}$$

Then see Figure 8.8:

$$\alpha = \arctan \frac{-\sin \theta_{se} \sin(\psi_0 - \lambda_s) + \phi \cos \theta_{se} - (\psi - \psi_0) \sin \theta_{se} \cos(\psi_0 - \lambda_s)}{\sin \theta_{se} \cos(\psi_0 - \lambda_s) - \theta \cos \theta_{se} - (\psi - \psi_0) \sin \theta_{se} \sin(\psi_0 - \lambda_s)}. \tag{8.133}$$

At normal state, $\phi, \theta, \psi - \psi_0$ are all small, $\alpha_0 = \psi_0 - \lambda_s$. The linearized equation becomes

$$\alpha = \alpha_0 - \frac{\cos \alpha_0}{\tan \theta_{se}}\phi + \frac{\sin \alpha_0}{\tan \theta_{se}}\theta + \psi - \psi_0. \tag{8.134}$$

The above equation is linear relation between α and ψ when the attitude angles are small.

8.5.1.2 Earth Sensor

A typical Earth sensor is an infrared scanning cone. See Figure 8.9. It is the frame of Earth sensor $\mathcal{F}_s$. In Figure 8.9, points A, M, E, C are all on the celestial sphere. The cone's symmetric axis is OA, which is on the axis $\hat{s}_1$, the Earth nadir is E, the intersection of the cone and the Earth is M, and $AM = \gamma$ is the cone angle of Earth sensor. $AE = \eta$ angle between the Earth nadir direction to the cone's axis. $ME = \rho = \arcsin(R_E/r_{sat})$, $\angle MAE = \mu/2$, $\angle CAE = \lambda$, where $\mu = \omega_{rot}(t_{out} - t_{in})$, $\lambda = \mu/2 - \omega_{rot}(t_{ref} - t_{in})$ Then the Earth nadir sensor frame is:

$$\mathbf{n}_s = \begin{bmatrix} \cos \eta \\ -\sin \eta \sin \lambda \\ \sin \eta \cos \lambda \end{bmatrix} \tag{8.135}$$

$$\cos \rho = \cos \gamma \cos \eta + \sin \gamma \sin \eta \cos \frac{\mu}{2}. \tag{8.136}$$

The nadir point is $[0, 0, 1]^T$

$$\mathbf{n}_s = \begin{bmatrix} \sin \theta \\ -\cos \theta \sin \phi \\ \cos \theta \cos \phi \end{bmatrix}. \tag{8.137}$$

For small $\theta = \pi/2 - \eta$, $\phi = \lambda$.

8.5.2 Attitude Determination

There are many methods to do attitude determination [4]. One well-known method is the triad algorithm, which is a deterministic solution using two measured and known vectors. It assumes that one vector is known to be more precise than another. Some of the information in the problem is neglected.

Assume two known unit vectors in the ECI frame, Sun's direction $\mathbf{s}_i$ and Earth nadir direction $\mathbf{n}_i$. Using Sun sensors and Earth sensors, these vectors in the Body frame can be measured: $\mathbf{s}_b$ for the Sun and $\mathbf{n}_b$ for the Earth's nadir. See Figure 8.10.

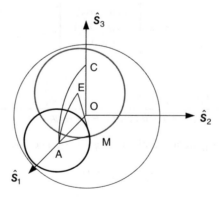

Figure 8.9 The frame of Earth sensor

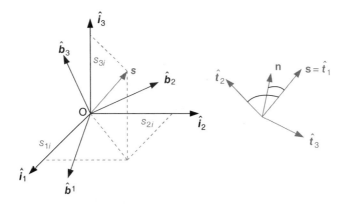

Figure 8.10 The geometry of triad algorithm

Define a new frame t, such that

$$t_1 = s, \quad t_{1i} = s_i, \quad t_{1b} = s_b, \tag{8.138}$$

$$t_2 = \frac{s \times n}{|s \times n|}, \quad t_{2i} = \frac{s_i \times n_i}{|s_i \times n_i|}, \quad t_{2b} = \frac{s_b \times n_b}{|s_b \times n_b|}, \tag{8.139}$$

$$t_3 = t_1 \times t_2, \quad t_{3i} = t_{1i} \times t_{2i}, \quad t_{3b} = t_{1b} \times t_{2b}. \tag{8.140}$$

Note that

$$\begin{bmatrix} t_{1i} \ t_{2i} \ t_{3i} \end{bmatrix} = \mathbf{R}^{it}\mathbf{I} = \mathbf{R}^{it} \tag{8.141}$$

$$\begin{bmatrix} t_{1b} \ t_{2b} \ t_{3b} \end{bmatrix} = \mathbf{R}^{bt}\mathbf{I} = \mathbf{R}^{bt} \tag{8.142}$$

or finally,

$$\mathbf{R}^{bi} = \mathbf{R}^{bt}\mathbf{R}^{ti} = \mathbf{R}^{bt} \left(\mathbf{R}^{it} \right)^{T} \tag{8.143}$$

$$\mathbf{R}^{bi} = \begin{bmatrix} t_{1b} \ t_{2b} \ t_{3b} \end{bmatrix} \begin{bmatrix} t_{1i} \ t_{2i} \ t_{3i} \end{bmatrix}^{T}. \tag{8.144}$$

Example

Suppose a spacecraft has two attitude sensors that provide the following measurements of the two vector v_1 and v_2:

$$v_{1b} = [0.8273, \ 0.5541, \ -0.0920]^{T} \tag{8.145}$$

$$v_{2b} = [-0.8285, \ 0.5522, \ -0.0955]^{T}. \tag{8.146}$$

These vectors have known inertial frame components of

$$v_{1i} = [-0.1517, \ -0.9669, \ 0.02050]^{T} \tag{8.147}$$

$$v_{2i} = [-0.8393, \ 0.4494, \ -0.3044]^{T}. \tag{8.148}$$

Applying the Triad algorithm, we construct the components of the vectors $t_j, j = 1, 2, 3$ in both the body and inertial frames:

$$t_{1b} = [0.8273, \ 0.5541, \ -0.0920]^T \tag{8.149}$$

$$t_{2b} = [-0.0023, \ 0.1671, \ 0.9859]^T \tag{8.150}$$

$$t_{3b} = [0.5617, \ -0.8155, \ 0.1395]^T \tag{8.151}$$

and

$$t_{1i} = [-0.1517, \ -0.9669, \ 0.2050]^T \tag{8.152}$$

$$t_{2i} = [0.2177, \ -0.2350, \ -0.9473]^T \tag{8.153}$$

$$t_{3i} = [0.9641, \ -0.0991, \ 0.2462]^T. \tag{8.154}$$

Using the above algorithm, we obtain the approximate rotation matrix:

$$\mathbf{R}^{bi} = \begin{bmatrix} 0.4156 & -0.8551 & 0.3100 \\ -0.8339 & -0.4943 & -0.2455 \\ 0.3631 & -0.1566 & -0.9185 \end{bmatrix}. \tag{8.155}$$

Then we can have other representations for the attitude:
quaternion: $q = [-0.8408, 0.5023, -0.2002, 0.0264]^T$,
Euler angle/axis: $\phi = 3.0887$, $a = [-0.8411, 0.5025, -0.2002]^T$,
Rodriguez parameters: $p = [-31.8031, 18.9991, -7.5711]^T$,
Modified Rodriguez parameters: $\sigma = [-0.8191, 0.4894, -0.1950]$,
Euler angle for 3-2-1 sequence: $\phi = -14.9614°$, $\theta = -18.0608°$, $\psi = 64.0834°$.

Applying this rotation matrix to v_{1i} gives v_{1b} exactly, because we used this condition in the formulation; however, applying it to v_{2i} does not gives v_{2b} exactly. If we know a priori that sensor 2 is more accurate than sensor 1, then we can use v_2 as the exact measurement, hopefully leading to a more accurate estimate of $\mathbf{R}^{bi}$.

In the above triad algorithm, it is assumed the Sun and the Earth direction are measured directly. And also the satellite is inertial oriented. Practically, most modern satellites are Earth-oriented, the Sun and the Earth directions can not be measured directly.

8.5.2.1 Earth-oriented Satellite

For the sake of clarity, we can redefine the angle in Equation (8.44) of the 3-2-1 sequence which is shown in Figure 8.7. The axis definition is also given at the beginning of Section 8.1. We assume that θ_1 is the yaw angle ψ for $\hat{z}$-axis, θ_2 is the pitch angle θ for $\hat{y}$-axis, θ_3 is roll ϕ for $\hat{x}$-axis, and rotation matrix $\mathbf{R}^{bo}$ is from orbital frame $\mathcal{F}_o$ to body frame $\mathcal{F}_b$, see Equation (8.44). The rotation from inertial form $\mathcal{F}_i$ to orbital frame $\mathcal{F}_o$ is $\mathbf{R}^{oi}$, see Equation (8.118). This time the rotation is from orbital frame $\mathcal{F}_o$ to body frame $\mathcal{F}_b$, still in 3-2-1 sequence.

$$\mathbf{R}^{bo} = \begin{bmatrix} r_{11} & r_{12} & r_{13} \\ r_{21} & r_{22} & r_{23} \\ r_{31} & r_{32} & r_{33} \end{bmatrix} \tag{8.156}$$

$$= \begin{bmatrix} c\psi c\theta & s\psi c\theta & -s\theta \\ -s\psi c\phi + c\psi s\theta s\phi & c\psi c\phi + s\psi s\theta s\phi & c\theta s\phi \\ s\psi s\phi + c\psi s\theta c\phi & s\psi s\theta c\phi - c\psi s\phi & c\theta c\phi \end{bmatrix}. \tag{8.157}$$

If a vector in $\mathcal{F}_i$ is $\mathbf{v}^i$, in $\mathcal{F}_b$ us $\mathbf{v}^b$, then

$$\mathbf{v}^b = \mathbf{R}^{bo}\mathbf{R}^{oi}\mathbf{v}^i. \tag{8.158}$$

If the orbit is a circular orbit, its angular orbital velocity is ω_o then $\boldsymbol{\omega}^{bi} = \boldsymbol{\omega}^{bo} + \mathbf{R}^{bo}\boldsymbol{\omega}^{oi}$, $\dot{\theta} = \mathbf{S}^{-1}\boldsymbol{\omega}^{bo}$. Then we have the relation between the Euler ϕ, θ, ψ and the angular velocity $\omega_x, \omega_y, \omega_z$.

$$\dot{\phi} = \omega_x + \sin\phi \tan\theta\, \omega_y + \cos\phi \tan\theta\, \omega_z + \sin\psi \sec\theta\, \omega_o \tag{8.159}$$

$$\dot{\theta} = \cos\phi\, \omega_y - \sin\phi\, \omega_z + \cos\psi\, \omega_o \tag{8.160}$$

$$\dot{\psi} = \sin\phi \sec\theta\, \omega_y + \cos\phi \sec\theta\, \omega_z + \tan\theta \sin\psi\, \omega_o. \tag{8.161}$$

8.5.2.2 Orbital Compass

When the attitudes ϕ, θ, ψ are small, Equations (8.159)–(8.161) become:

$$\dot{\phi} = \omega_x + \psi\, \omega_o \tag{8.162}$$

$$\dot{\theta} = \omega_y + \omega_o \tag{8.163}$$

$$\dot{\psi} = \omega_z - \phi\, \omega_o. \tag{8.164}$$

From Equations (8.162)–(8.164), θ is decoupled, and ϕ, ψ are coupled. Note that if $\boldsymbol{\omega} = [\omega_x, \omega_y, \omega_z]^T = \mathbf{0}$, and the satellite is in circular orbit, ω_o is a constant, then the $\hat{y}$-axis of the body is inertially fixed space.

If the satellite initial time is at t_0, its roll and yaw angles are $\phi(t_0), \psi(t_0)$

$$\phi(t) = \phi(t_0)\cos(\omega_o(t - t_0)) + \psi(t_0)\sin(\omega_o(t - t_0)) \tag{8.165}$$

$$\psi(t) = \psi(t_0)\cos(\omega_o(t - t_0)) - \phi(t_0)\sin(\omega_o(t - t_0)). \tag{8.166}$$

When $(t - t_0)\omega_o = \pi/2$, i.e., the orbit rotates 1/4 period, then $\phi(t) = \psi(t_0), \psi(t) = -\phi(t_0)$. This means the roll angle and yaw angle are exchanged. From the view of measurement, with one sensor in the roll-yaw channel, the roll-yaw system is observable. In modern control theory, $C = [1\ 0], A = [0\ 1; -1\ 0], \text{rank}[C;CA] = 2$.

In general cases, $\boldsymbol{\omega}$ is not zero. We have three gyros in each axis of the body, it gives the angular velocities along each axis $\omega_x, \omega_y, \omega_z$ in $\mathcal{F}_b$. The gyro model is

$$g_x = \omega_x + N_{gx} \tag{8.167}$$

$$g_y = \omega_y + N_{gy} \tag{8.168}$$

$$g_z = \omega_z + N_{gz}, \tag{8.169}$$

where N_{gx}, N_{gy}, N_{gz} are measurement error, i.e. gyro noise and constant drift. By submitting the gyro model into Equations (8.162)–(8.164),

$$\dot{\phi} = \psi\,\omega_o + g_x - N_{gx} \tag{8.170}$$

$$\dot{\theta} = \omega_o + g_y - N_{gx} \tag{8.171}$$

$$\dot{\psi} = -\phi\,\omega_o + g_z - N_{gx}. \tag{8.172}$$

If we have two Earth sensors along the body $\hat{x}$ and $\hat{y}$ axis each, its measurement is denoted ϕ_H, ψ_H, then

$$\phi_H = \phi + N_{\phi H} \tag{8.173}$$

$$\theta_H = \theta + N_{\theta H}, \tag{8.174}$$

where $N_{\phi H}, N_{\theta H}$ are measurement noise of the Earth sensors.

From the view of modern control theory, Equations (8.170)–(8.172) are system state equations, where ϕ, θ, ψ are state variable, g_x, g_y, g_z are input variables, N_{gx}, N_{gy}, N_{gz} are system noises. Equations (8.173)–(8.174) are measurement equations, ϕ_H, θ_H are the outputs. The estimator can be designed:

$$\dot{\bar{\phi}} = \bar{\psi}\,\omega_o + g_x + K_\phi(\phi_H - \bar{\phi}) \tag{8.175}$$

$$\dot{\bar{\theta}} = \omega_o + g_y + K_\theta(\theta_H - \bar{\theta}) \tag{8.176}$$

$$\dot{\bar{\psi}} = -\bar{\phi}\,\omega_o + g_z + K_\psi(\phi_H - \bar{\phi}), \tag{8.177}$$

where $\bar{\phi}, \bar{\theta}, \bar{\psi}$ are estimates of the state, K_ϕ, K_θ, K_ψ are gains of the estimator. Normally they are small, about $0.005 - 0.05(1/s)$. This set of equation is called "orbital compass," as the yaw angle is not directly measured. Its state transfer function is

$$\bar{\phi} - \phi = \frac{s}{\Delta(s)}N_{gx} + \frac{\omega_o}{\Delta(s)}N_{gz} + \frac{K_\phi s + \omega_o K_\psi}{\Delta(s)}N_{\phi H} \tag{8.178}$$

$$\bar{\theta} - \theta = \frac{1}{s + K_\theta}N_{gy} + \frac{K_\theta}{s + K_\theta}N_{\theta H} \tag{8.179}$$

$$\bar{\psi} - \psi = -\frac{K_\psi + \omega_o}{\Delta(s)}N_{gx} + \frac{s + K_\phi}{\Delta(s)}N_{gz} + \frac{K_\psi s - \omega_o K_\phi}{\Delta(s)}N_{\phi H}, \tag{8.180}$$

where $\Delta(s)$ is the characteristic polynomial,

$$\Delta(s) = s^2 + K_\phi s + \omega_o(K_\psi + \omega_o), \tag{8.181}$$

by proper choice of $K_\phi, K_\psi, K_\theta, \Delta(s)$ and $s + K_\theta$ can be stable. If N_{gx}, N_{gy}, N_{gz} are constant drifts of gyros, the steady state errors are

$$(\bar{\phi} - \phi)_{t\to\infty} = \frac{1}{K_\psi + \omega_o}N_{gz} \tag{8.182}$$

$$(\bar{\theta} - \theta)_{t\to\infty} = \frac{1}{K_\theta}N_{gy} \tag{8.183}$$

$$(\bar{\psi} - \psi)_{t\to\infty} = -\frac{1}{\omega_o}N_{gx} + \frac{K_\phi}{\omega_o(K_\psi + \omega_o)}N_{gz}. \tag{8.184}$$

Large K_θ, K_ψ can reduce the steady state error, but is not effective in refraining the noise.

8.5.2.3 Attitude Capture

Another kind of Sun sensor is the "0/1" Sun sensor which is used for Sun capture. See Figure 8.11. It is in the body frame $\mathcal{F}_b$. If the Sun is in the region IV, let the satellite rotate around $\hat{x}$-axis; in the region II, rotate about -$\hat{x}$-axis; region III and V, about -$\hat{y}$ axis. At last the Sun is in region I near -$\hat{z}$-axis. If we use two "0/1" Sun sensors, the control logic is

$$S_x = L_{14} + L_{22}, \quad S_{-x} = L_{12} + L_{24}, \tag{8.185}$$

$$S_y = L_{23} + L_{25}, \quad S_{-y} = L_{15} + L_{13}. \tag{8.186}$$

The meaning, for example, is that if the Sun is in region IV of the front sensor and region II in the back sensor, i.e. $L_{14} = L_{22} = 1$ while others are zero, the satellite should rotate about the $\hat{x}$-axis, etc. If the Sun is in the direction of -$\hat{z}$-axis in $\mathcal{F}_b$, and $r_{ij} \in \mathbf{R}^{bo}$ in

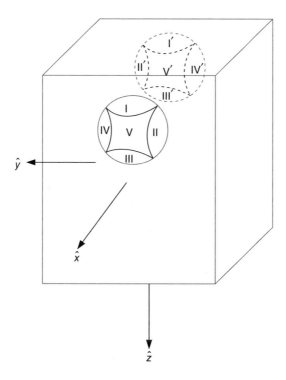

Figure 8.11 The "1/0" Sun sensor adapted from book [3]. The front sensor in the $\hat{x}$-direction includes $L_{11}, L_{12}, L_{13}, L_{14}, L_{15}$; whereas the back sensor in the $-\hat{x}$-direction includes $L_{21}, L_{22}, L_{23}, L_{24}, L_{25}$. They are all in the body frame $\mathcal{F}_b$

Equation (8.156), the Sun's vector in $\mathcal{F}_o$ is $\mathbf{s}_o = [s_{ox}, s_{oy}, s_{oz}]^T$, and $\mathbf{s}_b = [0, 0, -1]^T$ then $[r_{31}, r_{32}, r_{33}]^T = -[s_{ox}, s_{oy}, s_{oz}]^T$, also by the two Earth sensors.

$$r_{13} = \frac{-\cos \phi_H \sin \theta_H}{\sqrt{1 - \sin^2 \phi_H \sin^2 \theta_H}} \tag{8.187}$$

$$r_{23} = \frac{\sin \phi_H \cos \theta_H}{\sqrt{1 - \sin^2 \phi_H \sin^2 \theta_H}}. \tag{8.188}$$

As also for example $r_{12} = -r_{21}r_{22} + r_{23}r_{31}$, $r_{21} = -r_{12}r_{33} + r_{13}r_{32}$, then

$$r_{21} = \frac{r_{13}r_{32} - r_{32}r_{31}r_{33}}{r_{13}^2 + r_{23}^2} \tag{8.189}$$

$$r_{11} = -\frac{r_{23}r_{32} + r_{12}r_{31}r_{33}}{r_{13}^2 + r_{23}^2}. \tag{8.190}$$

More discussion about these models are given in reference [3].

8.6   Summary and Keywords

This chapter introduces satellite attitude notions and their representations by different methods and also describes a few attitude determination algorithms and their realizations.

Keywords: attitude kinematics, Euler angle, direct cosine matrix (DCM), quaternion, Euler theorem, satellite sensor, attitude determination, triad algorithm, orbital compass, attitude capture.

Problems

8.1   What are the 6 constraints on the 9 elements of a rotation matrix $\mathbf{R}$? List some other relations such as $r_{21} = -r_{12}r_{33} + r_{13}r_{32}$ and prove one of them.

8.2   Consider a unit vector $\hat{a}$, with components in a particular reference frame: $\hat{a} = [a_1, a_2, a_3]^T$. Show that $\hat{a}$ satisfies the following equations:
(a) $(\hat{a}\hat{a}^T)^2 = \hat{a}\hat{a}^T$.
(b) $\hat{a}^\times \hat{a} = 0$.
(c) $\hat{a}\hat{a}^T \hat{a}^\times = 0$.
(d) $\hat{a}^\times \hat{a}^\times = -\hat{a}^T \hat{a} \mathbf{1} + \hat{a}\hat{a}^T$.

8.3   Roll, pitch and yaw are Euler angles and are sometimes defined as a 3-2-1 sequence and sometimes defined as a 1-2-3 sequence What's the difference?

8.4   For Euler angle what happens when $\theta \to n\pi/2$, for odd n? What happens when the Euler angles and their rates are "small" ?

8.5 Using the relationship between the elements of the quaternion and the Euler angle and axis, verify that the expressions for **R**, in terms of $\bar{q}$ and **a**, Φ are equivalent.

8.6 Develop $\mathbf{S}(\theta)$ for a (2-3-1) rotation from $\mathcal{F}_i$ to $\mathcal{F}_b$, so that $\omega_b^{ib} = \mathbf{S}(\theta)\dot{\theta}$. Where is $\mathbf{S}(\theta)$ singular?

8.7 Consider the following rotation matrix $\mathbf{R}^{ab}$ that transforms vectors from $\mathcal{F}_b$ to $\mathcal{F}_a$:

$$\mathbf{R}^{ab} = \begin{bmatrix} 0.45457972 & 0.43387382 & -0.77788868 \\ -0.34766601 & 0.89049359 & 0.29351236 \\ 0.82005221 & 0.13702069 & 0.55564350 \end{bmatrix}. \tag{8.191}$$

Use at least two different properties of rotation matrices to convince yourself that $\mathbf{R}^{ab}$ is indeed a rotation matrix. Remark on any discrepancies you notice. If it is 1-2-3 sequence, what are its Euler angles?

8.8 Determine the Euler axis **a** and Euler principal angle Φ directly from $\mathbf{R}^{ab}$. Verify your results using the formula for $\mathbf{R}(\mathbf{a}, \Phi)$. Verify that $\mathbf{Ra} = \mathbf{a}$.

8.9 Determine the components of the quaternion $\bar{q}$, directly from $\mathbf{R}^{ab}$. Verify your results using the formula for $\mathbf{R}(\bar{q})$. Verify your results using the relationship between $\bar{q}$ and $(\mathbf{a}, \Phi)$.

8.10 Derive the formula for a (3-1-3) rotation. Determine the Euler angles for a (2-3-1) rotation, directly from $\mathbf{R}^{ab}$. Verify your results using the formula you derived.

8.11 What is the expression of small angle DCM, **R** for the $(\mathbf{a}, \Phi)$ representation? If $\phi = 0$, what do you do about the Euler axis/angle method?

8.12 Write a short paragraph discussing the relative merits of the three different representations of **R**.

8.13 This problem requires numerical integration of the kinematics equations of motion. You should have a look at the MATLAB® reference if needed. Suppose that $\mathcal{F}_b$ and $\mathcal{F}_a$ are initially aligned, so that $\mathbf{R}^{ba}(0) = \mathbf{1}$. At $t = 0$, $\mathcal{F}_b$ begins to rotate with angular velocity $\omega_b = e^{-4t}[\sin t, \sin 2t, \sin 3t]^T$ with respect to $\mathcal{F}_b$.
1. Use a (1-2-3) Euler angle sequence and make a plot of the three Euler angles vs. time for $t = 0$ to 10 s. Use a sufficiently small step size so that the resulting plot is "smooth."
2. Use the quaternion representation and make a plot of the four components of the quaternion vs. time for t = 0 to 10 s. Use a sufficiently small step size so that the resulting plot is "smooth."
3. Use the results of the two integrations to determine the rotation matrix at t = 10 s. Do this using the expression for $\mathbf{R}(\theta)$ and for $\mathbf{R}(\bar{q})$. Compare the results.

8.14 Prove Equations (8.81)–(8.83).

References

[1] Wertz, J.R. (1985) *Spacecraft Attitude Determination and Control*. Boston: D. Reidel.

[2] Schaub, H. and Junkins, J. (2003) *Analytical Mechanics of Space Systems*, AIAA Education Series. Reston, VA: AIAA.

[3] Tu, S. (2001) *Satellite Attitude Dynamics and Control* (in Chinese). Beijing: China Astronautic Publishing House.

[4] Liu, C. and Hu, W. (2014) Relative pose estimation for cylinder shaped spacecraft using a single image, *IEEE Transactions on Aero and Electronic Systems*, **50**(4): 3036–56.

9

Attitude Dynamics

This chapter introduces the definition of rigid body, linear momentum, extension of Newton's Second Law, angular momentum and development of Euler's law, Euler's equations, moments of inertia, coupling between rotational and translational motion. Special cases are analyzed such as axis symmetric torque-free, asymmetric torque-free, etc.

9.1 Rigid Body Models

Recall the fundamental dynamics Equation(8.1): $\mathbf{f} = d\mathbf{p}/dt$, $\mathbf{g} = d\mathbf{h}/dt$. We need to develop these equations for finite, rigid, bodies.

Particle model is an idealization that is useful for describing vehicle motion for many applications, for example, aircraft performance, spacecraft orbits, when only the translational motion is of interest.

Rigid body model is an idealization that is useful for describing more complicated vehicle motions, for example, aircraft roll, pitch, and yaw, spacecraft pointing, marine vessel motion when both translational and rotational motion are of interest.

A rigid body is an idealized model of a solid body of finite dimension in which deformation is neglected. Alternatively, the distance between any two points in the body remains constant in any motion. The motion of a rigid body is described by the position and velocity of any point in the body with respect to an inertial origin, and the orientation and angular velocity of a body frame with respect to an inertial frame.

Defining rigid body models takes two distinct approaches in the literature, both of which arrive at the basic equations of motion, that is, Equation(8.2): $\dot{\mathbf{r}} = \mathbf{p}/m$, $\dot{\mathbf{p}} = \mathbf{f}$.

One approach: begin with a system of n particles and take the limit as $n \to \infty$ to arrive at a continuous body. Newton's Second Law naturally extends to the system of particles and the resulting limit of a continuous body. Euler's Law results from application of Newton's Second Law.

Second approach: begin with a continuous body and assert that Newton's Second Law and Euler's Law are independent principles that apply to finite bodies, whether rigid or deformable.

Fundamental Spacecraft Dynamics and Control, First Edition. Weiduo Hu.
© 2015 John Wiley & Sons (Asia) Pte Ltd. Published 2015 by John Wiley & Sons (Asia) Pte Ltd.

9.1.1 Particle Model

It is easy to model the rigid body as a system of n mass particles of mass m_i, $i = 1, ..., n$. See Figure 2.1. The velocity of each particle is $\mathbf{v}_i$, and the linear momentum of each particle is $\mathbf{p}_i = m_i \mathbf{v}_i$. The system linear momentum is then

$$\mathbf{p} = \sum_{i=1}^{n} m_i \mathbf{v}_i. \tag{9.1}$$

Define the angular momentum of the system of particles, about the point O, as

$$\mathbf{H}^O = \sum_{i=1}^{n} \mathbf{r}_i \times m_i \mathbf{v}_i. \tag{9.2}$$

O is an inertial origin of $\mathcal{F}_i$ which is an inertial frame.

The system linear momentum satisfies:

$$\dot{\mathbf{p}} = \sum_{i=1}^{n} m_i \dot{\mathbf{v}}_i = \sum_{i=1}^{n} \dot{\mathbf{p}}_i = \sum_{i=1}^{n} \mathbf{f}_i \tag{9.3}$$

$$= \sum_{i=1}^{n} \mathbf{f}_i^{\text{internal}} + \sum_{i=1}^{n} \mathbf{f}_i^{\text{external}} = \sum_{i=1}^{n} \mathbf{f}_i^{\text{external}} = \mathbf{f}. \tag{9.4}$$

$\mathbf{f}$ is the total external force.

So Newton's Second Law (combined with the Third Law) allows the extension to systems of particles. The angular momentum of the system of particles satisfies:

$$\dot{\mathbf{H}}^O = \sum_{i=1}^{n} \frac{d}{dt} \left[\mathbf{r}_i \times m_i \mathbf{v}_i \right] = \sum_{i=1}^{n} \left[\dot{\mathbf{r}}_i \times m_i \mathbf{v}_i + \mathbf{r}_i \times m_i \dot{\mathbf{v}}_i \right] \tag{9.5}$$

$$= \sum_{i=1}^{n} \left[\mathbf{r}_i \times \mathbf{f}_i \right] = \sum_{i=1}^{n} \left[\mathbf{r}_i \times \mathbf{f}_i^{\text{internal}} \right] + \sum_{i=1}^{n} \left[\mathbf{r}_i \times \mathbf{f}_i^{\text{external}} \right] \tag{9.6}$$

$$= \sum_{i=1}^{n} \mathbf{g}_i^{O,\text{external}} = \mathbf{g}^{O,\text{external}}. \tag{9.7}$$

9.1.2 Continuous Model

Model the rigid body as a continuum, $\mathcal{B}$, that satisfies Newton's Second Law and Euler's Law. See Figure 9.1. The system linear momentum is

$$\mathbf{p} = \int_{B} \mathbf{v} \, dm. \tag{9.8}$$

The angular momentum about the point o is

$$\mathbf{h}^o = \int_{B} \mathbf{r} \times \mathbf{v} \, dm. \tag{9.9}$$

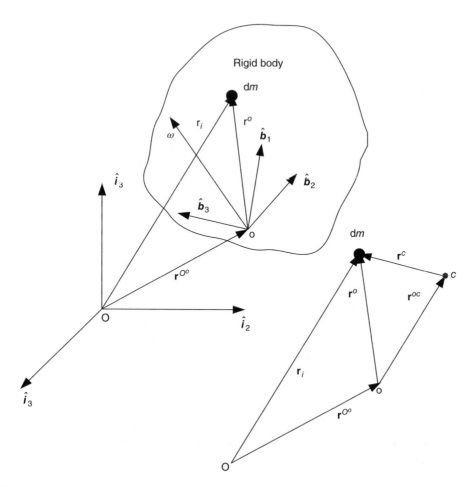

Figure 9.1 A rigid body in inertial frame $\mathcal{F}_i$ whose origin is at O and body frame $\mathcal{F}_b$ whose origin is at o in Section 8.1

These two vector quantities obey the linear and angular momentum principles as expressed by Newton and Euler.

$$\dot{\mathbf{p}} = \mathbf{f}, \quad \text{and} \quad \dot{\mathbf{h}}^o = \mathbf{g}^o. \tag{9.10}$$

Let's consider a uniform rectangle example first. See Figure 9.2. It defines a three-dimensional volume integral over the volume of the continuous body $\mathcal{B}$; in general:

$$\int_{\mathcal{B}} f(\mathbf{r}) dV = \int_{r_{1min}}^{r_{1max}} \int_{r_{2min}(r_1)}^{r_{2max}(r_1)} \int_{r_{3min}(r_1,r_2)}^{r_{3max}(r_1,r_2)} f(\mathbf{r}) dr_3 dr_2 dr_1. \tag{9.11}$$

Three types of integrals arise in the development of rigid body equations of motion: scalar, vector, and tensor. Alternatively, these three mathematical objects can be called zeroth-rank, first-rank, and second-rank tensors.

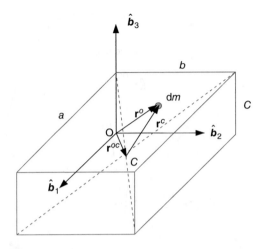

Figure 9.2   A rectangle example

For example, the total mass, m, is the integral over the body of the density, $\mu(\mathbf{r}, t)$, which can in general vary from point to point within the body, or even with time.

We usually assume that the density is constant. The rectangular prism has sides of length a, b, and c. Note that the origin of the body frame shown is not the mass center.

$$m = \int_B \mu dV = \int_B dm = \int_0^a \int_0^b \int_0^c \mu dr_3 dr_2 dr_1 = \mu abc. \tag{9.12}$$

Mass is a scalar, or zeroth-rank tensor, and is sometimes called the zeroth moment of inertia.

The first moment of inertia about o, $\mathbf{c}^o$, is the integral over the body of the position vector from o to each differential mass element dm.

The vector is denoted $\mathbf{r}^o$, where the superscript indicates that it is the vector from o to dm. Then the first moment of inertia is

$$\mathbf{c}^o = \int_B \mathbf{r}^o dm, \quad \text{or} \tag{9.13}$$

$$\mathbf{c}_b^o = \int_B \mathbf{r}_b^o \mu dV \tag{9.14}$$

$$= \int_0^a \int_0^b \int_0^c \mu [r_1, \ r_2, \ r_3]^T dr_3 dr_2 dr_1 \tag{9.15}$$

$$= \frac{\mu}{2} [a^2 bc, \ ab^2 c, \ abc^2]^T. \tag{9.16}$$

The first moment of inertia can be used to find the mass center c, defined as the point about which $\mathbf{c}$ vanishes.

$$\mathbf{c}^c = \int_B \mathbf{r}^c dm = \mathbf{0}, \quad \text{or} \tag{9.17}$$

$$\mathbf{c}_b^c = \int_B \mathbf{r}_b^c \mu dV = \mathbf{0}. \tag{9.18}$$

The position vector can be written as, (see Figure 9.2):

$$\mathbf{r}^o = \mathbf{r}^{oc} + \mathbf{r}^c. \tag{9.19}$$

Thus

$$\mathbf{c}^o = \int_B (\mathbf{r}^{oc} + \mathbf{r}^c) dm = \int_B \mathbf{r}^{oc} dm = \mathbf{r}^{oc} \int_B dm = M\mathbf{r}^{oc}. \tag{9.20}$$

For the rectangle $\mathbf{r}^{oc} = [a/2, b/2, c/2]^T$.

The linear momentum has already been defined as $\mathbf{p} = \int_B \mathbf{v} dm$ where $\mathbf{v}$ is the time derivative of the position vector $\mathbf{r}$ of a differential mass element $dm = \mu dV$. This position vector must be measured from an inertial origin, and the derivative is taken with respect to inertial space.

The position vector can be written as $\mathbf{r}^{Oo} + \mathbf{r}^o$, where $\mathbf{r}^{Oo}$ is the position vector from the inertial origin O to o and $\mathbf{r}^o$ is the vector from o to a point in the body. See Figure 9.1. Then we can have the velocity $\mathbf{v}$ of a differential mass element.

The body frame motion has two velocities: the velocity of its origin with respect to the inertial origin, $\mathbf{v}^{Oo}$, and the angular velocity of the frame with respect to an inertial frame, $\boldsymbol{\omega}^{bi}$. The velocity of a mass element dm is then

$$\mathbf{v} = \mathbf{v}^{Oo} + \boldsymbol{\omega}^{bi} \times \mathbf{r}^o. \tag{9.21}$$

Thus the linear momentum is

$$\mathbf{p} = \int_B \left(\mathbf{v}^{Oo} + \boldsymbol{\omega}^{bi} \times \mathbf{r}^o \right) dm. \tag{9.22}$$

The velocity of the origin and the angular velocity of the frame do not depend on the particular differential mass element, so they are constant with respect to the integration, so that the momentum becomes

$$\mathbf{p} = m\mathbf{v}^{Oo} + \boldsymbol{\omega}^{bi} \times \int_B \mathbf{r}^o dm. \tag{9.23}$$

The right term is the first moment of inertia about the origin, so that

$$\mathbf{p} = m\mathbf{v}^{Oo} + \boldsymbol{\omega}^{bi} \times \mathbf{c}^o. \tag{9.24}$$

Linear momentum principle: $\mathbf{f} = \dot{\mathbf{p}}$.

We can also express these equations in a general rotating reference frame. Recall the formula for differentiating a vector as expressed in a rotating frame $\mathcal{F}_b$:

$$\frac{d}{dt} \left[\{\hat{b}\}^T \mathbf{a} \right] = \{\hat{b}\}^T [\dot{\mathbf{a}}] + \boldsymbol{\omega}^\times \mathbf{a}, \tag{9.25}$$

where $\boldsymbol{\omega} = \boldsymbol{\omega}^{bi}$.

Applying the formula to $\mathbf{p}$ and to $\mathbf{v} = \dot{\mathbf{r}}$, we obtain the translational equations of motion in vector form, which are:

$$\mathbf{p} = m \left[\dot{\mathbf{r}} + \boldsymbol{\omega}^\times \mathbf{r} \right] + \boldsymbol{\omega}^\times \mathbf{c} \tag{9.26}$$

$$\mathbf{f} = \dot{\mathbf{p}} + \boldsymbol{\omega}^\times \mathbf{p}, \tag{9.27}$$

which we can rearrange in the usual $\dot{x} = f(x)$, form as the matrix translational equations of motion, expressed in a rotating frame:

$$\dot{\mathbf{r}} = \frac{1}{m} \left[\mathbf{p} - \boldsymbol{\omega}^\times \mathbf{c} \right] - \boldsymbol{\omega}^\times \mathbf{r} \tag{9.28}$$

$$\dot{\mathbf{p}} = -\boldsymbol{\omega}^{\times}\mathbf{p} + \mathbf{f}. \tag{9.29}$$

The $\dot{\mathbf{r}}$ equation is the kinematics equation, and is frequently written in an inertial frame instead of in a rotating frame.

Note that $\boldsymbol{\omega}$ is generally obtained from the rotational equations of motion, and that $\mathbf{f}$ may depend on the rotational motion, that is: i.e.

$$\mathbf{f} = \mathbf{f}(\mathbf{r}, \dot{\mathbf{r}}, \mathbf{R}, \boldsymbol{\omega}) \quad \text{or} \quad \mathbf{f} = \mathbf{f}(\mathbf{r}, \dot{\mathbf{r}}, \mathbf{q}, \boldsymbol{\omega}). \tag{9.30}$$

That is, the force generally depends on the dynamic state of the body: its position, velocity, attitude, and angular velocity.

9.1.3 Angular Momentum

There are two forms of angular momentum in common use: moment of momentum, and angular momentum. See Figure 9.1. The quantity $\mathbf{v}dm$ is in fact the momentum of the differential mass element, and by taking the cross product with the moment arm $\mathbf{r}^o$, we are forming the moment of momentum about o.

The *absolute angular momentum*, $\mathbf{H}^o$ about o is

$$\mathbf{H}^o = \int_B \mathbf{r}^o \times \mathbf{v}dm = \int_B \mathbf{r}^o \times \dot{\mathbf{r}}_i dm \tag{9.31}$$

$$= \int_B \mathbf{r}^o \times (\dot{\mathbf{r}}^{Oo} + \dot{\mathbf{r}}^o)dm \tag{9.32}$$

$$= \int_B \mathbf{r}^o \times \dot{\mathbf{r}}^{Oo}dm + \int_B \mathbf{r}^o \times \dot{\mathbf{r}}^o dm \tag{9.33}$$

$$= \int_B (\mathbf{r}^{oc} + \mathbf{r}^c) \times \dot{\mathbf{r}}^{Oo}dm + \int_B \mathbf{r}^o \times \dot{\mathbf{r}}^o dm \tag{9.34}$$

$$= m \cdot \mathbf{r}^{oc} \times \dot{\mathbf{r}}^{Oo} + \mathbf{h}^o, \tag{9.35}$$

where $\mathbf{r}^o$ is the position vector of a differential mass element from point o, and $\mathbf{v}$ is the velocity of the mass element with respect to inertial space. Usually choose o as c or a fixed point, then $\mathbf{H}^o = \mathbf{h}^o$. Here, $\mathbf{h}^o = \int_B \mathbf{r}^o \times \dot{\mathbf{r}}^o dm$ is the *relative angular momentum* about o, where $\mathbf{r}^o$ is the position vector of a differential mass element from point o, and $\dot{\mathbf{r}}^o$ is the velocity of the mass element with respect to o. See Figure 9.1. The difference between the two definitions is the velocity terms that are used.

The quantity $\dot{\mathbf{r}}^o dm$ is not the momentum of the differential mass element, and by taking the cross product with the moment arm $\mathbf{r}^o$, we are forming the moment of a momentum-like quantity about o.

We will work with the *angular momentum*, $\mathbf{h}^o$ about o is

$$\mathbf{h}^o = \int_B \mathbf{r}^o \times \dot{\mathbf{r}}^o dm. \tag{9.36}$$

The velocity term is

$$\dot{\mathbf{r}}^o = \boldsymbol{\omega} \times \mathbf{r}^o, \tag{9.37}$$

where $\omega = \omega^{bi}$. Thus

$$\mathbf{h}^o = \int_B \mathbf{r}^o \times [\omega \times \mathbf{r}^o] dm. \tag{9.38}$$

Expand the integrand using the identity:

$$\mathbf{a} \times (\mathbf{b} \times \mathbf{c}) = \mathbf{b}(\mathbf{a} \cdot \mathbf{c}) - \mathbf{c}(\mathbf{a} \cdot \mathbf{b}), \tag{9.39}$$

then

$$\mathbf{h}^o = \int_B \mathbf{r}^o \times [\omega \times \mathbf{r}^o] dm \tag{9.40}$$

$$= \int_B \left[\omega(\mathbf{r}^o \cdot \mathbf{r}^o) - \mathbf{r}^o(\omega \cdot \mathbf{r}^o) \right] dm. \tag{9.41}$$

Rewrite the two terms in the integrand as

$$\omega(\mathbf{r}^o \cdot \mathbf{r}^o) = \mathbf{r}^o \cdot \mathbf{r}^o \omega \tag{9.42}$$

$$\mathbf{r}^o(\omega \cdot \mathbf{r}^o) = \mathbf{r}^o \mathbf{r}^o \cdot \omega. \tag{9.43}$$

Whenever we write three vectors in the form $\mathbf{ab} \cdot \mathbf{c}$, parentheses are implied around the two vectors with the $\cdot$ between them. Thus, $\mathbf{ab} \cdot \mathbf{c} = \mathbf{a}(\mathbf{b} \cdot \mathbf{c})$ and $\mathbf{a} \cdot \mathbf{bc} = (\mathbf{a} \cdot \mathbf{b})\mathbf{c}$.

Introduce a new mathematical object, the identity tensor, $\mathbf{1}$, defined by its operation on vectors:

$$\mathbf{1} \cdot \mathbf{v} = \mathbf{v} \cdot \mathbf{1} = \mathbf{v}. \tag{9.44}$$

Use the identity tensor to rewrite the first term in the integrand as

$$\mathbf{r}^o \mathbf{r}^o \cdot \omega = \mathbf{r}^o \cdot \mathbf{r}^o \mathbf{1} \cdot \omega = (r^o)^2 \mathbf{1} \cdot \omega. \tag{9.45}$$

Combining all these expressions, we write $\mathbf{h}^o$ as

$$\mathbf{h}^o = \int_B \left[\omega(\mathbf{r}^o \cdot \mathbf{r}^o) - \mathbf{r}^o(\omega \cdot \mathbf{r}^o) \right] dm \tag{9.46}$$

$$= \int_B \left[(r^o)^2 \mathbf{1} \cdot \omega - \mathbf{r}^o \mathbf{r}^o \cdot \omega \right] dm. \tag{9.47}$$

The angular velocity vector is independent of the position vector to each individual mass element, so

$$\mathbf{h}^o = \int_B \left[(r^o)^2 \mathbf{1} \cdot \omega - \mathbf{r}^o \mathbf{r}^o \cdot \omega \right] dm \tag{9.48}$$

$$= \int_B \left[(r^o)^2 \mathbf{1} - \mathbf{r}^o \mathbf{r}^{oT} \right] dm \cdot \omega. \tag{9.49}$$

The integral is the moment of inertia tensor, and is constant, just as the total mass and first moment integrals are:

$$\mathbf{I}^o = \int_B \left[(r^o)^2 \mathbf{1} - \mathbf{r}^o \mathbf{r}^{oT} \right] dm. \tag{9.50}$$

If the point o is clear, omit the superscript:

$$I = \int_B \left[r^2 1 - rr^T \right] dm. \tag{9.51}$$

The angular momentum is then:

$$h^o = I \cdot \omega, \tag{9.52}$$

where

$$I = \int_B \left[r^2 1 - rr^T \right] dm. \tag{9.53}$$

9.2   Inertia Tensor

Recall angular momentum about o: $h^o = \int_B r^o \times \dot{r}^o dm$, $h^o = I \cdot \omega$, where $I = \int_B \left[r^2 1 - rr \right] dm$. The superscript o denotes that the angular momentum, and the inertia tensor, are taken about the point o.

Express angular momentum vector, angular velocity vector, and moment of inertia tensor in $\mathcal{F}_b$:

$$h_b^o = h_b^o \cdot \{\hat{b}\} \tag{9.54}$$

$$\omega_b^{bi} = \omega^{bi} \cdot \{\hat{b}\} \tag{9.55}$$

$$I_b^o = \{\hat{b}\} \cdot I \cdot \{\hat{b}\}^T \tag{9.56}$$

$$h_b^o = I_b^o \omega_b^{bi}. \tag{9.57}$$

Note that the angular velocity here is the angular velocity of the body-fixed frame with respect to an inertial frame.

9.2.1   Calculation of Inertia Tensor (I)

Think of I as you think of vectors: it is an abstract mathematical object that has numbers associated with it when you choose a particular reference frame.

We have computed body integrals involving $r dm$, and we had to express r in a body-fixed frame:

$$r = r^T \{\hat{b}\} = \{\hat{b}\}^T r. \tag{9.58}$$

Substitute these expressions into the terms in I:

$$r^2 = r \cdot r = r^T \{\hat{b}\} \cdot \{\hat{b}\}^T r = r^T 1 r = r^T r \tag{9.59}$$

$$rr^T = \{\hat{b}\}^T rr^T \{\hat{b}\}, \tag{9.60}$$

then

$$I = \int_B \left[r^2 1 - rr^T \right] dm \tag{9.61}$$

$$= \int_B \left[\mathbf{r}^T \mathbf{r} \{\hat{b}\}^T \{\hat{b}\} - \{\hat{b}\}^T \mathbf{r} \mathbf{r}^T \{\hat{b}\} \right] dm \tag{9.62}$$

$$= \{\hat{b}\}^T \int_B \left[\mathbf{r}^T \mathbf{r} \mathbf{1} - \mathbf{r} \mathbf{r}^T \right] dm \{\hat{b}\}. \tag{9.63}$$

Recall that $\{\hat{b}\} \cdot \{\hat{b}\}^T = 1$, and take the dot product of both sides with $\{\hat{b}\}$, from the left and $\{\hat{b}\}^T$ from the right.

Taking the dot products gives:

$$\{\hat{b}\}\mathbf{I}\{\hat{b}\}^T = \int_B \left[\mathbf{r}^T \mathbf{r} \mathbf{1} - \mathbf{r} \mathbf{r}^T \right] dm. \tag{9.64}$$

In the same way that taking dot products of a vector $\mathbf{v}$ with the unit vectors gives the components of the vector in the given frame, the expression on the left is the components of the inertia tensor in $\mathcal{F}_b$:

$$v_b = \mathbf{v} \cdot \{\hat{b}\} \tag{9.65}$$

$$\mathbf{I}_b = \{\hat{b}\}\mathbf{I}\{\hat{b}\}^T. \tag{9.66}$$

The matrix version of the moment of inertia tensor is computed in similar fashion as the first moment of inertia vector:

$$\mathbf{I}_b^o = \int_B \left[\mathbf{r}^T \mathbf{r} \mathbf{1} - \mathbf{r} \mathbf{r}^T \right] \mu dV. \tag{9.67}$$

The term in brackets is easily shown to be:

$$\mathbf{I} \equiv \begin{bmatrix} I_{11} & -I_{12} & -I_{13} \\ -I_{12} & I_{22} & -I_{23} \\ -I_{13} & -I_{23} & I_{33} \end{bmatrix} = \begin{bmatrix} r_2^2 + r_3^2 & -r_1 r_2 & -r_1 r_3 \\ -r_1 r_2 & r_1^2 + r_3^2 & -r_2 r_3 \\ -r_1 r_3 & -r_2 r_3 & r_1^2 + r_2^2 \end{bmatrix}. \tag{9.68}$$

For a rectangle, the required integration is:

$$\int_0^a \int_0^b \int_0^c \left[\mathbf{r}^T \mathbf{r} \mathbf{1} - \mathbf{r} \mathbf{r}^T \right] \mu dr_3 dr_2 dr_1. \tag{9.69}$$

Each element of the matrix is integrated independently of the rest. For example,

$$I_{11} = \frac{\mu abc}{3}(b^2 + c^2) = \frac{m}{3}(b^2 + c^2). \tag{9.70}$$

Carrying out the required integrations leads to

$$\mathbf{I}_b^o = m \begin{bmatrix} \frac{1}{3}(b^2 + c^2) & -\frac{1}{4}ab & -\frac{1}{4}ac \\ -\frac{1}{4}ab & \frac{1}{3}(a^2 + c^2) & -\frac{1}{4}bc \\ -\frac{1}{4}ac & -\frac{1}{4}bc & \frac{1}{3}(a^2 + b^2) \end{bmatrix}. \tag{9.71}$$

Note the pattern, especially that $\mathbf{I}^T = \mathbf{I}$.

Also, we have computed the moment of inertia matrix about point o, and with respect to the $\hat{b}$ frame. We need to be able to translate the origin, and rotate the frame.

9.2.2   Parallel Axis Theorem (PAT)

Suppose we know the moment of inertia matrix about the mass center, c, expressed in a body fixed reference frame, and we want the moment of inertia matrix about a different point o, defined by $\mathbf{r}_b^{co}$, expressed in the same frame, but with translated origin.

The former is denoted $\mathbf{I}_b^c$ and the latter is denoted $\mathbf{I}_b^o$ We can safely omit the subscript b, since the reference frame is the same for all the matrices.

Technically, the two frames are the same, since their unit vectors are in the same directions. Their origins, however, are displaced; hence the term parallel axes.

$$\mathbf{I}^c = \int_B \left[\mathbf{r}^{cT} \mathbf{r}^c \mathbf{1} - \mathbf{r}^c \mathbf{r}^{cT} \right] dm. \tag{9.72}$$

We write $\mathbf{r}^c$ as:

$$\mathbf{r}^c = \mathbf{r}^{co} + \mathbf{r}^o \tag{9.73}$$

and obtain:

$$\mathbf{I}^c = \int_B \left[\mathbf{r}^{oT} \mathbf{r}^o \mathbf{1} - \mathbf{r}^o \mathbf{r}^{oT} \right] dm - m \left[\mathbf{r}^{coT} \mathbf{r}^{co} \mathbf{1} - \mathbf{r}^{co} \mathbf{r}^{coT} \right]. \tag{9.74}$$

The integral in this expression is the moment of inertia matrix about the point o (expressed in $\mathcal{F}_b$). Parallel axis theorem:

$$\mathbf{I}^o = \mathbf{I}^c + m \left[\mathbf{r}^{coT} \mathbf{r}^{co} \mathbf{1} - \mathbf{r}^{co} \mathbf{r}^{coT} \right]. \tag{9.75}$$

Alternatively:

$$\mathbf{I}^o = \mathbf{I}^c - m \mathbf{r}^{co\times} \mathbf{r}^{co\times}. \tag{9.76}$$

Let's see an application of Parallel Axis Theorem. A cylindrical spacecraft has a rectangular solar panel. The cylinder has height $h = 4$ m, radius $r = 1.5$ m, and mass $m_c = 100$ kg. The solar panel is in the $\hat{b}_2$-$\hat{b}_3$ plane. It has length $c = 3$ m, height $b = 2$ m, and mass $m_p = 10$ kg. See Figure 9.3.

Look up the moments of inertia for a cylinder and a plate, which are usually given for principal axes, and about the mass center.

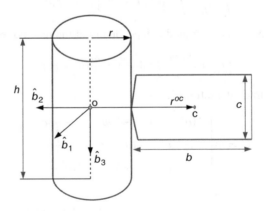

Figure 9.3   A cylindrical spacecraft with a rectangular solar panel

9.2.2.1 Cylinder

$$\mathbf{I}^c_{cyl} = m_c \begin{bmatrix} (3r^2 + h^2)/12 & 0 & 0 \\ 0 & (3r^2 + h^2)/12 & 0 \\ 0 & 0 & r^2/2 \end{bmatrix}. \tag{9.77}$$

9.2.2.2 Panel

$$\mathbf{I}^c_{pan} = m_p \begin{bmatrix} (b^2 + c^2)/12 & 0 & 0 \\ 0 & c^2/12 & 0 \\ 0 & 0 & b^2/12 \end{bmatrix}. \tag{9.78}$$

To add them, we have to decide what origin to use; some possibilities: center of cylinder, center of panel, or center of composite body. Compute the moment of inertia of the composite body about the mass center of the cylinder. Apply PAT to panel:

$$\mathbf{r}^{oc} = -\mathbf{r}^{co} = -(b/2 + r)\hat{b}_2. \tag{9.79}$$

Based on the panel moment of inertia and parallel axis then:

$$\mathbf{r}^{co} = [0 \ (b/2 + r) \ 0]^T \tag{9.80}$$

$$\mathbf{r}^{coT}\mathbf{r}^{co} = (b/2 + r)^2 \tag{9.81}$$

$$\mathbf{r}^{co}\mathbf{r}^{coT} = \begin{bmatrix} 0 & 0 & 0 \\ 0 & (b/2 + r)^2 & 0 \\ 0 & 0 & 0 \end{bmatrix} \tag{9.82}$$

$$\left[\mathbf{r}^{coT}\mathbf{r}^{co}\mathbf{1} - \mathbf{r}^{co}\mathbf{r}^{coT}\right] = \begin{bmatrix} (b/2 + r)^2 & 0 & 0 \\ 0 & 0 & 0 \\ 0 & 0 & (b/2 + r)^2 \end{bmatrix}. \tag{9.83}$$

The moment of inertia of the panel about the cylinder mass center is clearly still diagonal, as there are no off-diagonal elements that arise in application of the PAT.

For the panel (where o is the mass center of the cylinder):

$$\mathbf{I}^c_{pan} = m_p \begin{bmatrix} (b^2 + c^2)/12 + (b/2 + r)^2 & 0 & 0 \\ 0 & c^2/12 & 0 \\ 0 & 0 & b^2/12 + (b/2 + r)^2 \end{bmatrix}. \tag{9.84}$$

When we add $\mathbf{I}^o_{pan}$ pan to $\mathbf{I}^o_{cyl}$ together, we get $\mathbf{I}^o_{comp}$, the moment of inertia matrix for the composite body, about the point o, which is the mass center of the cylinder. But it is not the mass center of the composite body. It is required to compute the mass center of the composite body first. Then we can calculate the moment of inertia of the composite body about its mass center. The applicable form of the PAT is

$$\mathbf{I}^c = \mathbf{I}^o - m \left[\mathbf{r}^{coT}\mathbf{r}^{co}\mathbf{1} - \mathbf{r}^{co}\mathbf{r}^{coT}\right]. \tag{9.85}$$

Note that we have not used any of the numbers that were given! Strategy: Use symbols until the final calculations are required.

9.2.3   *Change of Vector Basis Theorem*

Suppose we know the moment of inertia matrix about some point, o, expressed in a body-fixed reference frame, $\mathcal{F}_a$, and we want the moment of inertia matrix about o, expressed in a different frame, $\mathcal{F}_b$. The former is denoted $\mathbf{I}_a^o$ and the latter is denoted $\mathbf{I}_b^o$. We cannot safely omit the subscripts a and b; however, since the origin o is the same for both inertia matrices, we can omit the superscript.

The moment of inertia tensor, $\mathbf{I}^o$ is the same mathematical object, but we are expressing it in a different frame, exactly as we might express a vector in two different frames.

We have:

$$\mathbf{I}_a = \int_B \left[\mathbf{r}_a^T \mathbf{r}_a \mathbf{1} - \mathbf{r}_a \mathbf{r}_a^T \right] dm. \tag{9.86}$$

The relationship between $\mathcal{F}_a$ and $\mathcal{F}_b$ is the rotation matrix $\mathbf{R}_{ba}$, and we use it to rotate $\mathbf{r}_a$ into $\mathcal{F}_b$:

$$\mathbf{r}_b = \mathbf{R}_{ba}\mathbf{r}_a \text{ or } \mathbf{r}_a = \mathbf{R}_{ab}\mathbf{r}_b. \tag{9.87}$$

What is $\mathbf{R}_{ab}$? Substitution into the integral for $\mathbf{I}_a$ leads to:

$$\mathbf{I}_a = \mathbf{R}_{ab} \int_B \left[\mathbf{r}_b^T \mathbf{r}_b \mathbf{1} - \mathbf{r}_b \mathbf{r}_b^T \right] dm \mathbf{R}_{ba}. \tag{9.88}$$

The integral is obviously the moment of inertia expressed in $\mathcal{F}_b$, which is what we want. Thus, pre-multiplying by $\mathbf{R}_{ba}$ and post-multiplying by $\mathbf{R}_{ab}$ leads to:

$$\mathbf{R}_{ba}\mathbf{I}_a\mathbf{R}_{ab} = \mathbf{I}_b. \tag{9.89}$$

This equation is the change of basis theorem, and is the equivalent of using a rotation matrix to rotate a vector from one frame to another. Note that

$$\mathbf{R}_{ab}\mathbf{I}_b\mathbf{R}_{ba} = \mathbf{I}_a. \tag{9.90}$$

9.2.3.1   Principal Axes

The most important application of the Change of Vector Basis Theorem is the principal axes for a rigid body.

Some Observations

The inertia tensor is symmetric:

$$\mathbf{u} \cdot \mathbf{I} \cdot \mathbf{v} = \mathbf{v} \cdot \mathbf{I} \cdot \mathbf{u}. \tag{9.91}$$

Exercise: convince yourself that the inertia matrix is symmetric:

$$\mathbf{I} = \mathbf{I}^T. \tag{9.92}$$

Fact: All eigenvalues of real, symmetric matrices are real.
Fact: All eigenvectors of real, symmetric matrices are real and are mutually orthogonal.

Conclusion: The eigenvectors of the inertia matrix comprise a set of base vectors for a body-fixed reference frame.

Compute the eigenvalues and eigenvectors of $\mathbf{I}$, expressed in a non-principal reference frame, $\mathcal{F}_a$.

The eigenvalue problem is stated as

$$\mathbf{I}_a\mathbf{x} = \lambda\mathbf{x}, \tag{9.93}$$

where λ and $\mathbf{x}$ are the unknowns.

If we find a $(\lambda, \mathbf{x})$ pair that solves the eigenvalue problem, then we have found an eigenvalue, λ, and an eigenvector, $\mathbf{x}$. For an inertia matrix, the eigenvalues are the principal moments of inertia, and the eigenvectors are the components of the principal frame base vectors, expressed in $\mathcal{F}_a$. In MATLAB®, for example, we can compute the eigensystem of a matrix $\mathbf{I}_a$, using

```
[V,D] = eig(Ia)
```

where D is a diagonal matrix containing the eigenvalues, and V is a matrix containing the eigenvectors Suppose we have calculated three eigenvalues and the associated eigenvectors: $\lambda_1, \lambda_2, \lambda_3, \mathbf{x}_1, \mathbf{x}_2, \mathbf{x}_3$. We can write the following:

$$\mathbf{I}_a[\mathbf{x}_1 \ \mathbf{x}_2 \ \mathbf{x}_3] = [\lambda_1\mathbf{x}_1, \lambda_2\mathbf{x}_2, \lambda_3\mathbf{x}_3] \tag{9.94}$$

```
I = % a fabricated example
1.9003 0.7171 1.0633
0.7171 1.7826 0.7806
1.0633 0.7806 1.6428

>> [V,D]=eig(I)
V =
 0.5837   0.5207 0.6230
 0.1809  -0.8314 0.5254
-0.7916   0.1940 0.5795

D =
0.6804 0          0
0          1.1513  0
0          0          3.4939

>> I*V
ans =
 0.3972   0.5995 2.1769
 0.1231  -0.9572 1.8357
-0.5386   0.2233 2.0246

>> V*D
ans =
 0.3972   0.5995 2.1769
 0.1231  -0.9572 1.8357
-0.5386   0.2233 2.0246
```

```
>> V'*I*V
ans =
  0.6804  0.0000  -0.0000
  0.0000  1.1513   0.0000
 -0.0000  0.0000   3.4939
```

$$\mathbf{I}_a \begin{bmatrix} \mathbf{x}_1, & \mathbf{x}_2, & \mathbf{x}_3 \end{bmatrix}^T = \begin{bmatrix} \lambda_1 \mathbf{x}_1 & \lambda_2 \mathbf{x}_2 & \lambda_3 \mathbf{x}_3 \end{bmatrix}^T \tag{9.95}$$

$$\mathbf{I}_a \mathbf{X} = \mathbf{X} \text{diag} \begin{bmatrix} \lambda_1 & \lambda_2 & \lambda_3 \end{bmatrix} \tag{9.96}$$

$$\mathbf{I}_a \mathbf{X} = \mathbf{X} \mathbf{\Lambda} \tag{9.97}$$

$$\mathbf{X}^{-1} \mathbf{I}_a \mathbf{X} = \mathbf{\Lambda} \tag{9.98}$$

$$\mathbf{R}^{ba} \mathbf{I}_a \mathbf{R}^{ab} = \mathbf{I}_b. \tag{9.99}$$

$\mathcal{F}_b$ is the principal frame.

9.2.4   Inertia with a Rotating Panel

A spacecraft is comprised of 3 rigid bodies: a cylinder, a rod, and a panel. The $\mathcal{F}_b$ frame is at the mass center of the cylinder with $\hat{b}_3$ axis parallel to the cylinder's symmetry axis, and the $\hat{b}_2$ axis parallel to the rod. See Figure 9.3. The cylinder diameter is 1 m, height is 2 m, and density is 100 kg/m^3. The rod length is 2 m and mass is 1 kg. The panel has length 2 m, width 0.5 m, and mass 4 kg. The panel is rotated about the rod through α_p; if $\alpha_p = 0$, then the panel is in the $\hat{b}_1 \hat{b}_2$ plane, and if $\alpha_p = 90^o$, then the panel is parallel to the $\hat{b}_2 \hat{b}_3$ plane.

Review the parallel axis theorem: $\mathbf{I}^o = \mathbf{I}^c + m \begin{bmatrix} \mathbf{r}^{coT} \mathbf{r}^{co} \mathbf{1} - \mathbf{r}^{co} \mathbf{r}^{coT} \end{bmatrix}$ Alternatively, $\mathbf{I}^o = \mathbf{I}^c - m \mathbf{r}^{cox} \mathbf{r}^{cox}$.

Change of vector basis theorem by $\mathbf{R}_{ba} \mathbf{I}_a \mathbf{R}_{ab} = \mathbf{I}_b$, $\mathbf{R}_{ab} \mathbf{I}_b \mathbf{R}_{ba} = \mathbf{I}_a$, which is used to compute principal axes and principal moments of inertia.

Steps to solution:

1. For each body, write $\mathbf{I}^c$ for the body in its own principal, centroidal reference frame.
2. For the panel, rotate the matrix from its principal frame to a frame parallel to $\mathcal{F}_b$.
3. For the panel and the rod, apply the parallel axis theorem to obtain $\mathbf{I}^o$ (where o is the mass center of the cylinder).
4. Add $\mathbf{I}^o_{pan}$ and $\mathbf{I}^o_{rod}$ to $\mathbf{I}^o_{cyl}$ to obtain $\mathbf{I}^o_{comp}$.
5. Compute first moment of inertia about o, $\mathbf{c}^o_{comp}$, and use this to find mass center location, $\mathbf{r}^{oc}$.
6. Apply parallel axis theorem to composite body to obtain $\mathbf{I}^c_{comp}$.
7. Compute eigensystem of $\mathbf{I}^c_{comp}$ to find principal moments of inertia and principal axes.

Actually $\mathbf{I}^c_{comp}$ is a time-variant due to the panel's rotation. We assume $\mathbf{I}^c_{cyl}$ is the moment of inertia about the c; this time we neglect rod (it can be regarded as part of panel), cp is mass center of the panel, the moment of inertia of panel about its mass center cp is $\mathbf{I}^{cp}_{pan}$,

$$\mathbf{I}^{cp}_{pan} = \begin{bmatrix} I_{xp} & 0 & 0 \\ 0 & I_{yp} & 0 \\ 0 & 0 & I_{zp}, \end{bmatrix} \tag{9.100}$$

where, $I_{zp} = I_{xp} + I_{yp}$. If α_p is the solar panel's angle (when the panel's normal vector pointing to $-\hat{z}$ axis $\alpha_p = 0$), rotation matrix is:

$$\mathbf{R}_p = \begin{bmatrix} \cos \alpha_p & 0 & \sin \alpha_p \\ 0 & 1 & 0 \\ -\sin \alpha & 0 & \cos \alpha_p \end{bmatrix}. \tag{9.101}$$

The panel inertial after rotation is $\mathbf{I}_{pan}^{cpt}$:

$$\mathbf{I}_{pan}^{cpt} = \mathbf{R}_p^T \mathbf{I}^{cp} \mathbf{R}_p = \begin{bmatrix} I_{xp} + I_{yp} \sin^2 \alpha_p & 0 & -I_{yp} \sin \alpha_p \cos \alpha_p \\ 0 & I_{yp} & 0 \\ -I_{yp} \sin \alpha_p \cos \alpha_p & 0 & I_{xp} + I_{yp} \cos^2 \alpha_p \end{bmatrix}. \tag{9.102}$$

If the m_p is the mass of the panel and $\mathbf{r}_{cp}$ is the vector from the total center c to the panel center cp, the moment of inertial after rotation is $\mathbf{I}_{comp}^c$

$$\mathbf{I}_{comp}^c = \mathbf{I}_{cyl}^c + \mathbf{I}_{pan}^{cpt} + m_p \mathbf{r}_{cp} \mathbf{r}_{cp}^T. \tag{9.103}$$

9.3 Equation of S/C Motion

9.3.1 Angular Momentum Principle

Going back to the definition $\mathbf{h}^o = \int_B \mathbf{r}^o \times \dot{\mathbf{r}}^o dm$, differentiate:

$$\dot{\mathbf{h}}^o = \int_B \mathbf{r}^o \times \ddot{\mathbf{r}}^o dm. \tag{9.104}$$

Recall Figure 9.1 that $\mathbf{r}^o = \mathbf{r}_i - \mathbf{r}^{Oo}$; therefore

$$\dot{\mathbf{h}}^o = \int_B \mathbf{r}^o \times \ddot{\mathbf{r}}_i dm - \int_B \mathbf{r}^o \times \ddot{\mathbf{r}}^{Oo} dm. \tag{9.105}$$

See Equation.(9.35) and define

$$\mathbf{g}^o = \int_B \mathbf{r}^o \times \mathbf{f} dm \Rightarrow \dot{\mathbf{h}}^o = \mathbf{g}^o - \mathbf{c}^o \times \ddot{\mathbf{r}}^{Oo}. \tag{9.106}$$

By Equation(9.20),

$$\dot{\mathbf{h}}^o = \mathbf{g}^o - m \mathbf{r}^{oc} \times \ddot{\mathbf{r}}^{Oo}, \tag{9.107}$$

then in more common form

$$\mathbf{I}^o \dot{\omega} + \omega^\times \mathbf{I}^o \omega + m \mathbf{r}^{ocx} \ddot{\mathbf{r}}^{Oo} = \mathbf{g}^o. \tag{9.108}$$

Observations:
If $o = c$, then $\mathbf{r}^{oc} = \mathbf{0}$, so $\dot{\mathbf{h}}^o = \mathbf{g}^o$; if o is not accelerating with respect to the inertial origin O, then $\ddot{\mathbf{r}}^{Oo} = \mathbf{0}$, so $\dot{\mathbf{h}}^o = \mathbf{g}^o$.

These observations motivate our usual selection of either an inertial origin or the mass center as the origin for the *angular momentum principle*.

By angular momentum principle, choose the origin at center of mass, $\mathbf{h}^c = \mathbf{I}^c \cdot \boldsymbol{\omega}^{bi}$, $\dot{\mathbf{h}}^c = \mathbf{g}^c$. Let's see how we express these equations in a rotating reference frame.

9.3.2 Rotational Equation of Motion

Recall the formula for differentiating a vector as expressed in a rotating frame $\mathcal{F}_b$:

$$\frac{d}{dt}\left[\{\hat{b}\}^T \mathbf{a}\right] = \{\hat{b}\}^T \left[\dot{\mathbf{a}} + \boldsymbol{\omega}^{\times}\mathbf{a}\right], \tag{9.109}$$

where $\boldsymbol{\omega} = \boldsymbol{\omega}^{bi}$. Note that $\boldsymbol{\omega}^{bi}$ plays two distinct roles in these equations.

Expressing the two equations in $\mathcal{F}_b$: gives:

$$\mathbf{h}_b^c = \mathbf{I}_b^c \boldsymbol{\omega}_b^{bi} \tag{9.110}$$

$$\dot{\mathbf{h}}_b^c + \boldsymbol{\omega}_b^{bi\times}\mathbf{h}_b^c = \mathbf{g}_b^c. \tag{9.111}$$

The latter is usually written as:

$$\dot{\mathbf{h}}_b^c = -\boldsymbol{\omega}_b^{bi\times}\mathbf{h}_b^c + \mathbf{g}_b^c. \tag{9.112}$$

Finally, assuming that the frame and origin are understood, we typically write:

$$\mathbf{h} = \mathbf{I}\boldsymbol{\omega} \tag{9.113}$$

$$\dot{\mathbf{h}} + \boldsymbol{\omega}^{\times}\mathbf{h} = \mathbf{g}. \tag{9.114}$$

Recall previously developed kinematics equations:

$$\dot{\bar{\mathbf{q}}} = \frac{1}{2}\begin{bmatrix} \mathbf{q}^{\times} + q_4\mathbf{1} \\ -\mathbf{q}^T \end{bmatrix} = \mathbf{Q}(\bar{\mathbf{q}})\boldsymbol{\omega} \quad \text{or} \tag{9.115}$$

$$\dot{\boldsymbol{\theta}} = \mathbf{S}^{-1}(\boldsymbol{\theta})\boldsymbol{\omega}. \tag{9.116}$$

Solve $\mathbf{h} = \mathbf{I}\boldsymbol{\omega}$ for $\boldsymbol{\omega}$, and substitute into the kinematics equations to obtain the system:

$$\boldsymbol{\omega} = \mathbf{I}^{-1}\mathbf{h} \tag{9.117}$$

$$\dot{\bar{\mathbf{q}}} = \mathbf{Q}(\bar{\mathbf{q}})\boldsymbol{\omega} \tag{9.118}$$

$$\dot{\mathbf{h}} = -\boldsymbol{\omega}^{\times}\mathbf{h} + \mathbf{g}. \tag{9.119}$$

We can numerically integrate the equations for $\bar{\mathbf{q}}$ and $\mathbf{h}$.

Recall that force generally depends on the dynamic state: position, velocity, attitude, and angular velocity:

$$\mathbf{f} = \mathbf{f}(\mathbf{r}, \dot{\mathbf{r}}, \mathbf{R}, \boldsymbol{\omega}). \tag{9.120}$$

The same applies to the torque:

$$\mathbf{g} = \mathbf{g}(\mathbf{r}, \mathbf{v}, \mathbf{R}, \boldsymbol{\omega}). \tag{9.121}$$

Note that we have omitted the overarrows, thus implying that $\mathbf{g}$ and its vector arguments are 3×1 matrices expressed in $\mathcal{F}_b$. $\mathbf{R}$ in Equation(9.20)–(9.21) is the rotation matrix.

9.3.3 Coupled Equations of Motion

The complete set of coupled translational and rotational equations of motion for a rigid body is summarized here:

$$\omega = \mathbf{I}^{-1}\mathbf{h}, \quad \mathbf{v} = \frac{1}{m}(\mathbf{p} - \omega^\times \mathbf{c}); \tag{9.122}$$

$$\mathbf{f} = \mathbf{f}(\mathbf{r}, \mathbf{v}, \mathbf{R}, \omega), \quad \mathbf{g} = \mathbf{g}(\mathbf{r}, \mathbf{v}, \mathbf{R}, \omega); \tag{9.123}$$

$$\dot{\mathbf{r}} = \frac{1}{m}\left[\mathbf{p} - \omega^\times \mathbf{c}\right] - \omega^\times \mathbf{r}, \quad \dot{\mathbf{p}} = -\omega^\times \mathbf{p} + \mathbf{f}, \tag{9.124}$$

$$\dot{\bar{\mathbf{q}}} = \mathbf{Q}(\bar{\mathbf{q}})\omega, \quad \dot{\mathbf{h}} = -\omega^\times \mathbf{h} + \mathbf{g}. \tag{9.125}$$

The first two equations provide velocities as functions of momenta.
The third and fourth equations define the force and moment in terms of the dynamic state.
The last four equations are the actual differential equations.

Note that there are 13 differential equations, so the state vector is $\mathbf{x} \in \mathbf{R}^{13}$ (a real vector of dimension 13).

9.4 Euler's Equation

A common approach is to substitute $\mathbf{h} = \mathbf{I}\omega$ into the angular momentum equation to obtain:

$$\mathbf{I}\dot{\omega} = -\omega^\times \mathbf{I}\omega + \mathbf{g} \tag{9.126}$$

$$\dot{\omega} = -\mathbf{I}^{-1}\omega^\times \mathbf{I}\omega + \mathbf{I}^{-1}\mathbf{g}. \tag{9.127}$$

The latter equation is the matrix form of Euler's equations.

If a nonprincipal reference frame is used, then $\mathbf{I}^{-1}$ is a bit complicated; however, for a principal frame, $\mathbf{I}^{-1}$ is simple.

For a principal frame, we write:

$$\mathbf{I} = \begin{bmatrix} I_1 & 0 & \\ 0 & I_2 & 0 \\ 0 & 0 & I_3 \end{bmatrix} = \mathrm{diag}\begin{bmatrix} I_1 & I_2 & I_3 \end{bmatrix}. \tag{9.128}$$

The inverse of a full-rank (em i.e., nonsingular) diagonal matrix is simply the diagonal matrix containing the scalar inverses: $\mathbf{I}^{-1} = \mathrm{diag}[1/I_1 \ 1/I_2 \ 1/I_3]$. Assuming a principal frame, Euler's equations may be written as the three scalar equations.

$$\dot{\omega}_1 = \frac{I_2 - I_3}{I_1}\omega_2\omega_3 + \frac{g_1}{I_1} \tag{9.129}$$

$$\dot{\omega}_2 = \frac{I_3 - I_1}{I_2}\omega_3\omega_1 + \frac{g_2}{I_2} \tag{9.130}$$

$$\dot{\omega}_3 = \frac{I_1 - I_2}{I_3}\omega_1\omega_2 + \frac{g_3}{I_3}. \tag{9.131}$$

Given initial conditions on ω and a means of computing the torque $\mathbf{g}$, these equations can be integrated to obtain $\omega(t)$, which can in turn be used to integrate the kinematics equations to obtain $\mathbf{R}(t)$.

As previously noted, the torque can depend on position, velocity, attitude, and angular velocity, so these equations are usually coupled with the translational dynamics and rotational kinematics equations.

Some useful insight into rotational dynamics is possible by considering two torque-free cases: axisymmetric ($I_1 = I_2$), asymmetric.

9.4.1   Axisymmetric, Torque-Free Rigid Body

See Figure 9.4. Setting $I_1 = I_2 = A$ and $I_2 = C$:

$$\dot{\omega}_1 = \frac{A - C}{A}\omega_2\omega_3 + \frac{g_1}{A} \tag{9.132}$$

$$\dot{\omega}_2 = \frac{C - A}{A}\omega_3\omega_1 + \frac{g_2}{A} \tag{9.133}$$

$$\dot{\omega}_3 = \frac{g_3}{C}. \tag{9.134}$$

Setting $\mathbf{g} = 0$, which means $\mathbf{h} = $ constant:

$$\dot{\omega}_1 = \frac{A - C}{A}\omega_2\omega_3 \tag{9.135}$$

$$\dot{\omega}_2 = \frac{C - A}{A}\omega_3\omega_1 \tag{9.136}$$

$$\dot{\omega}_3 = 0. \tag{9.137}$$

Integrate $\dot{\omega}_3$,

$$\dot{\omega}_3 = 0 \Rightarrow \omega_3 = \omega_3(t_0) \equiv \Omega. \tag{9.138}$$

The angular velocity about the symmetry axis is a constant, $\omega_3 = \Omega$, and the other two components of $\boldsymbol{\omega}$ satisfy:

$$\dot{\omega}_1 = \frac{A - C}{A}\Omega\omega_2 \tag{9.139}$$

$$\dot{\omega}_2 = \frac{C - A}{A}\Omega\omega_1. \tag{9.140}$$

Since A, C, and Ω are constants, these two equations comprise a linear, constant-coefficient, system of ordinary differential equations:

$$\dot{\mathbf{x}} = \mathbf{A}\mathbf{x} \Rightarrow \mathbf{x}(t) = e^{\mathbf{A}(t-t_0)}\mathbf{x}(t_0). \tag{9.141}$$

Since this system is so simple, we can solve it without resorting to the matrix exponential. Define:

$$\hat{\Omega} = \Omega(A - C)/A \tag{9.142}$$

$$\dot{\omega}_1 = \hat{\Omega}\omega_2 \tag{9.143}$$

$$\dot{\omega}_2 = -\hat{\Omega}\omega_1. \tag{9.144}$$

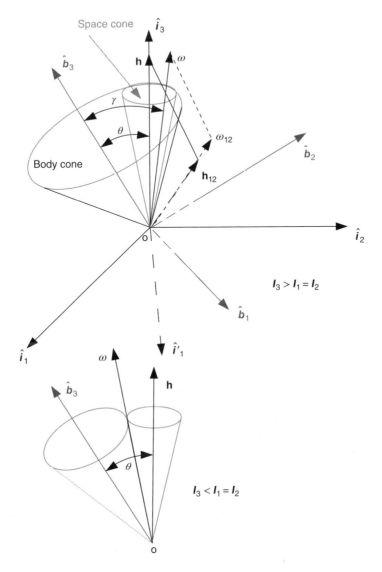

Figure 9.4 Body cone and space cone in $\mathcal{F}_b$ and $\mathcal{F}_i$ frame

Differentiate the first and substitute in from the second:

$$\ddot{\omega}_1 + \hat{\Omega}^2 \omega_1 = 0. \tag{9.145}$$

Since $\hat{\Omega}$ is a constant, this equation is a linear, constant coefficient, second-order, ordinary differential equation.

The solution to this equation can be written as:

$$\omega_1 = c_1 \cos \hat{\Omega} t + c_2 \sin \hat{\Omega} t, \tag{9.146}$$

where the two c_i's depend on initial conditions, i.e. $c_1 = \omega_1(t_0)$, $c_2 = \dot{\omega}(t_0)/\hat{\Omega} = \omega_2(t_0)$.

This can also be written as

$$\omega_1 = \omega_{12} \sin\left(\hat{\Omega}t + \varphi\right),$$
(9.147)

where $\omega_{12} = \pm\sqrt{c_1^2 + c_2^2} = \sqrt{\omega_1^2(t_0) + \omega_2^2(t_0)}$ is the amplitude and φ is the phase. Determination of φ in terms of c_1 and c_2 is left as an exercise.

We can always choose φ so that $a > 0$.

We have integrated the kinetics equations for rotational motion of an axisymmetric rigid body with zero torque:

$$\omega_1 = \omega_{12} \sin\left(\hat{\Omega}t + \varphi\right)$$
(9.148)

$$\omega_2 = \omega_{12} \cos\left(\hat{\Omega}t + \varphi\right)$$
(9.149)

$$\omega_3 = \Omega.$$
(9.150)

Alternatively,

$$\omega_1 = c_1 \cos\hat{\Omega}t + c_2 \sin\hat{\Omega}t$$
(9.151)

$$\omega_2 = -c_1 \sin\hat{\Omega}t + c_2 \cos\hat{\Omega}t$$
(9.152)

$$\omega_3 = \Omega.$$
(9.153)

Knowing the angular velocity as a function of time, we can integrate the kinematics equations (either $\dot{\boldsymbol{\theta}}$ or $\dot{\bar{\mathbf{q}}}$, or any other kinematics representation).

As in the exercise, where I gave you $\boldsymbol{\omega}(t)$, here we have $\boldsymbol{\omega}(t)$ derived as a solution for the axisymmetric torque-free rigid body.

You could take this $\boldsymbol{\omega}(t)$ and substitute it into your simulation code and get the attitude motion of an axisymmetric rigid body.

With the solution Equations(9.151)–(9.153) we can integrate the kinematics, expressed as 3-1-3 Euler angles, in closed-form as follows. See Figure 9.5.

We follow the classical definitions here, with the Euler angles defined as precession $\phi = \theta_1$, nutation $\theta = \theta_2$, self-spin $\psi = \theta_3$.

For this set of Euler angles, the relationship $\boldsymbol{\omega} = \mathbf{S}(\boldsymbol{\theta})\dot{\boldsymbol{\theta}}$ reduces to:

$$
\begin{bmatrix} \omega_1 \\ \omega_2 \\ \omega_3 \end{bmatrix} = \begin{bmatrix} 0 \\ 0 \\ \dot{\psi} \end{bmatrix} + \begin{bmatrix} \cos\psi & \sin\psi & 0 \\ -\sin\psi & \cos\psi & 0 \\ 0 & 0 & 1 \end{bmatrix} \begin{bmatrix} \dot{\theta} \\ 0 \\ 0 \end{bmatrix}
$$
(9.154)

$$
+ \begin{bmatrix} \cos\psi & \sin\psi & 0 \\ -\sin\psi & \cos\psi & 0 \\ 0 & 0 & 1 \end{bmatrix} \begin{bmatrix} 1 & 0 & 0 \\ 0 & \cos\theta & \sin\theta \\ 0 & -\sin\theta & \cos\theta \end{bmatrix} \begin{bmatrix} 0 \\ 0 \\ \dot{\phi} \end{bmatrix}.
$$
(9.155)

Applying the solution for $\boldsymbol{\omega}$, Equations(9.151)–(9.153), $\dot{\theta} = 0$, that is, combining the kinetics solution and the kinematics ODEs gave:

$$\omega_{12} \sin\left(\hat{\Omega}t + \varphi\right) = \dot{\phi} \sin\theta \sin\psi$$
(9.156)

$$\omega_{12} \cos \left(\hat{\Omega} t + \varphi \right) = \dot{\phi} \sin \theta \cos \psi \tag{9.157}$$

$$\Omega = \dot{\psi} + \dot{\phi} \cos \theta. \tag{9.158}$$

Study this set of equations long enough and some patterns begin to emerge.

Here ω_{12} and Ω are both > 0, by choice of phase φ (for $\omega_{12} > 0$), and choice of axis direction (for $\Omega > 0$).

Also note that $0 < \theta < \pi/2$, so that both $\sin \theta$ and $\cos \theta$ are > 0.

Furthermore, θ is constant.

Assume $\mathbf{h}$ (const) is aligned with $\hat{i}_3$, and

$$\mathbf{h} = \left[A\omega_1 \ A\omega_2 \ C\Omega \right]^T. \tag{9.159}$$

Component along $\hat{b}_3$ is

$$\mathbf{h}^T \hat{b}_3 = h \cos \theta = C\Omega. \tag{9.160}$$

Solve for θ(const):

$$\cos \theta = \frac{h_3}{h} = \frac{C\Omega}{\sqrt{A^2(\omega_1^2 + \omega_2^2) + C^2\Omega^2}} \tag{9.161}$$

$$= \frac{C\Omega}{\sqrt{A^2\omega_{12}^2 + C^2\Omega^2}}, \tag{9.162}$$

which proves that θ is constant.

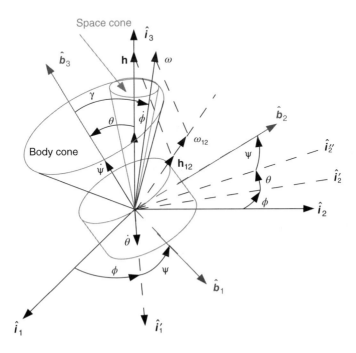

Figure 9.5 Free rigid body motion with 3-1-3 Euler angles in $\mathcal{F}_b$ and $\mathcal{F}_i$ frame

The first two kinematics equations:

$$\omega_{12} \sin\left(\hat{\Omega}t + \varphi\right) = \dot{\phi} \sin\theta \sin\psi \tag{9.163}$$

$$\omega_{12} \cos\left(\hat{\Omega}t + \varphi\right) = \dot{\phi} \sin\theta \cos\psi. \tag{9.164}$$

Divide the first by the second:

$$\tan\left(\hat{\Omega}t + \varphi\right) = \tan\psi. \tag{9.165}$$

Thus, either (1)$\psi = \hat{\Omega}t + \varphi$ or (2)$\psi = \pi + \hat{\Omega}t + \varphi$. In either case, the spin rate is:

$$\dot{\psi} = \hat{\Omega} = \Omega(A - C)/A. \tag{9.166}$$

Recall that $\sin(\pi + x) = -\sin x$, $\cos(\pi + x) = -\cos x$. If (1) is true, then $\omega_{12} = \dot{\phi}\sin\theta$. If (2) is true, then $\omega_{12} = -\dot{\phi}\sin\theta$. The leads to a contradiction, since $\Omega - \dot{\psi} > 0$. Thus (1) must be true, and since $\sin\theta > 0$, $\dot{\phi} = \omega_{12}/\sin\theta > 0$.

9.4.2 Axisymmetric, Torque-Free Rigid Body Summary

Solution of Euler's equations:

$$\omega_1 = \omega_{12} \sin\left(\hat{\Omega}t + \varphi\right) \tag{9.167}$$

$$\omega_2 = \omega_{12} \cos\left(\hat{\Omega}t + \varphi\right) \tag{9.168}$$

$$\omega_3 = \Omega, \tag{9.169}$$

where

$$\omega_{12} = \sqrt{\omega_1^2(0) + \omega_2^2(0)} > 0 \tag{9.170}$$

$$\varphi = \tan^{-1}\frac{\omega_1(0)}{\omega_2(0)} \tag{9.171}$$

$$\Omega = \omega(3) > 0 \tag{9.172}$$

$$\hat{\Omega} = \Omega(A - C)/A. \tag{9.173}$$

Solution of kinematics equations, see Figure 9.5:

$$\dot{\phi} = \omega_{12}/\sin\theta = \text{const.} \tag{9.174}$$

$$\dot{\theta} = 0 \tag{9.175}$$

$$\dot{\psi} = \hat{\Omega} \tag{9.176}$$

$$\cos\theta = \frac{C\Omega}{\sqrt{A^2\omega_{12}^2 + C^2\Omega^2}}. \tag{9.177}$$

Precession rate $\dot{\phi}$ is a positive constant. Nutation angle θ is also a constant. Spin rate sign depends on whether the body is minor or major axis spinner. See Figure 9.4, So for the space and body cone, $\mathbf{h}$ is a constant vector. $\mathbf{h}$, $\boldsymbol{\omega}$ and $\hat{b}_3$ are coplanar, and this plane rotates about $\mathbf{h}$ at the rate $\hat{\Omega}$ with respect to the body, which is called self-spin rate. The path of $\boldsymbol{\omega}$ creates the space cone in inertial space, and the $\boldsymbol{\omega}$'s trajectory in body frame shapes the body cone, where the body cone's symmetric axis coincides with the major axis. The rotation can be simply described as the body cone rolling on the space cone with $\boldsymbol{\omega}$ corresponding to the line of tangency without slipping between the body and space cone. More relations about the nutation angle θ are:

$$\tan\theta = \frac{h_{12}}{h_3} = \frac{I_1\omega_{12}}{I_3\omega_3} \tag{9.178}$$

$$\tan\gamma = \frac{\omega_{12}}{\omega_3} = \frac{\omega_{12}}{\Omega} \tag{9.179}$$

$$\tan\theta = \frac{I_1}{I_3}\tan\gamma = \text{constant.} \tag{9.180}$$

9.4.3 Asymmetric, Torque-Free Rigid Body

Begin with Equations(9.129)–(9.131). Unlike the axisymmetric case, these equations remain coupled and nonlinear.

When $\mathbf{g} = 0$, which leads to $\mathbf{h}$ constant in inertial space:

$$\dot{\omega}_1 = \frac{I_2 - I_3}{I_1}\omega_2\omega_3 \tag{9.181}$$

$$\dot{\omega}_2 = \frac{I_3 - I_1}{I_2}\omega_3\omega_1 \tag{9.182}$$

$$\dot{\omega}_3 = \frac{I_1 - I_2}{I_3}\omega_1\omega_2. \tag{9.183}$$

Or, in matrix form:

$$\dot{\boldsymbol{\omega}} = -\mathbf{I}^{-1}\boldsymbol{\omega}^\times\mathbf{I}\boldsymbol{\omega} \tag{9.184}$$

The rotational kinetic energy is:

$$T = \int_B \dot{\mathbf{r}} \cdot \dot{\mathbf{r}}dm = \frac{1}{2}\boldsymbol{\omega}^T\mathbf{I}\boldsymbol{\omega}. \tag{9.185}$$

Exercise: Develop T and show that T is a constant:

$$\frac{dT}{dt} = \frac{1}{2}\frac{d}{dt}\left(\boldsymbol{\omega}^T\mathbf{I}\boldsymbol{\omega}\right) = \frac{1}{2}\left[\frac{d}{dt}\left(\boldsymbol{\omega}^T\right)\mathbf{I}\boldsymbol{\omega} + \boldsymbol{\omega}^T\mathbf{I}\frac{d}{dt}\left(\boldsymbol{\omega}\right)\right] \tag{9.186}$$

$$= \frac{1}{2}\left[\dot{\boldsymbol{\omega}}^T\mathbf{I}\boldsymbol{\omega} + \boldsymbol{\omega}^T\mathbf{I}\dot{\boldsymbol{\omega}}\right] = \boldsymbol{\omega}^T\mathbf{I}\dot{\boldsymbol{\omega}} = -\boldsymbol{\omega}^T\mathbf{I}\mathbf{I}^{-1}\boldsymbol{\omega}^\times(\mathbf{I}\boldsymbol{\omega}) = 0. \tag{9.187}$$

In scalar form, the two conserved quantities T and h^2 are:

$$2T = I_1\omega_1^2 + I_2\omega_2^2 + I_3\omega_2^2 \tag{9.188}$$

$$h^2 = I_1^2\omega_1^2 + I_2^2\omega_2^2 + I_3^2\omega_2^2. \tag{9.189}$$

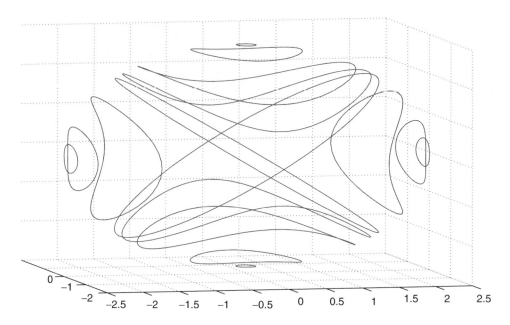

Figure 9.6   Numerical calculation of the ellipsoid of $\boldsymbol{\omega} = [\omega_1, \omega_2, \omega_3]^T$ for a fixed T

Figure 9.6 shows $\omega_1, \omega_2, \omega_3$'s relation for different given T and h. So, we have two conserved quantities (aka first integrals) and three first-order differential equations; therefore, we are two-thirds 2/3 of the way towards integrating Euler's equations.

Exercise: What is the distinction between stating $\mathbf{h} = $ const and $h = $ const? Which statement is true for the torque-free rigid body?

Assume, without loss of generality, that $I_1 > I_2 > I_3$, Thus, $2I_1 T - h^2 > 0$ and $2I_3 T - h^2 < 0$.

We can solve for ω_1^2 and ω_3^2 in terms of ω_2^2, T, h^2, and $\mathbf{I}$ and substitute into the differential equation for $\dot{\omega}_2$ to obtain:

$$\dot{\omega}_2 = f(\omega_2; \mathbf{I}, T, h^2). \tag{9.190}$$

The resulting differential equation can be integrated to obtain:

$$\omega_2 = \omega_{12}\, \text{sn}(\tau; k). \tag{9.191}$$

where a is a constant amplitude, and $\tau = \kappa(t - t_0)$.

The constants κ and k depend on $\mathbf{I}, T$, and h^2.

The solution for ω_2 involves the Jacobian elliptic function sn; the parameter $k \in [0, 1]$.

Similar solutions can be obtained for ω_1 and ω_2, and involve the three elliptic functions: sn, cn, and dn.

When k = 0, sn = sin, cn = cos, and dn = 1.

When k = 1, sn = tanh, cn = sech, and dn = sech.

In MATLAB®, elliptic functions are computed using the function call

`[sn,cn,dn]=ellipj(tau,m)`

where m = k². We have the general solution.

9.5   Summary and Keywords

This chapter introduces rigid body dynamics, including Euler's equation and moment of inertia, the coupling motion between rotation and translation; and also a few special cases are described, such as torque-free motion.

Keywords: attitude dynamics, kinetics, inertia tensor, free rigid body motion, space cone, body cone, precession, nutation, parallel axis theorem, absolute and relative angular momentum.

Problems

9.1   Determine φ in terms of initial conditions in Equation(9.147).

9.2   A rigid body S/C is at an orientation of $\psi = 5°$, $\theta = 10°$, $\phi = -3°$ has a total mass 200 kg and an inertia matrix:

$$\mathbf{I} = \begin{bmatrix} 66 & 2 & 10 \\ 2 & 31 & 7 \\ 10 & 7 & 19 \end{bmatrix} kg \cdot m^2.$$

Its orbital velocity at 5 km/s, and the (3-2-1) Euler angle yaw, pitch and roll rates are $\dot{\psi} = -2$ deg/s, $\dot{\theta} = 1$ deg/s and $\dot{\phi} = 3$ deg/s. Find the total kinetic energy of this S/C. What is its principle moment of inertia?

9.3   Consider the free rotational motion of an axially symmetric rigid body with $I_1 = I_2 = 0.5I_3$.
(a)   What is the largest possible value of the angle between ω and $\mathbf{h}$?

(b)   Find the critical value of kinetic energy that results in the largest angle between ω and $\mathbf{h}$.

9.4   Carry out the intermediate steps in proof of the Parallel Axis Theorem (PAT), making use of the fact that $\mathbf{c}^c = 0$ in Equation(9.17). Does it matter whether we use $\mathbf{r}^{co}$ or $\mathbf{r}^{oc}$ in PAT?

9.5   Show the two expressions are equivalent for a vector $\mathbf{a}$, $-\mathbf{a}^\times\mathbf{a}^\times = \mathbf{a}^T\mathbf{a}\mathbf{1} - \mathbf{a}\mathbf{a}^T$.

9.6   Why $\mathbf{I}_{pan}(2, 2) = c^2/12$ is the same in Equation(9.78) and Equation(9.84). The cylinder mass center is not the mass center of the composite body; compute the mass center for total composite body.

9.7   Why do we like to choose the origin of the frame to be at the center of mass for a spacecraft? Which equation can explain this very well?

9.8   At what conditions is the space cone inside or outside the body cone?

References

[1]   Greenwood, D.T. (1988) *Principles of Dynamics* 2nd edn. New York: Prentice-Hall.
[2]   Kaplan. M., (1976) *Modern Spacecraft Dynamics and Control*. New York: John Wiley & Sons, Inc.

10

Attitude Stabilization and Control

In this chapter a few satellite attitude stabilization and control methods are introduced. Table 10.1 shows a summary of the most frequently used methods. Then some attitude models based on rigid body models are described.

10.1 Free Body Motion Stability

Are there any special solutions for a free body motion? Again, consider Equation (9.181)–(9.183)

$$\dot{\omega}_1 = \frac{I_2 - I_3}{I_1}\omega_2\omega_3, \quad \dot{\omega}_2 = \frac{I_3 - I_1}{I_2}\omega_3\omega_1, \quad \dot{\omega}_3 = \frac{I_1 - I_2}{I_3}\omega_1\omega_2.$$

Stare at the equations long enough and see that if any two of the $\omega_j(j = 1, 2, 3)$ are zero, then all three $\dot{\omega}_j(j = 1, 2, 3)$ are zero.

Thus steady spin about either principal axis is the only steady spin that is possible.

We want to determine the stability of these steady spins.

Choice:

Prescribe an ordering of the principal moments of inertia (i.e., $I_1 > I_2 > I_3$), and check all three possible steady spins for stability; or, Select a single steady spin, and determine the conditions on the moments of inertia for stability.

Select $\boldsymbol{\omega} = [0\ \Omega\ 0]^T, \Omega > 0$.

One way to investigate stability is to linearize about the motion in question. Here we wish to linearize about $\boldsymbol{\omega} = [0\ \Omega\ 0]^T$.

By linearization, we mean adding a small perturbation to the motion in question, expand the equations of motion in a Taylor series, and determine whether the assumed small motion remains small.

Thus, we assume $\boldsymbol{\omega} = [0\ \Omega\ 0]^T + \delta\boldsymbol{\omega}$.

$$\omega_1 = 0 + \delta\omega_1 \tag{10.1}$$

$$\omega_2 = \Omega + \delta\omega_2 \tag{10.2}$$

$$\omega_3 = 0 + \delta\omega_3 \tag{10.3}$$

Fundamental Spacecraft Dynamics and Control, First Edition. Weiduo Hu.
© 2015 John Wiley & Sons (Asia) Pte Ltd. Published 2015 by John Wiley & Sons (Asia) Pte Ltd.

Table 10.1 A comparison of attitude control methods

Method	Accuracy(deg)	Axes	Characteristic
Spin stabilization	0.1–1.0	1, 2	Passive, simple, low cost, inertially oriented, need slip rings
Gravity gradient	1–5	2	Passive, simple, low cost, central body oriented
Jet RCS	0.01–1	3	Expensive, quick response, consumable required
Magnetic torquers	1–2	2	Cheap, slow, lightweight LEO only
Momentum wheel	0.1–1	2	Expensive, similar to dual spin
Reaction wheels	0.001–1	3	Expensive, precise, high power faster slew, internal torque
Control momentum Gyro (CMG)	0.001–1	3	Expensive, heavy, quick for fast slew, requires other momentum

Substitute these expressions into Euler's equations:

$$\delta\dot{\omega}_1 = \frac{I_2 - I_3}{I_1}(\Omega + \delta\omega_2)\delta\omega_3 \tag{10.4}$$

$$\delta\dot{\omega}_2 = \frac{I_3 - I_1}{I_2}\delta\omega_3\delta\omega_1 \tag{10.5}$$

$$\delta\dot{\omega}_3 = \frac{I_1 - I_2}{I_3}\delta\omega_1(\Omega + \omega_2). \tag{10.6}$$

Expand and ignore the products of $\delta\omega_j$ terms:

$$\delta\dot{\omega}_1 = \frac{I_2 - I_3}{I_1}\Omega\delta\omega_3 \tag{10.7}$$

$$\delta\dot{\omega}_2 = 0 \tag{10.8}$$

$$\delta\dot{\omega}_3 = \frac{I_1 - I_2}{I_3}\Omega\delta\omega_1. \tag{10.9}$$

These equations are linear, constant-coefficient, coupled, ordinary differential equations.

You should compare these equations with the Euler's equations for an axisymmetric rigid body.

One difference is that the constants are not quite the same as they were in the axisymmetric case (because $I_1 \neq I_3$ here).

Differentiate $\delta\dot{\omega}_1$ and substitute $\delta\dot{\omega}_3$ to obtain:

$$\delta\ddot{\omega}_1 + \frac{(I_2 - I_1)(I_2 - I_3)}{I_1 I_3}\Omega^2\delta\omega_1 = 0. \tag{10.10}$$

Clearly this is a second-order, linear, constant-coefficient, ordinary differential equation in the form:

$$\ddot{x} + kx = 0 \tag{10.11}$$

(mass-on-a-spring) $k > 0$ is required for stability, which translates to $I_2 > \{I_1, I_3\}$ or $I_2 < \{I_1, I_3\}$.

10.2 Dual-Spin Stabilization

Assume the angular velocity of the flywheel is:

$$\boldsymbol{\omega}^{wi} = \boldsymbol{\omega} + \boldsymbol{\omega}^{wb}, \tag{10.12}$$

where

$$\boldsymbol{\omega}^{wb} = \Omega_w \hat{b}_3 = \Omega_w [0, 0, 1]^T. \tag{10.13}$$

The total angular momentum:

$$\mathbf{h} = \mathbf{I}\boldsymbol{\omega} + I_w \boldsymbol{\omega}^{wb}, \tag{10.14}$$

then

$$\dot{\omega}_1 = \frac{I_2 - I_3}{I_1} \omega_2 \omega_3 - \frac{I_w}{I_1} \Omega_w \omega_2 + \frac{g_1}{I_1} \tag{10.15}$$

$$\dot{\omega}_2 = \frac{I_3 - I_1}{I_2} \omega_3 \omega_1 + \frac{I_w}{I_2} \Omega_w \omega_3 + \frac{g_2}{I_2} \tag{10.16}$$

$$\dot{\omega}_3 = \frac{I_1 - I_2}{I_3} \omega_1 \omega_2 + \frac{g_3}{I_3}. \tag{10.17}$$

If the steady state is $[0, 0, \Omega]^T$, variation equation about it is:

$$\delta\dot{\omega}_1 = \frac{I_2 - I_3}{I_1} \Omega \delta\omega_2 - \frac{I_w}{I_1} \Omega_w \delta\omega_2 \tag{10.18}$$

$$\delta\dot{\omega}_2 = \frac{I_3 - I_1}{I_2} \Omega \delta\omega_1 + \frac{I_w}{I_2} \Omega_w \delta\omega_1 \tag{10.19}$$

$$\delta\dot{\omega}_3 = 0, \tag{10.20}$$

then the characterization equation is:

$$\delta\ddot{\omega}_1 + \left(\frac{I_3 - I_2}{I_1} \Omega + \frac{I_w}{I_1} \Omega_w \right) \left(\frac{I_3 - I_1}{I_2} \Omega + \frac{I_w}{I_2} \Omega_w \right) \delta\omega_1 \tag{10.21}$$

thus the stability condition is:

$$I_3 > I_2 - I_w \Omega_w / \Omega \quad \text{and} \quad I_3 > I_1 - I_w \Omega_w / \Omega \tag{10.22}$$

or

$$I_3 < I_2 - I_w \Omega_w / \Omega \quad \text{and} \quad I_3 < I_1 - I_w \Omega_w / \Omega. \tag{10.23}$$

10.3   Gravity Gradient Stabilization

See Figure 10.1. For any differential mass element of S/C:

$$\mathrm{d}\mathbf{f}_g = -\frac{\mu}{r^2}\hat{e}_r \mathrm{d}m. \tag{10.24}$$

Any position from mass center of S/C is:

$$\mathbf{r}_c = \mathbf{r} - \mathbf{r}_{oc}, \tag{10.25}$$

we know $|\mathbf{r}_{oc}| \gg |\mathbf{r}_c|$. The torque about the mass center of S/C:

$$\mathbf{g}_g^c = \int_B \mathbf{r}_c \times \mathrm{d}\mathbf{f}_g = -\mu \int_B \frac{\mathbf{r}_c \times \mathbf{r}}{r^3} \mathrm{d}m \tag{10.26}$$

$$\approx -\mu \int_B \mathbf{r}_c \times (\mathbf{r}_{oc} + \mathbf{r}_c) \frac{1 - 3(\mathbf{r}_c \cdot \mathbf{r}_{oc})/r_{oc}^2}{r_{oc}^3} \mathrm{d}m \tag{10.27}$$

$$= \frac{\mu}{r_{oc}^5} \mathbf{r}_{oc} \times \int_B \mathbf{r}_c(\mathbf{r}_{oc} - 3\mathbf{r}_c)\mathrm{d}m \cdot \mathbf{r}_{oc} \tag{10.28}$$

$$= -\frac{3\mu}{r_{oc}^5} \mathbf{r}_{oc} \times \int_B (\mathbf{r}_c \mathbf{r}_c)\mathrm{d}m \cdot \mathbf{r}_{oc} \tag{10.29}$$

$$= \frac{3\mu}{r_{oc}^3} \frac{\mathbf{r}_{oc}}{r_{oc}} \times \int_B (\mathbf{r}_c^T \mathbf{r}_c \mathbf{1} - \mathbf{r}_c \mathbf{r}_c^T)\mathrm{d}m \cdot \frac{\mathbf{r}_{oc}}{r_{oc}} \tag{10.30}$$

$$= \frac{3\mu}{r_{oc}^3} \hat{o}_3 \times \int_B (\mathbf{r}_c^T \mathbf{r}_c \mathbf{1} - \mathbf{r}_c \mathbf{r}_c^T)\mathrm{d}m \cdot \hat{o}_3. \tag{10.31}$$

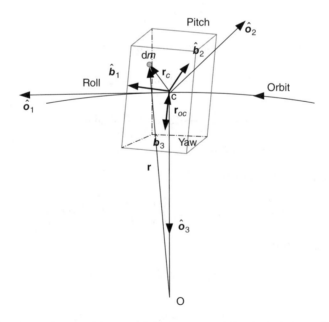

Figure 10.1   Gravity gradient model, where O is the center of the Earth, $\mathcal{F}_b$ is the body frame whose origin is at C, the mass center of the rigid body, $\mathcal{F}_o$ is the orbital frame in Section 8.1

Above derivation is similar to the derivation of CW-equation, see Section 4.1. We give detailed explanation here as gravity gradient is a typical problem that orbit and attitude are coupled.

The torque can be simplified as:

$$\mathbf{g}_g^c = 3\frac{\mu}{r_{oc}^3}\hat{o}_3 \times \mathbf{I} \cdot \hat{o}_3 = 3\omega_o^2\hat{o}_3 \times \mathbf{I} \cdot \hat{o}_3 \tag{10.32}$$

$$= 3\omega_o^2\begin{bmatrix} (I_3 - I_2)r_{23}r_{33} \\ (I_1 - I_3)r_{13}r_{33} \\ (I_2 - I_1)r_{13}r_{23} \end{bmatrix}. \tag{10.33}$$

Note that $\hat{o}_3$ here is the third axis of orbital frame expressed in body frame, that is $\hat{o}_3^b = [r_{13}, r_{23}, r_{33}]^T$ of is the third column of the rotation matrix $\mathbf{R}^{bo}$, see Equation (10.39). ω_o is circular orbit angular velocity.

The basic problem involves the rotational motion of a rigid satellite in a Keplerian orbit. The circular orbit assumption is well-known and is useful for many satellite applications.

This analysis is about the rotational motion of a rigid satellite in a circular orbit, with the body frame "close" to the orbital frame; that is roll, pitch, and yaw angles and rates are small.

Newton:

$$\ddot{\mathbf{r}} + \frac{\mu}{r^3}\mathbf{r} = \mathbf{f}, \tag{10.34}$$

Euler:

$$\dot{\mathbf{h}} = 3\frac{\mu}{r^3}\hat{o}_3 \times \mathbf{I} \cdot \hat{o}_3. \tag{10.35}$$

This assumes size of satellite $\ll$ size of orbit. $\hat{o}_3$ is the nadir vector of the orbital reference frame.

Expressing the rotational equations in the body frame gives:

$$\mathbf{I}\dot{\omega} = -\omega^\times\mathbf{I}\omega + 3\frac{\mu}{r^3}\hat{o}_3^\times\mathbf{I}\hat{o}_3. \tag{10.36}$$

An equilibrium of these equations is for the principal body frame to be aligned with the orbital frame, in which case:

$$\omega^\times\mathbf{I}\omega = 0 \tag{10.37}$$

$$\hat{o}_3^\times\mathbf{I}\hat{o}_3 = \mathbf{0}. \tag{10.38}$$

Use a 1-2-3 (roll-pitch-yaw), i.e. $\theta_1 - \theta_2 - \theta_3$ Euler angle sequence to relate $\mathcal{F}_b$ to $\mathcal{F}_o$:

$$\mathbf{R}^{bo} = \begin{bmatrix} c\theta_2c\theta_3 & s\theta_1s\theta_2c\theta_3 + c\theta_1s\theta_3 & s\theta_1s\theta_3 - c\theta_1s\theta_2c\theta_3 \\ -c\theta_2s\theta_3 & c\theta_1c\theta_3 - s\theta_1s\theta_2s\theta_3 & s\theta_1c\theta_3 + c\theta_1s\theta_2s\theta_3 \\ s\theta_2 & -s\theta_1c\theta_2 & c\theta_1c\theta_2 \end{bmatrix}. \tag{10.39}$$

For small Euler angles,

$$\mathbf{R}^{bo} = \mathbf{1} - \boldsymbol{\theta}^\times \Rightarrow \hat{o}_3 = [-\theta_2, \theta_1, 1]^T. \tag{10.40}$$

Thus $\hat{o}_3 = [-\theta_2, \theta_1, 1]^T$, the gravity gradient torque is:

$$\mathbf{g}_{gg} = 3\omega_o^2\hat{o}_3^\times\mathbf{I}\hat{o}_3 = 3\omega_o^2\begin{bmatrix} (I_3 - I_2)\theta_1 \\ (I_3 - I_1)\theta_2 \\ 0 \end{bmatrix}, \tag{10.41}$$

where ω_o is orbital angular velocity.

Write angular velocity in terms of Euler angles and their rates in $\mathcal{F}_b$ frame:

$$\omega^{bi} = \omega^{bo} + \omega^{oi} \tag{10.42}$$

$$= \begin{bmatrix} \dot{\theta}_1 \\ \dot{\theta}_2 \\ \dot{\theta}_3 \end{bmatrix} + \begin{bmatrix} 1 & \theta_3 & -\theta_2 \\ -\theta_3 & 1 & \theta_1 \\ \theta_2 & -\theta_1 & 1 \end{bmatrix} \begin{bmatrix} 0 \\ -\omega_o \\ 0 \end{bmatrix} = \begin{bmatrix} \dot{\theta}_1 - \omega_o\theta_3 \\ \dot{\theta}_2 - \omega_o \\ \dot{\theta}_3 + \omega_o\theta_1 \end{bmatrix}. \tag{10.43}$$

Drop the superscript bi and differentiate:

$$\dot{\omega} = \begin{bmatrix} \ddot{\theta}_1 - \omega_o\dot{\theta}_3 \\ \ddot{\theta}_2 \\ \ddot{\theta}_3 + \omega_o\dot{\theta}_1 \end{bmatrix}. \tag{10.44}$$

Now we have $\hat{o}_3$, ω and $\dot{\omega}$ for Euler's equations, that is Equation (10.36).

$$I_1\ddot{\theta}_1 + (I_2 - I_3 - I_1)\omega_o\dot{\theta}_3 - 4(I_3 - I_2)\omega_o^2\theta_1 = 0 \tag{10.45}$$

$$I_2\ddot{\theta}_2 + 3(I_1 - I_3)\omega_o^2\theta_2 = 0 \tag{10.46}$$

$$I_3\ddot{\theta}_1 + (I_3 + I_1 - I_2)\omega_o\dot{\theta}_1 + (I_2 - I_1)\omega_o^2\theta_3 = 0. \tag{10.47}$$

Note that the pitch equation (θ_2) is decoupled from the roll (θ_1) and yaw (θ_3) equations, and is of the form $\ddot{x} + kx = 0$, so that the pitch motion is stable if $I_1 > I_3$.

The Earth-pointing axis, $\hat{b}_3$, cannot be the major axis. If the other two equations of motion are unstable, then the pitch motion does not necessarily remain small, since the nonlinear terms are no longer negligible.

In matrix second-order form:

$$\mathbf{M}\ddot{\mathbf{x}} + \mathbf{G}\dot{\mathbf{x}} + \mathbf{K}\mathbf{x} = 0. \tag{10.48}$$

To analyze the stability, we seek solution when $\mathbf{x} = e^{\lambda t}\mathbf{c}_0$. So

$$e^{\lambda t}\left[\lambda^2\mathbf{M} + \lambda\mathbf{G} + \mathbf{K}\right]\mathbf{c}_0 = \mathbf{0}, \tag{10.49}$$

where

$$\mathbf{M} = \begin{bmatrix} I_1 & 0 \\ 0 & I_3 \end{bmatrix}, \tag{10.50}$$

$$\mathbf{G} = (I_1 + I_3 - I_2)\omega_o\begin{bmatrix} 0 & -1 \\ 1 & 0 \end{bmatrix}, \tag{10.51}$$

$$\mathbf{K} = \omega_o^2\begin{bmatrix} 4(I_2 - I_3) & 0 \\ 0 & (I_2 - I_1) \end{bmatrix}. \tag{10.52}$$

To find the characteristic polynomial, we set the determinant:

$$\begin{vmatrix} \lambda^2 I_1 + 4\omega_o^2(I_2 - I_3) & -\lambda\omega_o(I_1 + I_3 - I_2) \\ \lambda\omega_o(I_1 + I_3 - I_2) & \lambda^2 I_3 + \omega_o^2(I_2 - I_1) \end{vmatrix} = 0. \tag{10.53}$$

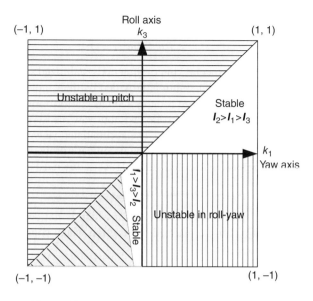

Figure 10.2   Region of stability for gravity gradient

If we assume $k_1 = (I_2 - I_3)/I_1, k_3 = (I_2 - I_1)/I_3$, then

$$(\lambda/\omega_o)^4 + (1 + 3k_1 + k_1 k_3)(\lambda/\omega_o)^2 + 4k_1 k_3 = 0, \tag{10.54}$$

if we assume $s = \lambda/\omega_o$ then

$$s^2 + b_1 s + b_0 = 0, \tag{10.55}$$

the condition for the λs are pure imaginary is $b_0 > 0, b_1 > 0, b_1^2 - 4b_0 > 0$. This is equivalent to:

$$k_1 > k_3 \tag{10.56}$$

$$k_1 k_3 > 0 \tag{10.57}$$

$$1 + 3k_1 + k_1 k_3 > 0 \tag{10.58}$$

$$(1 + 3k_1 + k_1 k_3)^2 - 16k_1 k_3 > 0. \tag{10.59}$$

See Figure 10.2 [3].

10.4   Three-axis Stabilization

The z-axis is pointing to the Earth; y-axis is pointing to the orbital normal; ω_o is the orbital angular velocity. The only momentum wheel is aligned on the body y-axis, h_{wy} and $\dot{y}_{wy}$. By linearizing the Euler equation and including the gravity gradient, the scalar form becomes:

$$I_x \ddot{\phi} + \left[4\omega_o^2(I_y - I_z) - \omega_o h_{wy}\right]\phi - [h_{wy} + \omega_o(I_x + I_z - I_2)]\dot{\psi} = T_{dx} + T_{cz} \tag{10.60}$$

$$I_z\ddot{\psi} + \left[\omega_o(I_x + I_z - I_y) + h_{wy}\right]\dot{\phi} + \left[\omega_o^2(I_y - I_x) - \omega_o h_{wy}\right]\psi = T_{dx} + T_{cx} \qquad (10.61)$$

$$I_y\ddot{\theta} + 3\omega_o^2(I_x - I_z)\theta + \dot{h}_{wy} = T_{dy} + T_{cy}. \qquad (10.62)$$

Momentum-biased satellites are a special kind of three-axis-stabilized satellite, with dual-spin property, whose constant angular momentum is provided by their momentum wheel, but not as the dual-spin satellites do using a large rotator, a small platform for mounting antenna or camera. These kinds of satellite can be controlled without yaw measurements, as the roll and yaw errors interchange every quarter of a orbit. The accumulated yaw error can be attenuated by the roll measurement and control. See the orbital compass problem in Section 8.5.2. To introduce the basic idea of control and stabilization, let's use the most simple model:

$$\ddot{\theta} = T_{cy}/I_y = t_{cy}. \qquad (10.63)$$

Actually any part in the left part of Equation (10.62) which is not desired can be included in the control T_{cy}. PID control can be realized by reaction wheel control. If we chose $t_{cy} = (-2\xi\omega_n\dot{\theta} - \omega_n^2\theta)$ the close-loop system is:

$$\ddot{\theta} + 2\xi\omega_n\dot{\theta} + \omega_n^2 = 0. \qquad (10.64)$$

This is a standard form in control theory, where θ is measured by the Earth sensor in pitch axis.

Another method of control is phase plane which is used for the jet engine control of S/C. For pitch channel Equation (10.63), most simple control logic is if $\theta \geq 0$, then $t_{cy} = -1$, and otherwise $t_{cy} = 1$; this is called "bang-bang control." The final state is a limit circle with two joint parabolic trajectories. See Figure 10.3. To remove the limit circle, the tilt switch line can be used.

$$t_{cy} = -1 \text{ when } \theta + \tau\dot{\theta} \geq 0, \qquad (10.65)$$

$$t_{cy} = +1 \text{ when } \theta + \tau\dot{\theta} < 0, \qquad (10.66)$$

where $\tau > 0$. This time there is no limit cycle, but chatter appears at the final stage along the switch line when approaching zero, see Figure 10.4. To remove the chatter, dead zone can be added. The control logic becomes:

$$t_{cy} = -1 \text{ when } \dot{\theta} \geq 0 \text{ and } \theta + \tau\dot{\theta} \geq \epsilon_2 \qquad (10.67)$$

$$\text{or } \dot{\theta} < 0 \text{ and } \theta + \tau\dot{\theta} \geq \epsilon_1 \qquad (10.68)$$

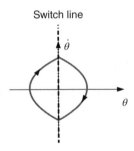

Figure 10.3   Phase plane of bang-bang control

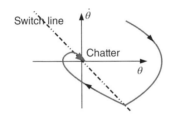

Figure 10.4   Phase plane of tilt bang-bang control

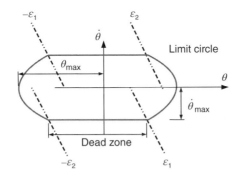

Figure 10.5   Phase plane of tilt bang-bang control with dead zones

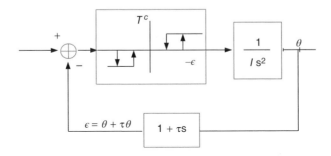

Figure 10.6   Control block diagram

$$t_{cy} = +1 \quad \text{when } \dot{\theta} \leq 0 \text{ and } \theta + \tau\dot{\theta} \leq -\epsilon_2 \tag{10.69}$$

$$\text{or } \dot{\theta} > 0 \text{ and } \theta + \tau\dot{\theta} < -\epsilon_1 \tag{10.70}$$

$$t_{cy} = 0 \quad \text{others.} \tag{10.71}$$

See Figure 10.5. With this control logic, a limit cycle can exist, but we can adjust the parameter ϵ_1 and ϵ_2 to change its size, i.e. $\theta_{\max}$ and $\dot{\theta}_{\max}$. For the control block diagram for this logic, see Figure 10.6. Practical phase plane design can be even more complicated as more factors are considered, such as delay, overshoot, disturbance, etc. See Figure 10.7, in which, in Region

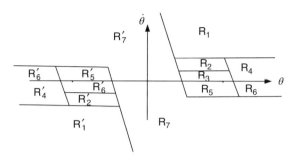

Figure 10.7   A practical phase plane design (not to scale) adapted from reference [2]

R_1 and R_1' large deceleration and acceleration are used. In Region R_7 and R_7' no propulsion is given. In different regions, different magnitudes and widths of jets are performed [2].

10.5   Complete Dynamical Modeling of Spacecraft

The augmented Euler equation of motion, which is an extension of Equations (9.122)–(9.125). Solar panel flexibility and liquid sloshing are included:

$$m_T \frac{d\mathbf{v}_T}{dt} + \mathbf{B}_t \ddot{\eta} + \mathbf{D}_t \ddot{\sigma} = \mathbf{F} \tag{10.72}$$

$$\mathbf{I}\dot{\omega} + \omega \times \left[\mathbf{I}\omega + \mathbf{B}_r \dot{\eta} + \mathbf{D}_r \dot{\sigma}\right] + \mathbf{B}_r \ddot{\eta} + \mathbf{D}_r \ddot{\sigma} = \mathbf{T}_c + \mathbf{T}_d \tag{10.73}$$

$$\mathbf{U}\ddot{\eta} + 2\zeta_\eta \mathbf{\Omega}_\eta \dot{\eta} + \mathbf{\Omega}_\eta^2 \eta + \mathbf{B}_r^T \dot{\omega} + \mathbf{B}_t^T \frac{d\mathbf{v}_T}{dt} = 0 \tag{10.74}$$

$$\mathbf{E}\ddot{\sigma} + 2\zeta_\sigma \mathbf{\Omega}_\sigma \dot{\sigma} + \mathbf{\Omega}_\sigma^2 \sigma + \mathbf{D}_r^T \dot{\omega} + \mathbf{D}_t^T \frac{d\mathbf{v}_T}{dt} = 0, \tag{10.75}$$

where $\mathbf{B} = \mathbf{K}\mathbf{C}^T$.

$$\mathbf{C} = \begin{bmatrix} \cos \alpha_p & 0 & -\sin \alpha_p \\ 0 & 1 & 0 \\ \sin \alpha_p & 0 & \cos \alpha_p \end{bmatrix}, \tag{10.76}$$

where α_p is solar panel rotation angle.
m_T is the total mass, $\mathbf{v}_T$ is the velocity;

$\mathbf{U}$: unit matrix $m \times m$;
ζ_η: flexible modes damping matrix, diagonal $(m \times m)$;
$\mathbf{\Omega}_\eta$: flexible models frequency matrix $(m \times m)$;
$\mathbf{B}_r$: flexible modes coupling matrix $(3 \times m)$ for rotation;
$\mathbf{B}_t$: flexible modes coupling matrix $(3 \times m)$ for translation;
$\mathbf{K}$: matrix of coupling coefficients $(m \times 3)$;

E: unit matrix $(n \times n)$;

ζ_σ: sloshing modes damping matrix, diagonal $(n \times n)$;

Ω_σ: sloshing modes frequency matrix, diagonal $(n \times n)$;

$\mathbf{D}_r$: sloshing modes coupling matrix $(3 \times n)$ for rotation.

$\mathbf{D}_t$: sloshing modes coupling matrix $(3 \times n)$ for translation.

And

$$\Omega_\eta = \mathrm{diag}\left[\Omega_{xp1}, \Omega_{xp2}, \cdots, \Omega_{yp1}, \Omega_{yp2}, \cdots, \Omega_{zp1}, \Omega_{zp2}, \cdots\right]^T, \tag{10.77}$$

$$\zeta_\eta = \mathrm{diag}\left[\zeta_{xp1}, \zeta_{xp2}, \cdots, \zeta_{yp1}, \zeta_{yp2}, \cdots, \zeta_{zp1}, \zeta_{zp2}, \cdots\right]^T, \tag{10.78}$$

$$\eta = \left[\eta_{xp1}, \eta_{xp2}, \cdots, \eta_{yp1}, \eta_{yp2}, \cdots, \eta_{zp1}, \eta_{zp2}, \cdots\right]^T, \tag{10.79}$$

$$\tag{10.80}$$

then

$$\mathbf{K}^T = \begin{bmatrix} k_{xp1}, k_{xp2}, \cdots & 0 & 0 \\ 0 & k_{yp1}, k_{yxp2}, \cdots & 0 \\ 0 & 0 & k_{zp1}, k_{zp2}, \cdots \end{bmatrix} \tag{10.81}$$

$$\mathbf{B} = \mathbf{KC}^T = \begin{bmatrix} k_{xp1} \cos \alpha_p & 0 & -k_{xp1} \sin \alpha_p \\ k_{xp1} \cos \alpha_p & 0 & -k_{xp1} \sin \alpha_p \\ \vdots & 0 & \vdots \\ 0 & k_{yp1} & 0 \\ 0 & k_{yp2} & 0 \\ 0 & \vdots & 0 \\ k_{zp1} \sin \alpha_p & 0 & k_{zp1} \cos \alpha_p \\ k_{zp1} \sin \alpha_p & 0 & k_{zp1} \cos \alpha_p \\ \vdots & 0 & \vdots \end{bmatrix}. \tag{10.82}$$

Equation (10.73) is a nonlinear equation of augmented Euler moment equation; Equation (10.74) and (10.75) are linearized equations of structural and sloshing dynamics, so they are correct only for small amplitude.

10.5.1 Flexible Panels Example

Normally, spacecraft (S/C) is composed of a rigid body plus two symmetric solar panels.

The solar panel can be modeled as two discrete mass m, positioned at a distance L from the center of mass (COM) of rigid body.

For the one axis model (θ) with one order flexibility, (u), such as the pitch axis:

$$(I_0 + 2mL^2)\ddot{\theta} + 2mL\ddot{u} = T \tag{10.83}$$

$$m\ddot{u} + K_d\dot{u} + Ku = -mL\ddot{\theta}, \tag{10.84}$$

where I_0 is the moment of inertia of the S/C main rigid body. $I_p = 2mL^2$ is the moment of inertia of one of the two (panel masses) about the cm. The transfer function is:

$$\frac{\theta(s)}{T(s)} = \frac{1}{I_0 s^2} \frac{s^2 + \frac{K_d}{m}s + \frac{k}{m}}{s^2 + \frac{I_0+2mL^2}{I_0}\frac{K_d}{m}s + \frac{I_0+2mL^2}{I_0}\frac{K}{m}} \tag{10.85}$$

$$= \frac{1}{I_0 s^2} \frac{s^2 + 2\xi\sigma s + \sigma^2}{s^2 + 2\xi_s\sigma_s s + \sigma_s}, \tag{10.86}$$

where $\sigma = \sqrt{K/m}$, the panel natural frequency; $\xi = K_d/(2\sqrt{mK})$ is its damping coefficient. The modal frequency of the entire system is $\sigma_s = \sqrt{(I_0 + 2mL^2)/I_0}\sigma$, and also the damping coefficient $\xi_s = \sqrt{(I_0 + 2mL^2)/I_0}\xi$.

For second order flexibility, each panel can be a model as two point mass m_1, m_2 with the distance to CM x_1, x_2, then the system model is:

$$(I_0 + 2m_1 x_1^2 + 2m_2 x_2^2)\ddot{\theta} + 2mx_1\ddot{u}_1 + 2mx_2\ddot{u}_2 = T \tag{10.87}$$

$$m_1\ddot{u}_1 + D_{11}\dot{u}_1 + D_{12}\dot{u}_2 + K_{11}u_1 + K_{12}u_2 = -\ddot{\theta}m_1 x_1 \tag{10.88}$$

$$m_2\ddot{u}_2 + D_{12}\dot{u}_2 + D_{22}\dot{u}_2 + K_{12}u_1 + K_{22}u_2 = -\ddot{\theta}m_2 x_2. \tag{10.89}$$

10.5.2 Liquid Sloshing Example

The mass of the non moving part of the fuel tank is M_o at the mass center of the whole system, with moment of inertial I_o thrust force f, thrust arm d, moving mass m; equivalent arm of m is b.

Two simplified models are used for liquid sloshing; the oscillation frequency is $\omega_{osc} = \sqrt{g/L}$, for pendulum model; $\omega_{osc} = \sqrt{k/m}$ for spring-mass model; equivalent arm length L; mass m and spring constant $k = mgL = m\omega_{osc}^2$.

$$I_o\ddot{\theta} + (bk + mg)x = df \tag{10.90}$$

$$m\ddot{x} + k(1 + m/M)x - mb\ddot{\theta} = -mf/M_o. \tag{10.91}$$

10.5.3 A Mass Spring S/C Example

A rigid body with a mass spring model is a foundation to analyze spacecraft with solar panel and liquid sloshing. Let us give a more practical discussion. The S/C body mass without mass-spring is m_C; origin is at O; the mass of the mass-spring is m_p; the spring coefficient is k_p; its position vector is $\mathbf{r}_p$; its motion is constrained in the plane $x = x_p$; its equilibrium point is $\mathbf{r}_p^0 = [x_p, 0, 0]^T$. The moment of inertia of body $\mathbf{I}_C$. Other variable $\mathbf{v}_0 = [v_x, v_y, y_z]^T$, $\boldsymbol{\omega} = [\omega_x, \omega_y, \omega_z]^T$, $\mathbf{r}_p = \mathbf{r}_p^0 + \mathbf{r}_p^r$, $\mathbf{r}_p^r = [0, y_p, z_p]^T$, $\dot{\mathbf{r}}_p = \dot{\mathbf{r}}_p^r = [0, \dot{y}_p, \dot{z}_p]^T$, $m = m_C + m_p$.

The total kinetic energy is:

$$T = \frac{1}{2}\mathbf{v}_0^T m_C \mathbf{v}_0 + \frac{1}{2}\boldsymbol{\omega}^T \mathbf{I}_C \boldsymbol{\omega} + \frac{1}{2}m_p(\mathbf{v}_0 + \boldsymbol{\omega}^\times \mathbf{r}_p + \dot{\mathbf{r}}_p)^T(\mathbf{v}_0 + \boldsymbol{\omega}^\times \mathbf{r}_p + \dot{\mathbf{r}}_p); \tag{10.92}$$

the system internal potential is:

$$V = \frac{1}{2}(\mathbf{r}_p^r)^T k_p \mathbf{r}_p^r; \tag{10.93}$$

the Lagrange function is $L = T - V = L(\mathbf{v}_0, \boldsymbol{\omega}, \dot{\mathbf{r}}_p^r, \mathbf{r}_p^r)$. If the external force $\mathbf{F}$ is along the x axis through the mass center, that is, there is no external torque. And if the $\mathbf{r}_p^r$ is perpendicular to the external force $\mathbf{F}$, that is, no general force, then

$$m\dot{\mathbf{v}}_0 + m\boldsymbol{\omega}^\times \mathbf{v}_0 + m_p \ddot{\mathbf{r}}_p + m_p \dot{\boldsymbol{\omega}}^\times \mathbf{r}_p^r + m_p \boldsymbol{\omega}^\times \boldsymbol{\omega}^\times \mathbf{r}_p^r + 2m_p \boldsymbol{\omega}^\times \dot{\mathbf{r}}_p = \mathbf{F} \tag{10.94}$$

$$\mathbf{I}\dot{\boldsymbol{\omega}} + \boldsymbol{\omega}^\times \mathbf{I}\boldsymbol{\omega} + \mathbf{I}_p^r \dot{\boldsymbol{\omega}} + \boldsymbol{\omega}^\times \mathbf{I}_p^r \boldsymbol{\omega} + \dot{\mathbf{I}}_p^r \boldsymbol{\omega} + m_p \mathbf{r}^\times \ddot{\mathbf{r}}_p + m_p \mathbf{r}_p^\times \dot{\mathbf{v}}_0$$

$$+ m_p \boldsymbol{\omega}^\times \mathbf{r}_p^\times \dot{\mathbf{r}}_p + m_p \boldsymbol{\omega}^\times \mathbf{r}_p'^\times \mathbf{v}_0 + m_p \mathbf{v}_0^\times \boldsymbol{\omega}^\times \mathbf{r}_p^r = 0 \tag{10.95}$$

$$m_p [\ddot{y}_p + \dot{v}_y + x_p \dot{\omega}_z - z_p \dot{\omega}_x - y_p(\omega_x^2 + \omega_z^2) - 2\omega_x \dot{z}_p + z_p \omega_y \omega_z$$

$$+ x_p \omega_x \omega_y + \omega_z v_x - \omega_x v_z] + k_p y_p = 0 \tag{10.96}$$

$$m_p [\ddot{z}_p + \dot{v}_z + y_p \dot{\omega}_x - x_p \dot{\omega}_y - z_p(\omega_x^2 + \omega_y^2) + 2\omega_x \dot{y}_p + y_p \omega_y \omega_z$$

$$+ x_p \omega_x \omega_z + \omega_x v_y - \omega_y v_x] + k_p z_p = 0, \tag{10.97}$$

where $\mathbf{I}_p$ and $\mathbf{I}_p^r$ are:

$$\mathbf{I}_p = \begin{bmatrix} 0 & 0 & 0 \\ 0 & m_p x_p^2 & 0 \\ 0 & 0 & m_p x_p^2 \end{bmatrix} \tag{10.98}$$

$$\mathbf{I}_p^r = \begin{bmatrix} m_p(y_p^2 + z_p^2) & -m_p x_p y_p & -m_p x_p z_p \\ -m_p x_p y_p & m_p z_p^2 & -m_p y_p z_p \\ -m_p x_p z_p & -m_p y_p z_p & m_p y_p^2 \end{bmatrix}. \tag{10.99}$$

More discussion about this model is given in reference [2].

10.6   Summary and Keywords

This chapter describes a few methods of satellite stabilization and introduces some attitude models including flexibility, liquid sloshing with examples.

Keywords: free body motion stability, duel-spin stabilization, gravity gradient stabilization, three-axis stabilization, phase plane control, solar panel flexibility, liquid sloshing.

Problems

10.1 Derive some relations between θ_{max}, $\dot{\theta}_{max}$, ϵ_1, ϵ_2 and τ in Figure 10.5.

10.2 If we ignore the roll and yaw motion of a rigid satellite in a circular orbit, we can develop a single second-order ordinary differential equation for the pitch angle θ about the $\hat{\mathbf{b}}_2$ axis. This equation is:

$$I_2 \ddot{\theta} + 3\omega_o^2(I_1 - I_3) \sin \theta \cos \theta = 0. \tag{10.100}$$

Tell me as much as you can about the equilibrium solutions of this equation. Questions you should address include: how many equilibria are there? and how do their stability properties depend on the moments of inertia?

Find its Jacobi integral, plot its phase plane.

All of your conclusions must be supported by your direct application of this differential equation; that is, you cannot simply state facts that were covered in class.

10.3 A spacecraft is in a circular, equatorial, prograde orbit with altitude 500 km. The sub-satellite point is on the Greenwich meridian ($L = 0$). You may assume that $\theta_g = 0$ at this time. The orientation of the spacecraft body frame F_b with respect to the orbital frame F_o is described by the quaternion:

$$\bar{q}^{bo} = [0.008\,628, 0.086\,293, 0.008\,628, 0.996\,195]^T. \tag{10.101}$$

What are the latitude δ? and longitude, L, of the "target" pointed to by a sensor whose axis is $\mathbf{q}_b = [0.00\,944, \ 0.00\,230, \ 0.99\,995]^T$? Give your answers in degrees.

10.4 Actually, not everything needed to this problem is given, some references are needed. A disk-shaped satellite is to be used in a gravity-gradient stabilized configuration, with the flat face of the disk pointing at the Earth. The disk has mass m, radius r, and negligible thickness.

(a) Compute the Smelt parameters k_1 and k_3 for this configuration.

(b) Is the satellite stable in this configuration? Please note that "yes" or "no" is not an acceptable answer!

The Acme Boom Company proposes the addition of a long boom to enhance the stability of the satellite. To save money and mass, we want to use the shortest boom possible.

(c) What is the minimum length boom we should use? Model the boom as a rigid rod with density per unit length.

In order to avoid a resonance condition, we want the pitch frequency to be greater than $2\omega_o$, where ω_o is the orbital angular velocity.

(d) Is it possible to do this using a boom? If no, why not and how might you modify the satellite to satisfy this requirement? If yes, how long would the boom have to be?

10.5 This problem involves linearization and stability analysis of a system of nonlinear ordinary differential equations. Consider the following system of differential equations:

$$\dot{\omega}_1 = \frac{I_2 - I_3 - I_r}{I_1}\omega_2\omega_3 + \frac{I_r}{I_1}\omega_2\omega_r \tag{10.102}$$

$$\dot{\omega}_2 = \frac{I_3 - I_2 - I_r}{I_2}\omega_1\omega_3 + \frac{I_r}{I_2}\omega_1\omega_r \tag{10.103}$$

$$\dot{\omega}_3 = \frac{I_1 - I_2}{I_3}\omega_1\omega_2, \tag{10.104}$$

where I_1, I_2, I_3 are the (constant) principal moments of inertia of a spacecraft containing a momentum wheel spinning about the "3" axis. The wheel has moment of inertia I_r about its spin axis, and is spinning with constant angular velocity I_r. The "3" axis is an intermediate axis; that is, $I_1 > I_3 > I_2$. Thus the "3" axis would be an unstable spin axis if this were a simple rigid body. However, the wheel can be used to stabilize the motion.

(a) Carefully show that a steady spin about the "3" axis, $\omega_{eq} = [0\ 0\ \Omega]^T$, is an equilibrium motion (steady spin) for any value of ω_r.

(b) Let $I_1 = 25, I_2 = 15, I_3 = 20, I_r = 3$ (kg m^2), and $\Omega = 1$ (rad/s). Show that this equilibrium is unstable if $\omega_r = 0$.

(c) Using the same numerical values as in (b), determine the critical value $\omega_r^* > 0$ (in rad/s) such that if $0 < \omega_r < \omega_r^*$ the motion is unstable and if $\omega_r > \omega_r^*$ the motion is (marginally) stable.

References

[1] Wertz, J.R. (1985) *Spacecraft Attitude Determination and Control*. Boston: D. Reidel.

[2] Tu, S. (2001) *Satellite Attitude Dynamics and Control* (in Chinese). Beijing: China Astronautic Publishing House.

[3] Junkins, J.L. and Turner, J.D. (1986) *Optimal Spacecraft Rotational Maneuvers*. Amsterdam: Elsevier.

[4] Sidi, Marcel J. (2000) *Spacecraft Dynamics and Control*. Cambridge: Cambridge University Press.

Appendix A

Math Review

Some math problems are reviewed in this chapter, which are often asked by students in orbital mechanics class.

A.1 Power, Taylor Series

Actually this is an approximation method. A power series is an infinite series of the form:

$$\sum_{n=0}^{\infty} a_n(x - x_0)^n = a_0 + a_1(x - x_0) + a_2(x - x_0)^2 + \cdots \tag{A.1}$$

Assume a_n, x, x_0 are real variables, the n'th partial sum is:

$$S_n(x) = \sum_{j=0}^{n} a_j(x - x_0)^j. \tag{A.2}$$

The remainder is:

$$R_n(x) = \sum_{j=n+1}^{\infty} a_j(x - x_0)^j. \tag{A.3}$$

If $\lim_{n \to \infty} S_n(x_1) = S(x_1)$, the series converges at $x = x_1$ and the sum is $S(x_1)$.

Note: $S(x_1) = S_n(x_1) + R_n(x_1)$. For convergence, $|R_n(x)| = |S(x) - S_n(x)| < \varepsilon$ when $n > N$. That is, for sufficiently large n, the remainder term is as small as desired.

Assume a continuously differentiable (analytical) function has the power series:

$$f(x) = \sum_{n=0}^{\infty} a_n(x - x_0)^n \tag{A.4}$$

$$f(x_0) = a_0 \tag{A.5}$$

$$f'(x) = a_1 + 2a_2(x - x_0) + 3a_3(x - x_0)^2 + \cdots \Rightarrow f'(x_0) = 1 \cdot a_1 \tag{A.6}$$

Fundamental Spacecraft Dynamics and Control, First Edition. Weiduo Hu.
© 2015 John Wiley & Sons (Asia) Pte Ltd. Published 2015 by John Wiley & Sons (Asia) Pte Ltd.

$$f''(x) = 2a_2 + 2 \cdot 3a_3(x - x_0) + \cdots \Rightarrow f''(x_0) = 1 \cdot 2a_2 \tag{A.7}$$

$$f'''(x) = 2 \cdot 3a_3 + 2 \cdot 3 \cdot 4a_4(x - x_0) + \cdots \Rightarrow f'''(x_0) = 1 \cdot 2 \cdot 3a_3. \tag{A.8}$$

In general,

$$f^{(n)}(x_0) = n!a_n \quad \text{or} \quad a_n = \frac{f^{(n)}(x_0)}{n!}. \tag{A.9}$$

Taylor Series of an analytical function:

$$f(x) = \sum_{n=0}^{\infty} \frac{f^{(n)}(x_0)}{n!}(x - x_0)^n. \tag{A.10}$$

A.1.0.1 Examples

$$\exp(x) = \sum_{n=0}^{\infty} \frac{x^n}{n!}, \quad \text{converges for all } x, \tag{A.11}$$

$$\ln(1 + x) = \sum_{n=0}^{\infty} (-1)^n \frac{x^{n+1}}{(n + 1)!}, \quad \text{converges for } |x| < 1, \tag{A.12}$$

$$\sin(x) = \sum_{n=0}^{\infty} (-1)^n \frac{x^{2n+1}}{(2n + 1)!}, \quad \text{converges for all } x, \tag{A.13}$$

$$\cos(x) = \sum_{n=0}^{\infty} (-1)^n \frac{x^{2n}}{(2n)!}, \quad \text{converges for all } x, \tag{A.14}$$

Notes: (1) Taylor Series for $\sin(x)$ shows the useful approximation for small x, $\sin(x) \approx x$. (2) Taylor Series for $\cos(x) \cong 1 - (1/2)x^2$. (3) Assume $0! = 1$ here. A typical example is $e^x|_{x=1} = 2.71\,828\,183$. Its Taylor Series is

$$e^x = \sum_{n=0}^{\infty} \frac{x^n}{n!}, \quad a_n = \frac{1}{n!}. \tag{A.15}$$

$n = 0$	$a_0 = 1$	$e^1 \approx 1$
$n = 1$	$a_1 = 1$	$e^1 \approx 2$

n	a_n	e^1
0	1	1
1	2	2
2	0.5	2.5
⋮	⋮	⋮
11	2.51e-08	2.71828183
12	2.09e-09	2.71828183

A.1.0.2 Solving $f(x)$

Orbital dynamics yields Kepler's Equation $M = E - e \cdot \sin(E)$, where M = mean anomaly, E = eccentric anomaly, e = eccentricity.
Given M and e, solve for E.

$$f(E) = E - e \cdot \sin(E) - M = 0. \tag{A.16}$$

Solve $f(E) = 0$, where f is a nonlinear function. Then, we can have:

$$E = M + e \cdot \sin(M) + e^2 \cdot \sin(M) \cdot \cos(M) - \frac{1}{8}e^3 \sin(M) + \frac{3}{8}e^3 \sin(3M) + O(e^4). \tag{A.17}$$

This is the basic idea of the so-called elliptic expansion. Similarly, high order and even other variables expansions can also be done such as radius r and true anomaly θ, see reference [2, 3].

$$
\begin{aligned}
\frac{r}{a} = 1 + \frac{1}{2}e^2 &+ \left(-e + \frac{3}{8}e^3 - \frac{5}{192}e^5 + \frac{7}{9216} \right) \cos M \\
&+ \left(-\frac{1}{2}e^2 + \frac{1}{3}e^4 - \frac{1}{16}e^6 \right) \cos 2M + \left(-\frac{3}{8}e^3 + \frac{45}{128}e^5 - \frac{567}{5120}e^7 \right) \cos 3M \\
&+ \left(-\frac{1}{3}e^4 + \frac{2}{5}e^6 \right) \cos 4M + \left(-\frac{125}{384}e^5 + \frac{4375}{9216}e^7 \right) \cos 5M \\
&- \frac{27}{80}e^6 \cos 6M - \frac{16807}{46080}e^7 \cos 7M + O(e^8),
\end{aligned}
\tag{A.18}
$$

$$
\begin{aligned}
\theta = M &+ \left(2e - \frac{1}{4}e^3 + \frac{5}{96}e^5 + \frac{107}{4608}e^7 \right) \sin M \\
&+ \left(\frac{5}{4}e^2 - \frac{11}{24}e^4 + \frac{17}{192}e^6 \right) \sin 2M + \left(\frac{13}{12}e^3 - \frac{43}{64}e^5 + \frac{95}{512}e^7 \right) \sin 3M \\
&+ \left(\frac{103}{96}e^4 - \frac{451}{480}e^6 \right) \sin 4M + \left(\frac{1097}{960}e^5 - \frac{5957}{4608}e^7 \right) \sin 5M \\
&+ \frac{1223}{960}e^6 \sin 6M + \frac{47273}{32256}e^7 \sin 7M + O(e^8).
\end{aligned}
\tag{A.19}
$$

A.1.0.3 Solving $f(x, y, y')$

To solve $dy/dx = x \cdot y + 1$, $y(0) = 1$, we can use numerical technique: Runge-Kutta 4 (RK4) which is a 4th order Taylor Series. Let's see the step from $x_0 = 0$, to endpoint x_n in steps of size h. Let $h = 0.1$, and $y_0 = 1.0$, $f(x, y) = x \cdot y + 1$.
RK4 Algorithm: Let $i = 0$,

$$k_1 = h \cdot f(x_i, y_i) \tag{A.20}$$

$$k_2 = h \cdot f(x_i + \frac{1}{2}h, y_i + \frac{1}{2}k_1) \tag{A.21}$$

$$k_3 = h \cdot f(x_i + \frac{1}{2}h, y_i + \frac{1}{2}k_2) \tag{A.22}$$

$$k_4 = h \cdot f(x_i + h, y_i + k_3), \tag{A.23}$$

then, $x_{i+1} = x_i + h$, and

$$y_{i+1} = y_i + \frac{1}{6}\left[k_1 + 2k_2 + 2k_3 + k_4\right]. \tag{A.24}$$

Repeat until $x_{i+1} = x_n$. For the 4th order Taylor Series method, 1st neglected term contains $(y - y_i)^5 \approx h^5$. Is this sufficiently accurate? Normally used values for h are 0.1 or 0.01. The answer is clear.

A.2 Differential Correction

See Section A.1 on the 1st order Taylor Series – single variable.

$$f(x) = f(x_0) + (x - x_0)f'(x_0) \Rightarrow f(x) - f(x_0) = (x - x_0)f'(x_0).$$

Then,

$$\Delta f = \Delta x \cdot f'.$$

Multi-variable:

$$f(x, y) = f(x_0, y_0) + \frac{\partial f(x_0, y_0)}{\partial x}(x - x_0) + \frac{\partial f(x_0, y_0)}{\partial y}(y - y_0).$$

Then,

$$\Delta f = \Delta x \frac{\partial f}{\partial x} + \Delta y \frac{\partial f}{\partial y}.$$

Given observation of $\mathbf{r}(t_0)$ and $\mathbf{v}(t_0)$ predict position and velocity at later time. Really predict ρ (See Figure 2.19) and $\dot{\rho}$
We want a prediction for the next 6 observations at different times $t_1, ..., t_6$
Let $\Delta\rho_{ij}$ = difference between observed and predicted value
of component i at time j:

$$\Delta\rho_{11} = \frac{\partial\rho_{11}}{\partial r_I}\Delta r_I + \frac{\partial\rho_{11}}{\partial r_J}\Delta r_J + \frac{\partial\rho_{11}}{\partial r_K}\Delta r_K + \frac{\partial\rho_{11}}{\partial v_I}\Delta v_I + \frac{\partial\rho_{11}}{\partial v_J}\Delta v_J + \frac{\partial\rho_{11}}{\partial v_K}\Delta v_K$$

$$\Delta\rho_{12} = \cdots$$

$$\cdots \tag{A.25}$$

$$\Delta\rho_{16} = \cdots$$

$$\Delta\rho_{21} = \cdots$$

$$\cdots$$

$$\Delta\rho_{36} = \cdots$$

36 equations if range rate is also included – must be linearly independent.
    Partial derivatives: Approximated by finite differences:

$$f'(x) \approx \frac{f(x + \Delta x) - f(x)}{\Delta x} \tag{A.26}$$

$$\frac{\partial\rho_{ij}}{\partial r_I} \approx \frac{\rho_{ij}(r_I + \Delta r_I, r_J, r_K, v_I, v_J, v_K) - \rho_{ij}(r_I, r_J, r_K, v_I, v_J, v_K)}{\Delta r_I}. \tag{A.27}$$

Corrected elements: Update original observation of $\mathbf{r}(t_0)$ and $\mathbf{v}(t_0)$, Then recompute predicted values and iterate until residuals $<$ tolerance. Its purpose is to develop $\mathbf{r}$ and $\mathbf{v}$ at time t_0 which correctly predicts the observations at a later time.

Algorithm Process

n is the number of observations and p is the number of parameters. Let: $\mathbf{W}$ be $n \times n$ diagonal matrix whose entries are square of confidence in observation measurements; $\mathbf{A}$ be $n \times p$ matrix of partial derivatives; $\mathbf{b}$ be $n \times 1$ matrix of residuals; $\mathbf{z}$ be $p \times 1$ matrix of computed corrections. By linear System Theory: when $p > n$, no unique solution; $p = n$, unique solution (linearly independent); $p < n$, no unique solution – develop least squares.

Example (for detailed discussion see reference [1])

Let $y = \alpha + \beta x$, x is an independent variable, y is a dependent variable of measurement, α and β are parameters interested. Initial estimate is $\alpha = 2$ and $\beta = 3$. Then, $x_1 = 2$, predict $y_1 = 8$, $x_2 = 3$, predict $y_2 = 11$. Observe that $x_1 = 2$, $y_1 = 1$ and $x_2 = 3$, $y_2 = 2$. It is 100% confidence in observation values:

$$\mathbf{W} = \begin{bmatrix} 1 & 0 \\ 0 & 1 \end{bmatrix}. \tag{A.28}$$

Have 2 equations and 2 unknowns:

$$1 = \alpha + 2\beta, 2 = \alpha + 3\beta \Rightarrow$$

$$\begin{bmatrix} 1 & 2 \\ 1 & 3 \end{bmatrix} \begin{bmatrix} \alpha \\ \beta \end{bmatrix} = \begin{bmatrix} 1 \\ 2 \end{bmatrix}$$

$$\begin{bmatrix} \alpha \\ \beta \end{bmatrix} = \begin{bmatrix} 3 & -2 \\ -1 & 1 \end{bmatrix} \begin{bmatrix} 1 \\ 2 \end{bmatrix} = \begin{bmatrix} -1 \\ 1 \end{bmatrix}.$$

Now have $y = -1 + x$. Check: $x_1 = 2$, $y_1 = 1$; $x_2 = 3$, $y_2 = 2$. Predictions equal observations as expected since the relation is linear. If $\mathbf{Y}(x)$ is nonlinear, for example 4 points and 2 parameters, use least-squares method.

$$\mathbf{Y} = \begin{bmatrix} y_1 \\ y_2 \\ y_3 \\ y_4 \end{bmatrix} = \beta \begin{bmatrix} x_1 \\ x_2 \\ x_3 \\ x_4 \end{bmatrix} + \alpha = \begin{bmatrix} 1 & x_1 \\ 1 & x_2 \\ 1 & x_3 \\ 1 & x_4 \end{bmatrix} \begin{bmatrix} \alpha \\ \beta \end{bmatrix} \equiv \mathbf{H} \begin{bmatrix} \alpha \\ \beta \end{bmatrix} \tag{A.29}$$

$$\Rightarrow \begin{bmatrix} \alpha \\ \beta \end{bmatrix} = [\mathbf{H}^T\mathbf{H}]^{-1} \mathbf{H}^T\mathbf{Y}. \tag{A.30}$$

In real-world, Kalman Filter is used (see Section 4.3), normally the method works with 6 elements, $\mathbf{W}$ is the variance/covariance matrix. Iterative process based on differential correction is applied.

A.3   Spherical Trigonometry [4]

Great circle is a circle formed by the intersection of a sphere and a plane through the center of the sphere.

Spherical Triangle is a set of curves consisting of three arcs of great circles that form the boundary of a portion of a spherical surface.

With reference to Figure 7.3, the point of Beijing could be point A, the point of Sydney could be point B, the point of an island on the equator in the Indian Ocean (west of Indonesia) could be point C, and the curve connecting A,B,C are great arcs. Then A, B, C are the vertices, and let a, b, c be their opposite sides, respectively, where $a = \widehat{CB}$, $b = \widehat{AC}$, $c = \widehat{AB}$. Also see Figure A.1, in which C is at the North Pole.

Law of Sines:

$$\frac{\sin a}{\sin A} = \frac{\sin b}{\sin B} = \frac{\sin c}{\sin C}. \tag{A.31}$$

Side Law of Cosines:

$$\cos a = \cos b \cdot \cos c + \sin b \cdot \sin c \cdot \cos A \tag{A.32}$$

$$\cos b = \cos a \cdot \cos c + \sin a \cdot \sin c \cdot \cos B \tag{A.33}$$

$$\cos c = \cos a \cdot \cos b + \sin a \cdot \sin b \cdot \cos C. \tag{A.34}$$

Angle Law of Cosines:

$$\cos A = -\cos B \cdot \cos C + \sin B \cdot \sin C \cdot \cos a \tag{A.35}$$

$$\cos B = -\cos A \cdot \cos C + \sin A \cdot \sin C \cdot \cos b \tag{A.36}$$

$$\cos C = -\cos A \cdot \cos B + \sin A \cdot \sin B \cdot \cos c. \tag{A.37}$$

There are some interesting special cases, such as when $C = 90°$, then $\cos A = \sin B \cos a$, $\tan b = \tan c \cos A$.

Note: $A + B + C \geq 180°$. Let $s = 1/2(a + b + c)$

$$\sin \frac{1}{2}A = \sqrt{\frac{\sin(s - b) \cdot \sin(s - c)}{\sin b \cdot \sin c}}, \cos \frac{1}{2}A = \sqrt{\frac{\sin s \cdot \sin(s - a)}{\sin b \cdot \sin c}} \tag{A.38}$$

$$\sin \frac{1}{2}B = \sqrt{\frac{\sin(s - a) \cdot \sin(s - c)}{\sin a \cdot \sin c}}, \cos \frac{1}{2}B = \sqrt{\frac{\sin s \cdot \sin(s - b)}{\sin a \cdot \sin c}} \tag{A.39}$$

$$\sin \frac{1}{2}C = \sqrt{\frac{\sin(s - a) \cdot \sin(s - b)}{\sin a \cdot \sin b}}, \cos \frac{1}{2}C = \sqrt{\frac{\sin s \cdot \sin(s - c)}{\sin a \cdot \sin b}}. \tag{A.40}$$

If we assume

$$r = \sqrt{\frac{\sin(s - a) \cdot \sin(s - b) \cdot \sin(s - c)}{\sin s}} \tag{A.41}$$

then

$$\tan \frac{1}{2}A = \frac{r}{\sin(s - a)}, \tan \frac{1}{2}B = \frac{r}{\sin(s - b)}, \tan \frac{1}{2}C = \frac{r}{\sin(s - c)}. \tag{A.42}$$

Example 1

Given 2 points: P_1 and P_2 on surface of sphere.
Find A, B, C and a, b, c.

See Figure A.1. Let $a = 90° - \text{Lat}(P_1), b = 90° - \text{Lat}(P_2)$, $C = \text{Long}(P_2) - \text{Long}(P_1)$. Based on Law of Cosines Equation (A.34), inverse the cosine $\Rightarrow$ obtain c. Then we have all the 3 sides a, b, c, let $s = (1/2)(a + b + c)$ and by Equation (A.41), then we have:

$$\tan \frac{1}{2}A = \frac{r}{\sin(s - a)} \Rightarrow A = 2 \tan^{-1} \left[\frac{r}{\sin(s - a)} \right] \qquad (A.43)$$

$$\tan \frac{1}{2}B = \frac{r}{\sin(s - b)} \Rightarrow B = 2 \tan^{-1} \left[\frac{r}{\sin(s - b)} \right]. \qquad (A.44)$$

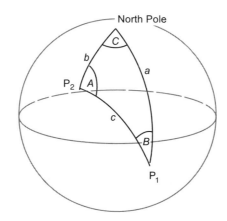

Figure A.1   Geometry of Example 1 and Example 3

Example 2

Know: B and a, given d.
Find latitude and longitude of P_3 which is d units from P_1 on great circle route.

See Figure A.2. Let λ = length of arc from North Pole to P_3, then $\cos \lambda = \cos a \cdot \cos d + \sin a \cdot \sin d \cdot \cos B$, Inverse the cosine gives λ. Then we have all 3 sides a, d, λ, $\text{Lat}(P_3) = 90° - \lambda$. Let $s = (1/2)(a + \lambda + d)$ and again by Equation (A.41). Then,

$$\tan \frac{1}{2}C = \frac{r}{\sin(s - d)} \Rightarrow C = 2 \tan^{-1} \left[\frac{r}{\sin(s - d)} \right].$$

$\text{Long}(P_3) = \text{Long}(P_1) \pm C$, $\pm$ depends on east or west.

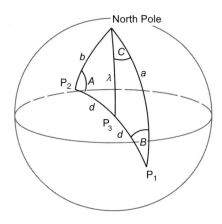

Figure A.2   Geometry of Example 2 and Example 4

Example 3

Sydney is at $151°12' = 2.6389E$, $-33°52' = -0.5911S$;
Beijing is at $116°23'17'' = 2.0314E$, $39°54'27'' = 0.6965N$;
Find great circle distance and heading from Sydney to Beijing. See Figure A.1.
As $\cos c = \cos a \cdot \cos b + \sin a \cdot \sin b \cdot \cos C$ as where $a = \pi/2 - (-0.5911) = 2.1619$, $b = \pi/2 - 0.6965 = 0.8743$, $C = 2.6389 - 2.0314 = 0.6076$, then $\cos c = 0.1645 \Rightarrow c = 1.4046 = 80.4777°$, $(1.4046 \text{ radians} = 8958.5 \text{ km} = 4837.2(\text{nm}))$. And as

$$\frac{\sin c}{\sin C} = \frac{\sin b}{\sin B} \Rightarrow \sin B = \frac{\sin b}{\sin c} \cdot \sin C \Rightarrow B = 0.4601 = 26.3615°.$$

Then the heading angle $= -26.3615°[333.6385°]$.

Example 4

Find the latitude and longitude of midway point in Example 3, that is, $d = c/2$. See Figure A.2.

$$\cos \lambda = \cos a \cdot \cos d + \sin a \cdot \sin d \cdot \cos B \tag{A.45}$$

$$= \cos(2.1619)\cos(0.7023) + \sin(2.1619)\sin(0.7023)\cos(0.4601) \tag{A.46}$$

$$= 0.0552 \Rightarrow \lambda = 1.5156, \tag{A.47}$$

then

$$s = \frac{1}{2}(a + \lambda + \frac{c}{2}) = 2.1899 \tag{A.48}$$

$$r = \sqrt{\frac{\sin(s-a) \cdot \sin(s-\lambda) \cdot \sin(s-c/2)}{\sin s}} = 0.1462 \qquad (A.49)$$

$$\tan \frac{1}{2}C = \frac{r}{\sin(s-c/2)} = 0.1467 \Rightarrow C = 0.2914. \qquad (A.50)$$

Then we have the midpoint at Latitude $= \pi/2\text{-}\lambda = 0.05\,552\,\text{N} = 3.1648°\text{N}$; Longitude $= 2.6389 - 0.2914 = 134.5053°\text{E}$.

The bottom line of this section is that each student should know that Spherical Trig is important and is used to analyze events near the Earth's surface and on the celestial sphere.

A.4   Summary and Keywords

This appendix introduces a few sections of background knowledge related to spacecraft dynamics, such as elliptic expansion and spherical trigonometry with application examples.

Keywords: Taylor series, elliptic expansion, Runge-Kutta method, differential correction, spherical side law of cosines, spherical angle law of cosine, spherical law of sine.

References

[1] Bate, R., Mueller, D., and White, J. *Fundamentals of Astrodynamics*. New York: Dover Publication, Inc., 1971.
[2] Danby, J.M.A. *Fundamentals of Celestial Mechanics*, 2nd edn. Richmond, VA: Willmann-Bell, 1992.
[3] Taff, L.G. *Celestial Mechanics: A Computational Guide for the Practitioner*. New York: John Wiley & sons, Inc., 1985.
[4] Harris, J.W. and Stocker, H. *"Spherical Geometry." §4.9 in Handbook of Mathematics and Computational Science*. New York: Springer-Verlag, pp. 108–113, 1998.

Index